Manufacturing Tribology

Edited by
Jibin T. Philip, Chacko Preno Koshy,
and M. D. Mathew

CRC Press
Taylor & Francis Group
Boca Raton London New York

CRC Press is an imprint of the
Taylor & Francis Group, an **informa** business

First edition published 2026
by CRC Press
2385 NW Executive Center Drive, Suite 320, Boca Raton FL 33431

and by CRC Press
4 Park Square, Milton Park, Abingdon, Oxon, OX14 4RN

CRC Press is an imprint of Taylor & Francis Group, LLC

ISBN: 9781032417028 (hbk)
ISBN: 9781032417035 (pbk)
ISBN: 9781003359333 (ebk)

DOI: 10.1201/9781003359333

Typeset in Times
by codeMantra

Contents

Preface

Tribology is the branch of science and engineering that deals with the study of friction, wear, and lubrication of interacting surfaces in relative motion. It influences every aspect of manufacturing, from the microscopic interactions in nanomanufacturing to the macroscopic dynamics of large-scale industrial operations. As the manufacturing industry continues to evolve, the importance of understanding and optimizing these tribological interactions becomes increasingly critical. Hence, there exists a phenomenal need to understand the fundamental influence of tribology on the discrete production methodologies and methods, such that they can be positively and productively monitored to achieve sustainable, economic, and efficient manufacturing of products and machines. Advancements in tribology, such as the development of novel lubricants, surface coatings, and material technologies, continue to enhance manufacturing processes. However, challenges remain in understanding the complex tribological interactions under varying conditions and scales.

This book contains eight chapters, each dealing with a distinct aspect of tribology in manufacturing. The first chapter, "Introduction to Tribology in Manufacturing," provides a foundational understanding of tribological principles and their relevance to manufacturing processes. It explores the fundamental concepts of friction, wear, and lubrication and highlights their critical role in increasing efficiency, reliability, and performance in manufacturing. This multidisciplinary field encompasses the study of friction, wear, and lubrication, which are fundamental to the performance and longevity of machinery and components. Friction is the resistance to motion that occurs when two surfaces slide against each other. Some level of friction is beneficial for processes like braking and gripping. However, excessive friction can lead to energy loss and heat generation, resulting in reduced efficiency and potential damage to components. Wear is the gradual removal of material from a surface due to mechanical action. Wear can lead to the degradation of tools, dies, and machine parts, affecting product quality and increasing production costs. Tribological studies help in identifying wear mechanisms and developing measures to reduce wear. Lubrication involves the application of substances (called lubricants) to reduce friction and wear between surfaces. Different types of lubricants, including oils, greases, and solid films, are used depending on the specific application and operating conditions. Advanced lubrication techniques, such as hydrodynamic and boundary lubrication, are employed to achieve better performance.

Chapter 2 on "Tribological Aspects in Nano/Micro-Manufacturing" discusses issues where components and parts are produced at microscopic scales and tribological considerations become even more critical. At these scales, the surface-to-volume ratio is significantly higher, and surface interactions dominate. Therefore, the traditional assumptions about friction, wear, and lubrication will not apply. Advanced lubricants, coatings, and surface treatments have been developed to address these unique challenges. Understanding and managing tribological aspects in nano and micro-manufacturing is essential for the development of high-performance miniaturized devices and systems, which are increasingly critical in fields such as electronics, biomedical engineering, and advanced materials science.

Chapter 3 is on “Nano-cutting Fluids for Machining: An Overview of the Tribological Aspects.” Nano-cutting fluids, enhanced with nanoparticles, represent a significant advancement in machining technology. These fluids improve machining performance by reducing friction, wear, and heat generation during cutting processes. The nanoparticles enhance the thermal conductivity and lubricating properties of the base fluid. This leads to more efficient heat dissipation and a smoother cutting action. The use of nano-cutting fluids is particularly beneficial in high-precision and high-speed machining operations, where traditional cutting fluids may fall short. As the demand for advanced manufacturing techniques grows, nano-cutting fluids are expected to play a crucial role in enhancing the efficiency and quality of machining processes.

Chapter 4 is on “Tribology of Metal Additively Manufactured Materials.” It discusses wear and friction challenges associated with additive manufacturing (AM) materials and presents strategies for optimizing their tribological performance. Unlike conventionally manufactured metals, AM materials exhibit unique microstructure, surface texture, and porosity due to their layer-by-layer fabrication process. These distinct attributes can significantly influence their tribological behavior. The tribological performance of AM materials can be optimized through postprocessing techniques such as heat treatment, surface finishing, and coating applications.

Chapter 5 is on “Tribo-behavioural Enhancement through Surface Texturing.” This chapter covers various surface texturing methods and their impact on reducing friction and wear in manufacturing applications. Surface texturing is an innovative technique for creating patterned micro- or nano-scale features on their surfaces. These textures can significantly reduce friction and wear by trapping lubricants, facilitating debris removal, and altering the contact mechanics between interacting surfaces. Surface texturing can improve the hydrodynamic lubrication regime, allowing a more stable lubricant film to form between the surfaces in relative motion. Advances in manufacturing technologies, such as laser surface texturing and micro-machining, have enabled the precise and reproducible creation of surface textures, making it a viable strategy for enhancing the tribo-behavior of a wide range of materials in various industrial applications.

Chapter 6 is on “Measurement and Control of Wettability and Surface Energy of Materials” and provides insights into the techniques used to measure and control the wettability and surface energy of materials. These properties are crucial for optimizing lubrication and adhesion in manufacturing processes. Controlling wettability and surface energy is essential in numerous fields, including coatings, adhesives, biomedical devices, and microfluidics. Creating superhydrophobic surfaces can lead to self-cleaning materials, while enhancing surface energy can improve adhesion in coating processes. Methods to modify these properties include surface treatments like plasma etching, chemical vapor deposition, and the application of functional coatings.

Chapter 7 is on “Modeling and Simulation of Tribological Responses Related to Specific Manufacturing Processes.” It emphasizes the role of computational models and simulations in predicting and optimizing tribological behavior. Advanced computational methods, such as finite element analysis, discrete element method, and molecular dynamics, allow for a detailed examination of tribological interactions at different scales, from macroscopic to atomic levels. These models incorporate material properties, surface roughness, contact mechanics, and environmental factors to simulate real-world manufacturing scenarios. By integrating tribological modeling and simulation into the design and planning stages, manufacturers can enhance process efficiency, extend tool life, and improve product quality, ultimately reducing costs and downtime.

This book concludes with Chapter 8 on "Manufacturing and Surface Treatment Strategies for Corrosive and Tribo-Corrosive Environments—A Case Study on Titanium Alloys for Biomedical Applications." This chapter presents a detailed case study on titanium alloys, exploring their tribological performance in biomedical applications and the surface treatment strategies employed to enhance their durability and resistance to corrosion and wear. Titanium alloys are widely used in biomedical applications due to their excellent mechanical properties, biocompatibility, and inherent corrosion resistance. However, when exposed to the harsh, corrosive, and tribo-corrosive environments of the human body, these materials can experience degradation that compromises their performance. Surface treatment methods, such as anodization, plasma spraying, surface texturing, and chemical vapor deposition, are applied to enhance the surface properties of titanium alloys. Case studies on titanium alloy implants have demonstrated that these combined strategies significantly improve their performance in corrosive and tribo-corrosive environments, leading to longer-lasting and more reliable biomedical devices.

By encompassing a broad spectrum of tribological topics and their applications in manufacturing, this book aims to equip readers with the knowledge and tools needed to address the complex challenges of modern manufacturing environments. We hope that "Tribology in Manufacturing" will serve as a valuable resource for advancing research, improving industrial practices, and fostering innovation in the field of tribology.

Dr. Jibin T. Philip
National Postdoctoral Fellow (N-PDF), SERB-DST India
Indian Institute of Science (IISc), Bangalore;
Assistant Professor
Amal Jyothi College of Engineering
Kerala, India

Dr. Chacko Preno Koshy
Associate Professor
Saintgits College of Engineering
Kerala, India

Dr. M. D. Mathew
Dean (Postgraduate Studies and Research)
Saintgits College of Engineering (Autonomous)
Kerala, India

About the Editors

Jibin T. Philip is currently serving as an Assistant Professor in the Department of Mechanical Engineering at Amal Jyothi College of Engineering, Kottayam, Kerala, India. Prior to this, he worked as a National Post-Doctoral Fellow (N-PDF), funded by the Science and Engineering Research Board (SERB), Department of Science and Technology (DST), Government of India, in the Department of Mechanical Engineering at the Indian Institute of Science (IISc), Bangalore, Karnataka, India. He has also served as a Senior Research Fellow (SRF) for a project funded by SERB-DST, Government of India. He completed his Doctor of Philosophy (Ph.D.) in Mechanical Engineering at the National Institute of Technology Mizoram, Aizawl, Mizoram, India, in collaboration with the National Institute of Technology Calicut, Kerala, India. He obtained his M.Tech. and B.Tech. degrees from APJ Abdul Kalam Technological University, Thiruvananthapuram, Kerala, India, and Mahatma Gandhi University, Kottayam, Kerala, India, respectively. He has published more than 25 international and national journal papers (SCI/Scopus), more than 28 international and national conference papers and has published four book chapters. He is the recipient of several awards and accolades at international and national levels, including those from Institute of Physics Publishing (IOP, UK), Centre des matériaux (Mines Paris—PSL, CNRS, Evry, France), The Institution of Engineers India (IEI), Indian National Academy of Engineering (INAE), Indian Society for Technical Education (ISTE), and International Travel Grant (SERB-DST India) for his excellence in research and academics. Philip is the lead editor for a book series and three book projects with the international publisher Taylor & Francis, in the areas of tribology, solid-state processing, and natural fiber composites. He is frequently called upon as a reviewer for many international peer-reviewed journals of publishers and as an invited speaker for seminars/webinars/FDPs/workshops. He is also a member of Board of Studies (BoS) at Saintgits College of Engineering (Autonomous), Kottayam, Kerala, India. His primary research interests include tribology, solid-state processing, process-structure-property correlation, advanced manufacturing/machining, surface engineering, and analysis/testing of materials (mechanical, metallurgical, morphological, corrosion, and biocompatibility properties).

Chacko Preno Koshy is currently a member of faculty at the School of Mechanical Sciences and Technology, APJ Abdul Kalam Technological University, Thiruvananthapuram, Kerala, India. He earned his B.Tech. in Mechanical Engineering and M.Tech. in Machine Design from Mahatma Gandhi University, Kottayam, and subsequently obtained his Ph.D. from the National Institute of Technology, Calicut. His research focuses on nanotribology, biobased nanolubricants and nanofuels, polymer-based nanocomposites, synthesis of nanoparticles for lubricant applications, and biomechanics.

His research was granted two Australian patents for innovations in 2021. He has published over 30 international peer-reviewed journal papers and presented in more than 35 international and national conferences. He has undertaken sponsored research funded by various agencies, such as AICTE, CERD APJAKTU, and KSCSTE. He has coordinated several FDPs/STTPs funded by AICTE, APJAKTU, and KSCSTE in the areas of biomimetics, nanotribology, nanofluidics, tribology, and surface engineering.

He is the first Secretary of the American Society of Mechanical Engineers (ASME) Kerala Professional Section for the terms 2024–2025 and 2025–2026, Joint Secretary of the Society of Automotive Engineers (SAE) Cochin Division (2022–2024 and 2024–2026), and Executive Committee Member of the Indian Society of Heat and Mass Transfer (ISHMT) (2023–2026). He is an approved research supervisor of APJAKTU and also serves as an academic auditor of the parent university. Currently, he is a reviewer of over ten prestigious technical journals.

He was a finalist of the Indian National Academy of Engineering (INAE) Innovative Doctoral Level Student Projects Award in Mechanical Engineering in 2019 and a mentor award of the INAE Innovative PG Student Projects Award in 2017. He received a mentor award of ISTE–G.S.F.C. National Award for the Best M.Tech. Thesis Award in Mechanical Engineering (first prize) in 2017 and also a mentor award of the ISTE National Award for the Best B.Tech. Project in 2017. He and his students received eight best papers/innovations awards in national and international conferences/contests. He was the Organizing Secretary of TRIBOINDIA 2021—an International Conference on Industrial Tribology under the Aegis of Tribology Society of India (TSI) during December 2021. He was the guest editor of a Special Issue of TRIBOINDIA 2021 in a prestigious journal.

He has delivered more than 40 invited talks in prestigious institutions like NIT Surat and NIT Calicut on various research fields related to nanofluidics, tribology, nanotribology, engine tribology, surface engineering, and life cycle analysis of lubricants. He also serves as a visiting faculty of Government Engineering Colleges of Kerala under the Visiting Faculty Scheme of the Government of Kerala. He is a member of the American Society of Mechanical Engineers (ASME) and a life member of professional bodies such as Indian Society for Technical Education (ISTE), Institution of Engineers (IE), Tribology Society of India (TSI), Indian Society of Heat and Mass Transfer (ISHMT), and Society of Automotive Engineers (SAE).

M. D. Mathew is currently Associate Provost and Dean of Research & International Relations at Saintgits College of Engineering (Autonomous), Kottayam, Kerala, India. He brings over three decades of experience in materials science and mechanical engineering, with specialization in high-temperature materials behavior, creep deformation, and small-scale mechanical testing techniques such as impression creep and ball indentation. During his research career at Indira Gandhi Centre for Atomic Research, under the Department of Atomic Energy, Government of India, spanning over 35 years, he made several major contributions to the development of fast breeder and fusion reactor technologies in India. The pioneering work on the creep life evaluation of ferritic-martensitic steels and austenitic stainless steels has influenced design methodologies for critical components in power and process industries. His research work has also led to the indigenous development of many advanced materials for critical applications in nuclear reactors.

Mathew holds a Bachelor's degree in Metallurgical Engineering, postgraduate degrees in Physics and Business Administration, and a Doctorate in Physics-Metallurgy interdisciplinary. He has published over 220 research papers in national and international journals, presented over 100 research papers in national and international conferences, and published several books and book chapters. He currently serves as an editor, key reader, or reviewer for 17 international journals.

Mathew received research fellowships from several countries, including Japan, South Korea, Germany, and USA, where he carried out research work at major nuclear research centers and universities. He has been ranked among the top 2% scientists in the world by Stanford University, USA, consecutively from 2020 to 2024. He received Asia's Outstanding Researcher Award, AIRA 2023, for his contributions to nuclear energy and its role in reducing global warming. He is also a program evaluator for the Accreditation Board for Engineering and Technology (ABET), USA. He is a fellow of The Indian Institute of Metals, The Institution of Engineers (India), and The Indian Institute of Welding; a member of The Minerals, Metals and Materials Society (USA) and the Global Engineering Deans Council (USA); and a life member of several professional bodies in India.

In addition to his research contributions, Mathew has played a key role in fostering international academic partnerships, promoting interdisciplinary research, and mentoring early-career researchers. His vision for academic excellence and global engagement continues to shape the research ecosystem at Saintgits and beyond.

List of Figures

List of Tables

Contributors

K. S. Abhimanyu
Amal Jyothi College of Engineering
Kerala, India

M. Gururaj Acharya
Indian Institute of Science (IISc)
Bangalore, Karnataka, India

Subrahmanyam Adabala
FEA Engineer
Regal Rexnord
Hyderabad, India

Alphene G. Alex
Amal Jyothi College of Engineering
Kanjirappally, Kerala, India

Vijayavarman Chakkravarthy
Cranfield University
Bedford, United Kingdom

Jitender Kumar Chaurasia
Liaisonec Engineering Pvt Ltd
Kanchipuram, Tamil Nadu, India

Deepak G. Dilip
Department of Mechanical Engineering
Mar Baselios College of Engineering and Technology
Trivandrum, Kerala, India

Sandeep Kumar Dwivedi
Mechanical Engineering Department
Bhopal Institute of Technology and Science
Bhopal, Madhya Pradesh, India

N. Anand Krishnan
Department of Mechatronics Engineering
Jyothi Engineering College
Thrissur, Kerala, India

Jinoop Arackal Narayanan
School of Computing, Engineering and Digital Technologies
Teesside University
Middlesbrough, United Kingdom

Nida Naveed
Faculty of Technology
University of Sunderland
Sunderland, United Kingdom

N. L. Parthasarathi
Materials Development and Technology Group
Metallurgy and Materials Group
Indira Gandhi Centre for Atomic Research
Kalpakkam, India

B. Depu Kumar Patro
National Institute of Technology Karnataka
Surathkal, Mangalore, India

Krishnkant Sahu
Mechatronics Engineering Department
Sharad Institute of Technology College of Engineering
Yadrav, Ichalkaranji, India

Dipti Samantaray
Materials Development and Technology Group
Metallurgy and Materials Group
Indira Gandhi Centre for Atomic Research
Kalpakkam, India

M. Ravi Sankar
Indian Institute of Technology Tirupati
Tirupati District, Andhra Pradesh, India

M. Shanmuka Srinivas
Indian Institute of Technology Tirupati
Tirupati District, Andhra Pradesh, India

P. S. Suvin
National Institute of Technology Karnataka
Surathkal, Mangalore, India

Adesh Kumar Tomar
Mechanical Engineering Department
Graphic Era (Deemed to be University)
Dehradun, Uttarakhand, India

M. Vasudevan
Materials Development and Technology Group
Metallurgy and Materials Group
Indira Gandhi Centre for Atomic Research
Kalpakkam, India

K. Vipindas
Department of Mechanical Engineering
Indian Institute of Information Technology Design and Manufacturing Kurnool
Kurnool, Andhra Pradesh, India

1 Introduction to Tribology in Manufacturing

N. L. Parthasarathi, Dipti Samantaray, and M. Vasudevan

1.1 INTRODUCTION

The interaction between two bodies in contact is a natural occurrence. This interaction becomes evident as a force when the equilibrium between the two bodies is disrupted, either to initiate or maintain motion at constant or varying speeds. This force is known as friction, and the energy used to overcome it is dissipated as heat. Consequently, in moving bodies, the kinetic energy lost as a result of friction (Röder et al., 2000) results in wear from the rubbing action between surfaces. A lubricant is introduced between the surfaces. The study of these interactions and their effects is called tribology (Jost, 1966).

Although the term "tribology" was coined by Jost in 1966, our awareness and application of its principles date back to ancient times (Dowson et al., 1979). Diverse subjects of human interest, including but not exclusive to engineering, medicine, biology, and physical geology, are concerned with the understanding of the origin and consequences of rubbing surfaces in contact. The scientific approach to tribology started with intriguing experiments of Leonardo da Vinci who pictorially recorded his observations on friction over the period of 20 years (da Vinci, 1493). These observations introduced the concept of the coefficient of friction. It is deduced from his drawings and notes that he understood friction and its laws two centuries before the laws of friction were enunciated by Guillaume Amontons (Amontons, 1699). Figure 1.1 gives the evolution of tribology over the years.

The application of concepts of tribology in manufacturing is as old as human knowledge of tool making. Historical evidences suggest that tools were made by rubbing stones together for hunting, and drills were made for drilling holes during the Palaeolithic period (Bharat, 2020). Nevertheless, the scientific approach of using tribology for manufacturing started with Leonardo da Vinci's application of his knowledge on the interplay between friction, wear, and lubrication to prepare conceptual designs of polishing machines and rolling mills.

In the present day, the role of tribology in manufacturing is well recognized since friction, lubrication, and wear have claimed a vital space in the field of manufacturing. Knowing friction is inevitable in conventional processes where moving bodies (in the form of workpiece, die, and tools) have to be in contact, every possible attempt is made to understand the interplay between influencing factors such as materials, lubricants, lubrication methods, surface properties, temperature, and type of contact.

DOI: 10.1201/9781003359333-1

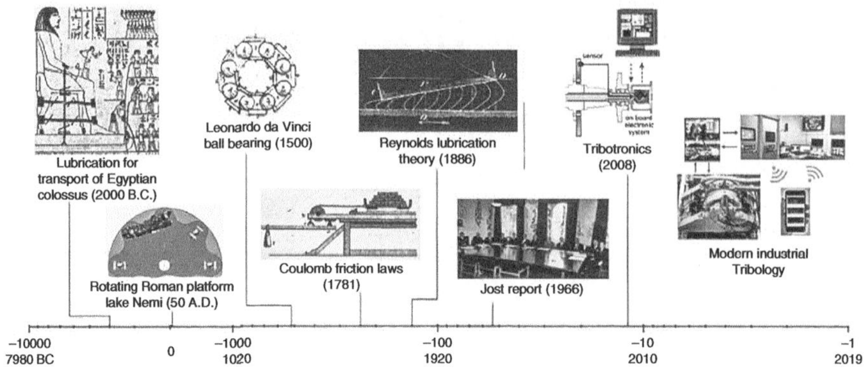

FIGURE 1.1 The evolution of tribology. (Ciulli, 2019.)

In manufacturing, the purpose of the study of tribology is dual: to minimize friction and wear where it is counterproductive and to use productive wear and friction to develop new technologies to improve quality and enhance production. Such demands justify the systematic understanding of friction, wear, and lubrication and their roles in various manufacturing processes.

1.2 ORIGIN OF FRICTION

Contrary to the appearance to the naked eye, smooth-looking polished surfaces are far from smooth. They are rough and have multiple sharp projections, referred to as "asperities." Some surfaces contain asperities across multiple scales. When two surfaces are placed together under load, asperities of both surfaces interact through various mechanisms. Adhesion is one of the mechanisms that offer resistance to motion, offered by virtue of topographical interaction. Due to a smaller area of contact, high pressure is developed at the tip of the asperities, and they fuse together. An effort to move the body breaks the joint formed between the asperities. The force required to break the asperity interaction is referred to as the frictional force.

Deformation is another mechanism through which the topographical interaction happens. Sometimes, locking rather than welding happens between the asperities. Under the force, the asperities undergo deformation. The resistance of the material to deformation affects the frictional force. Small-scale distortion of the asperities takes place. Nevertheless, bulk deformation happens in specific situations. This phenomenon is known as ploughing. The displacement of material that happens through ploughing is much higher and is of the order of magnitude than that which happens through deformation of asperities.

The above interactions are also responsible for the creation of new topography for further interaction. The continuous process involves the formation and breakdown of asperities on moving surfaces in contact due to phenomena like surface adhesion followed by tearing, gouging, and ploughing. The resistive force offered by the asperity interaction or topographical interaction may vary under different atmospheres, loading conditions, and at different rubbing speeds. In certain circumstances, chemical interaction may happen between the surfaces. Chemical interactions between the surfaces can amplify or reduce the resistive force offered by asperities interaction (Chen et al., 2019).

1.3 FRICTION TYPES

Static and dynamic (or kinetic) friction are the two general categories into which friction can be divided. Bodies in motion experience dynamic or kinetic friction, while static bodies feel friction when at rest. The friction can be further classified on the basis of contacts, motions, and use of lubricants (ASM Handbook, 1992). Kinetic friction involves power loss and induces heating, stress, cracking, fatigue, and morphological changes in the surfaces. Figure 1.2 depicts a friction curve plotted against sliding displacement. The peak friction value occurs shortly after the initial sliding begins.

1.4 FRICTION LAWS AND MODELS

The observations on rubbing bodies made by Amontons and Leonardo da Vinci can be summed up as follows: "The friction or the force of friction acts between two sliding surfaces" (Amontons, 1699). The apparent contact area between surfaces has no bearing on the frictional force, which is precisely proportional to the load-sustaining contact between them. "The frictional force is independent of the sliding speed," as noted by Coulomb. The fundamental rules of friction are these. The main law of friction can be stated mathematically as

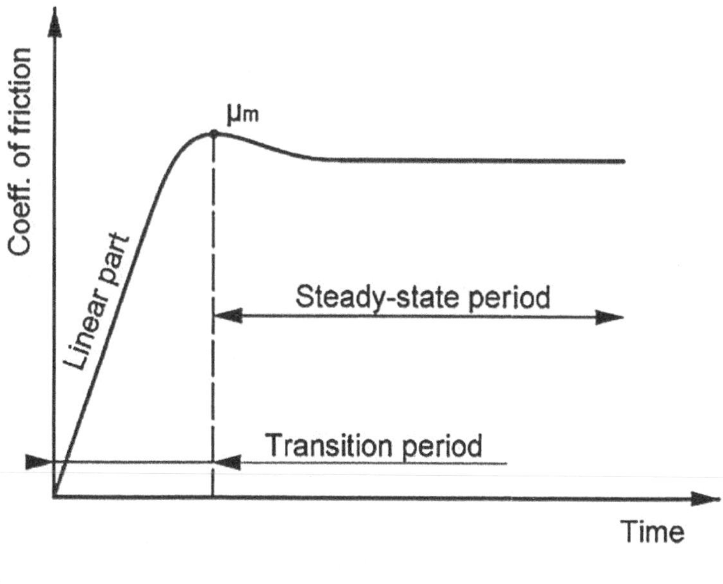

FIGURE 1.2 Changes in the friction coefficient over the sliding distance. (Krbata et al., 2020.)

$$F = \mu N \tag{1.1}$$

F denotes the frictional force. μ represents the proportionality constant and is referred to as the Coulomb factor or coefficient of friction. To measure the static coefficient of friction (Figure 1.3), gradually increase the angle of the inclined plane, α, until the sled just begins to slip. At this point, the tangent of the angle (Figure 1.3) represents the static coefficient of friction.

Numerous factors, including temperature, the type of motion, the composition and lubrication of the material, the nature of the contact, and the size of the normal force N, can affect the coefficient of friction. The friction force fluctuates according to the coefficient of friction when the normal load remains constant. When the sliding velocity $v=0$, the static friction force F_s is linked to the static coefficient of friction (μ_s). At greater sliding velocities, the kinetic coefficient of friction (μ_k) is applied, and this corresponds to the kinetic friction force F_k. The rolling friction force F_r and the rolling coefficient of friction (μ_r) are connected in rolling motion. The actual contact area, the deformation mode, the normal force, and the material qualities are some of the variables that affect this coefficient.

Later research revealed that Coulomb's friction model is only applicable in two situations: (1) the normal load must be within the materials' elastic limit and (2) the surface asperities in contact must be sufficiently apart to prevent interaction between their stress fields. This suggests that the apparent contact area is far larger than the real contact area. This law is used in manufacturing to examine and develop fixtures and jaws. Asperities deform and flatten under loading in several plastic deformation manufacturing processes. The real contact area gets closer to the apparent contact area as the weight increases. Differentiating between the real area of contact from a microscopic perspective and the apparent area of touch from a macroscopic perspective is necessary to understand the interaction between surfaces. Therefore, under these circumstances, the linear relationship between friction force and normal load as reported by Leonardo, Amontons, and Coulomb is not valid (Amontons, 1699). According to Li et al. (2014) and Michael (2010), sliding speed, or the difference in

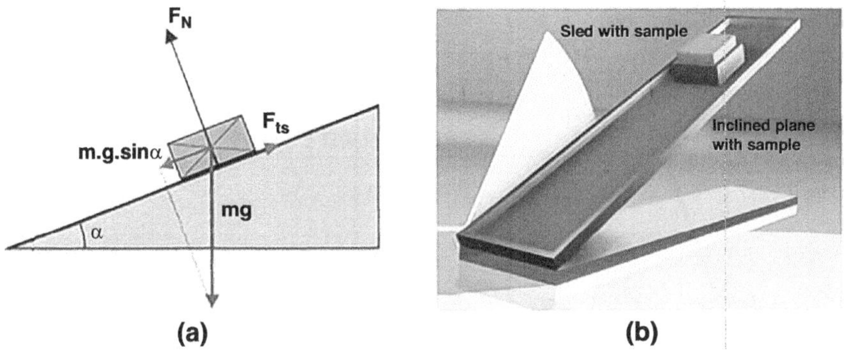

FIGURE 1.3 (a) Technique of using an inclined plane to measure the static coefficient of friction. (b) Setup of the device used to measure friction on an inclined plane. (Kelemenová et al., 2020.)

surface velocities, and the area of contact also affect friction. Tresca's friction model is widely employed in situations where lubrication is not used.

$$F = mkA \tag{1.2}$$

where F denotes the frictional force, A represents the area of contact, k denotes the shear strength of the material which is velocity dependent, and m is the friction factor. The friction factor depends on adhesion and ploughing. This model focuses solely on adhesion in metal forming processes. Tresca's friction model is not widely applicable to processes like sheet metal forming or lubricated bulk forming. A different model was proposed for sheet metal forming by Wilson, Hsu, and Huang in 1995, which takes additional factors such as ploughing and lubrication into account. According to Wilson's model, the stress resulting from friction is characterized as

$$\tau_f = \tau_a A + \tau_p A + \tau_h (1 - A) \tag{1.3}$$

where A represents the contact area fraction and τ_a, τ_p, and τ_h denote the stress generated due to adhesion ploughing and hydrodynamic phenomenon, respectively.

$$\tau_a = mk \tag{1.4}$$

and

$$\tau_p = \theta_p kH \tag{1.5}$$

where m represents the adhesion factor and k denotes the shear strength. θ_p represents the ploughing coefficient and H denotes the dimensionless surface hardness. In metal cutting applications, the frictional shear stress at the interface is described as

$$\tau_c = \mu \tau_n \tag{1.6}$$

Here, τ_n represents the interface normal stress. μ is the frictional coefficient at the tool–chip boundary, which reflects a weighted average of contributing factors like adhesion (μ_{ad}), abrasion (μ_{ab}), and deformation (μ_d), among others.

$$\mu = k_1\mu_{ad} + k_2\mu_{ab} + k_3\mu_d \tag{1.7}$$

where k_1, k_2, and k_3 represent the probability of each phenomenon.

Since shearing happens at higher strain rates than those seen in metal forming operations, to accommodate the plastic strain rate, a law was proposed.

$$\mu = \mu_0 \left(1 + a\dot{\varepsilon}_p\right) \tag{1.8}$$

where μ_0 is the frictional coefficient when the objects behave elastically and $\dot{\varepsilon}_p$ represents the plastic strain. This can be regarded as the kinetic frictional coefficient between the tool and specimen when the anticipated area is greater than the apparent area. In this context, "a" is a coefficient approximately equal to the ratio of the undeformed chip thickness to the cutting speed.

1.5 WEAR

The progressive damage of rubbing surfaces of the bodies is known as wear. The gradual damage may cause displacement of material from the surface without any loss of material or may cause loss of material from at least one component in contact. Wear is of special interest in manufacturing. While it is a matter of concern for the longevity of tools and dies, it is beneficially exploited for cutting and finishing operations.

Wear mechanisms fall into various categories: Adhesive wear, abrasive wear, corrosive wear, fatigue wear, fretting wear, and erosion are the first six types of wear. Understanding and characterizing wear is crucial for estimating the lifespan of tools and moulds. Wear is influenced by various factors including the geometry and movement of the bodies, their chemical compositions, the amount of force or load applied, the movement between the bodies, sliding speed, sliding distance, temperature, the testing environment (such as vacuum, air, argon, water, oil, or liquid metal), and the presence of debris. Wear progressively damages surfaces and changes the geometry of the bodies. In manufacturing, the changes in the shapes of dies and tools have a significant effect on surface quality and dimensional attributes of the products. Hence, understanding of wear draws significant attention from tool dies and tool makers.

1.6 FRICTION AND WEAR MEASUREMENT

Numerous tests are designed to assess the friction factor and rate of wear. The pin-on-disc (POD) test is particularly prevalent. Figure 1.4 shows a basic setup for a POD test, which includes a pin and a disc that together constitute the testing pair for friction. In this configuration, a load is applied to press the pin against the disc. To create relative motion, the disc is rotated on its axis. At the point of contact, both the normal and tangential forces are noted. This straightforward configuration allows for the measurement of how load and velocity affect the frictional force.

In the POD test, it is possible to use either conforming or nonconforming test geometries by altering the shape of the pin. POD testing equipment can be designed with the pin and disc axes arranged horizontally, vertically, or at an angle, depending on the researcher's interest in the debris generated at the interface. Some POD testing machines include heating capabilities to examine friction and wear at higher temperatures. Additionally, some machines are equipped with specialized chambers that enable variation of the testing atmosphere. In this test, a ring- or disc-shaped specimen is placed inside a testing apparatus and rotated along its axis to drive a pin against it. Standardized protocols have been established for performing POD testing (ASTM, G-99, 2017).

The POD test machine is an asymmetric system. For a dissimilar pair, the interchange in materials for pin and disc generates different data. To avoid these differences, symmetric systems are chosen. In symmetric systems, the geometry of the

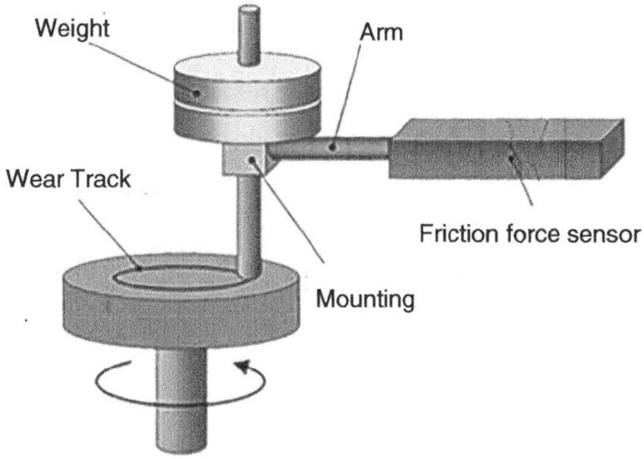

FIGURE 1.4 Diagrammatic representation of a pin-on-disc test setup. (Trzepiecinski and Lemu, 2020.)

rubbing bodies is chosen to be similar. Mostly, a pair of cylinders of similar diameter or a pair of discs of similar radius is chosen as the rubbing pairs.

Sliding under higher loads develops a very large shear strain at the interface. Transmission electron microscopic studies have shown that the microstructure generated at the interface has a resemblance with the microstructure achieved by deformation processes such as rolling, extrusion, and wire-drawing (Rigney, 1981).

Wear is typically measured by the reduction in mass and volume or changes in dimensions. It can also be evaluated based on the extent of surface damage, such as the dimensions of the wear scar, or by quantifying the amount of wear debris produced. The rate of wear is commonly defined as the mass lost or volume over a given period of time.

1.6.1 Wear and Friction Study on AISI 316 LN using a POD Tribometer

Austenitic stainless steel of AISI type 316L (N) is widely utilized in fast reactor applications as a structural material. Table 1.1 provides a list of this material's chemical makeup. POD studies have been performed to evaluate wear and friction properties over a wide range of temperatures. Table 1.2 provides specifics on the parameters used in these wear testing.

The ASTM standard G99-05 was followed in conducting the studies. The detailed wear characterization in relation to sliding temperature is shown in Figure 1.5. Gradual and uniform wear is shown by the wear patterns, which show a continuously rising trend with a positive slope as sliding distance rises. Parthasarathi et al.

TABLE 1.1
AISI Type 316 L (N) Austenitic Stainless Steel's Chemical Makeup

Material	C	Cr	Ni	Mo	Mn	Si	P	S	B	Co	Fe
316LN SS	0.024	17.7	10.8	2.0	1.2	0.4	0.04	0.01	-	0.16	Balance

TABLE 1.2
Materials and Experimental Parameters of the Dry Sling Wear Test.

Pin Material	Disc Material	Load	Sliding Speed	Temperature
AISI type 316L (N)	AISI type 316L (N)	20N	0.8 m/s	RT – 550°C (in steps of 50°C)

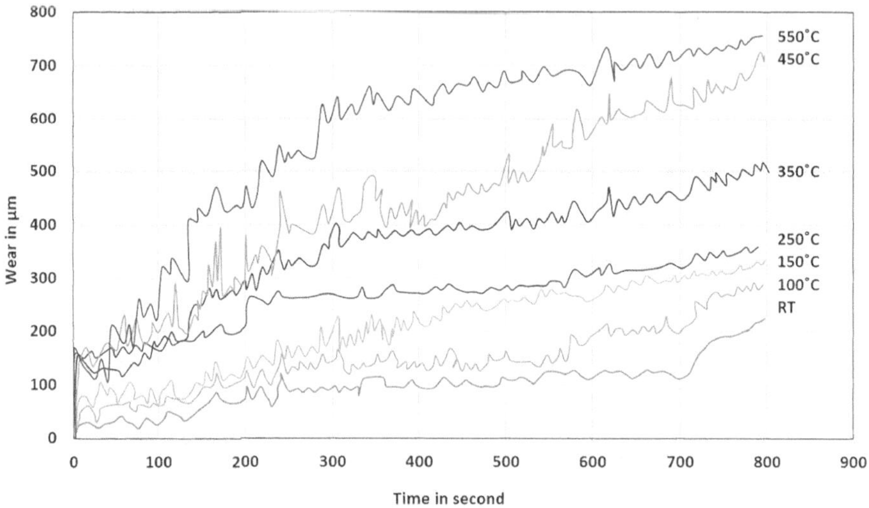

FIGURE 1.5 Composite graph showing the change in wear over time and temperature. (Adapted and redrawn from Parthasarathi et al., 2013a.)

(2012) have reported that there is a corresponding increase in wear severity with rising temperatures. The material softens as a result of the temperature rise, increasing wear. When AISI 316L (N) is heated from ambient temperature to 550°C, its elastic modulus drops by almost 23%.

The elastic contact between a sphere and a flat surface is described by the Hertz contact theory, which states that the indentation's width and depth are inversely correlated with the contacting materials' elastic modulus. Thus, as temperature rises, indentation width and depth increase. According to Ledbetter (1982), the maximum wear value increased from roughly 200 µm at ambient temperature to roughly 800 µm at 550°C. Furthermore, as the test temperature rises, the wear track's breadth and frictional force both rise, increasing the amount of material lost during sliding wear.

A combined perspective of the temperature-dependent coefficient of friction during dry sliding wear is shown in Figure 1.6, which suggests that the temperature increases the contact area between the disc and pin. At higher temperatures, there is more material loss because the wear debris functions as an abrasive substance, changing the wear mechanism from a two-body to a three-body abrasive mechanism. As the temperature rises, more and more worn-out layers adhere to the wear track, increasing surface roughness (Figure 1.7).

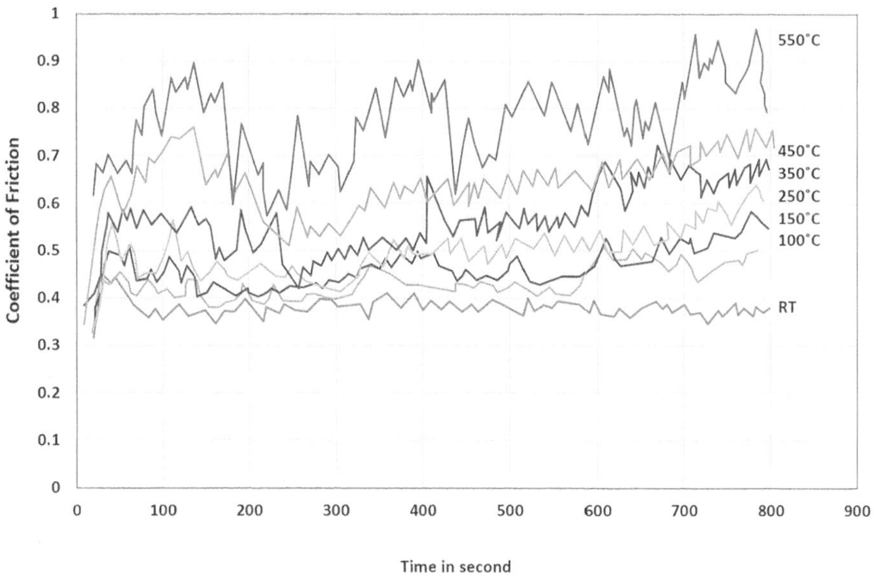

FIGURE 1.6 Changes in the coefficient of friction across various temperatures. (Adapted and redrawn from Parthasarathi et al, 2013b.)

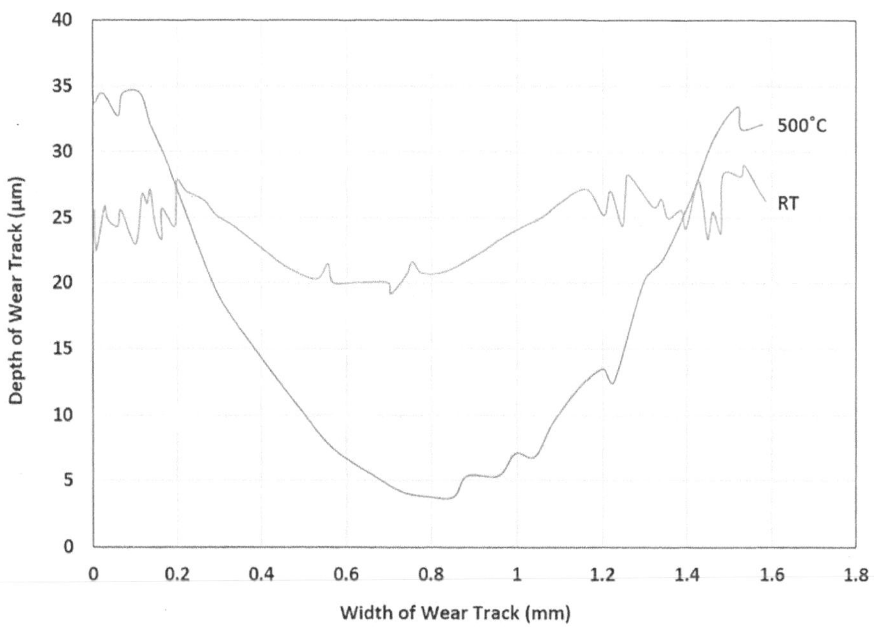

FIGURE 1.7 Profile of the worn surface at (a) room temperature and (b) 550°C. (Adapted from Parthasarathi et al., 2013a.)

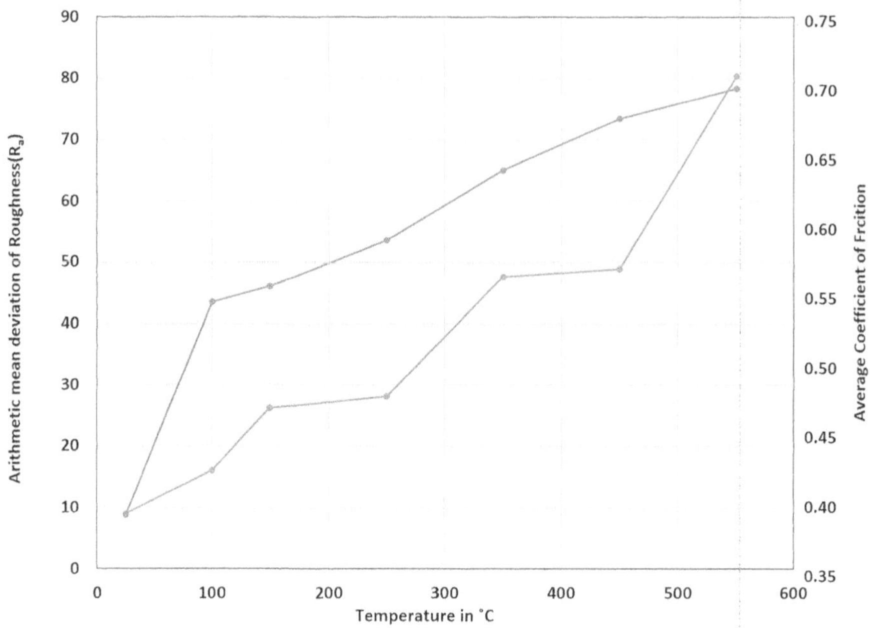

FIGURE 1.8 Changes in surface roughness (R_a) relative to temperature and friction coefficient. (Adapted and redrawn from Parthasarathi et al., 2013b.)

Figure 1.8 shows an increasing trend in the arithmetic mean deviation of surface roughness (R_a) during dry sliding wear as operating temperatures rise. The coefficient of friction data computed during the sliding wear tests on AISI 316L (N) stainless steel at different temperatures was confirmed by the line profile measurements on the wear tracks. The relationship between surface roughness and coefficient of friction was found by the profilometric feature analysis of wear tracks.

1.6.2 Determining the Friction Factor through Ring Compression Testing

The ring compression test (RCT) is an established method for evaluating the adhesion friction factor through a unique testing approach. This method is widely recognized for its straightforward and effective determination of the adhesion friction factor during plastic deformation processes. The test involves a ring-shaped specimen, typically dimensioned in a 6:3:1 or 6:3:2 OD: ID ratio as depicted in Figure 1.9. During the test, axial deformation alters the specimen's geometry based on the applied strain and the friction between the tool and specimen material.

Under well-lubricated conditions, where friction is negligible, the ring behaves similarly to a solid disk, with material flowing outward and the internal diameter increasing. Conversely, in poorly lubricated scenarios, high friction forces the material to move both inward and outward, decreasing the internal diameter. The friction coefficient is then derived from changes in the internal diameter and the specimen's height or thickness.

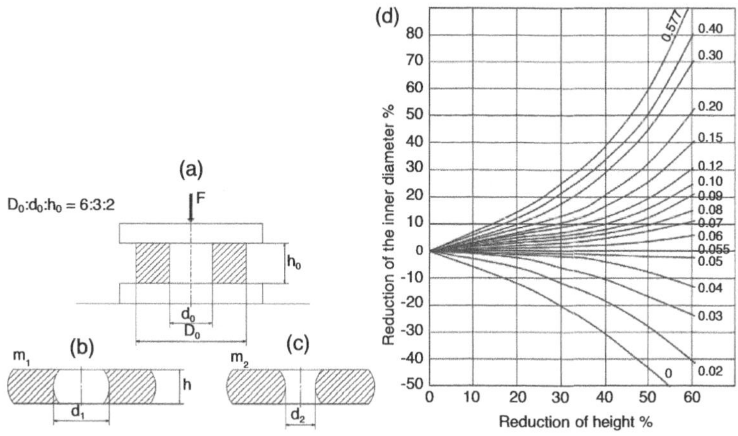

FIGURE 1.9 View of (a) an illustration depicting the ring upsetting procedure along with the variations in sample dimensions post-upsetting for various friction coefficient values. (b) Instances of inadequate lubrication indicated by $d_1 > d_0$. (c) Examples of effective lubrication with $8d_2 < d_0$ and $m_2 > m_1$. (d) A graphical representation showing friction coefficient measurements. (Dudkiewicz et al., 2023.)

One of the benefits of the RCT is that it bypasses the need for a separate tribometer and boasts high reproducibility. Despite these advantages, the RCT has not yet been standardized for friction and wear characterization but remains useful for estimating friction in metal forming operations such as forging and drawing. Friction factors are determined using calibration curves developed through numerical simulations. A typical calibration curve, along with room temperature RCT results, is shown in Figure 1.9. It is important to note that the friction coefficients obtained via RCT are average values, reflecting the overall interaction between the specimen and tool throughout the deformation process.

1.6.3 Friction Studies on AISI 316L (N) using Ring Compression Tests

RCTs were performed on a 100 kN servo-hydraulic machine under computer control using ring-shaped specimens with parameters of 12 mm outer diameter, 6 mm inner diameter, and 4 mm height. Figure 1.10 shows the specimen that was utilized in the RCT. At 900°C, 1000°C, 1100°C, and 1200°C, these rings were compressed to 50% of their original height while, in dry conditions, retaining a consistent actual strain rate of 0.01 s^{-1}. Post-experimentation, measurements of the inner diameter, outer diameter, and height of each compressed specimen were taken. Changes in the inner diameter percentage were calculated and juxtaposed on a calibration curve to determine friction factors under different conditions, as shown in Figure 1.11. The friction factors for forging AISI 316L (N), derived from these tests, are listed in Table 1.3. It was observed that the friction factors did not directly correlate with temperature, being affected by variables such as tool surface roughness, material thermal softening, and oxide layer presence.

FIGURE 1.10 Initial ring specimen. (Bose et al., 2018.)

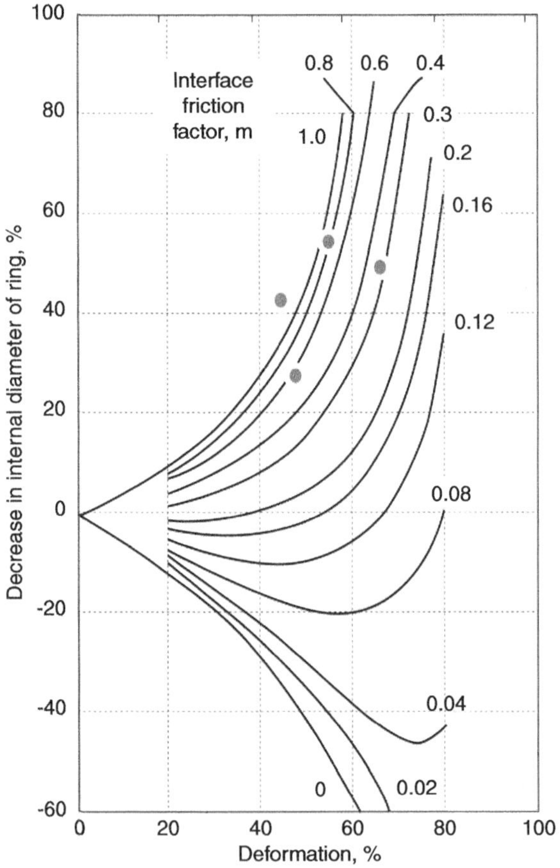

FIGURE 1.11 Friction factor calibration curve. (Bose et al., 2018.)

Simulations of these experiments were conducted using ABAQUS/Standard, considering the dies as analytical solids with a reference point on their outer periphery. An axisymmetric setup employed CAX4RT elements for the RCT, with a fine mesh

TABLE 1.3
Coefficient of Friction Predicted by RCT

Temperature (°C)	Condition	Coefficient of Friction
900	Dry	0.39
1000	Dry	0.46
1100	Dry	0.41
1200	Dry	0.28

Source: Bose et al. (2018).

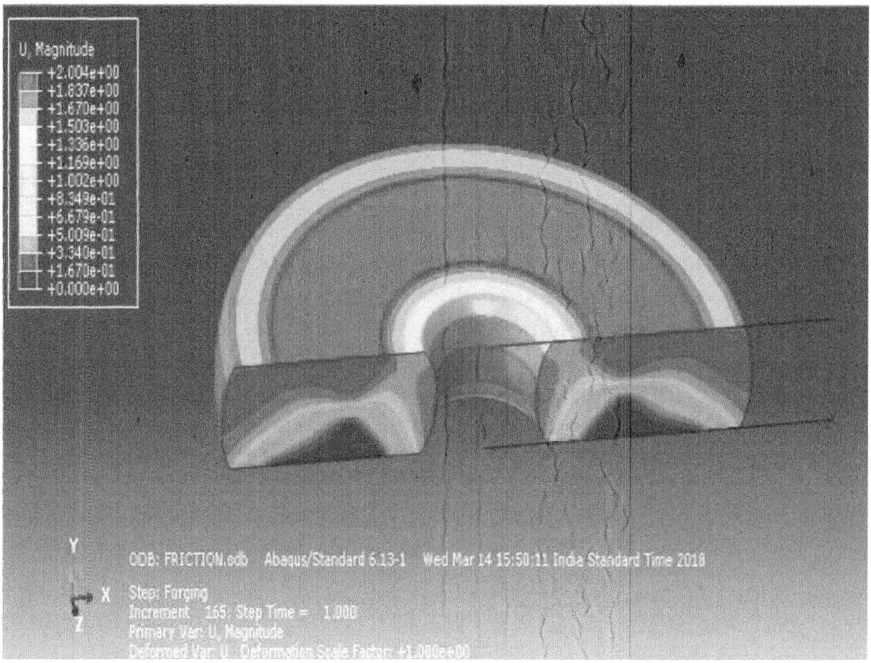

FIGURE 1.12 Coefficient of friction predicted by RCT. (Bose et al., 2018.)

size of 0.5 mm to ensure solution convergence. The bottom die's movements were restricted, while the top die was subjected to a downward vertical displacement of 2 mm. The ring was compressed to half of its original height. Figure 1.12 illustrates the material displacement within the ring specimen during testing.

1.7 FRICTIONAL HEATING

Friction force work is calculated by multiplying the frictional force with the distance it acts over, transforming this energy into heat. This generated heat is partly dissipated into the environment directly from the moving surface, while another portion is conducted through the body of the object. The energy conducted can cause physical and structural alterations within the body; some of this energy may also be conducted to other parts of the object and then dissipated to the surroundings. The energy that remains at the interface can lead to local temperature increases. In addition to external friction between tool/die surfaces and the workpiece, internal friction from layer-to-layer rubbing during the deformation of the workpiece also contributes to heat generation, which is particularly significant in high-speed metal forming.

In manufacturing, certain processes like friction stir welding and friction welding harness this frictional heat to effectively join metals. However, in machining, excessive heat can negatively impact both the tool and the workpiece. Excessive surface temperatures can reduce the material's hardness by exceeding its softening point, diminishing machining efficiency. This localized heating can alter the colour of some metals and alloys, change the local structure, and affect material properties, potentially leading to premature failure of the component. Similarly, in metal forming, frictional heat may cause localized defects and uneven material flow in the workpiece. While frictional heating is inevitable in high-speed manufacturing processes, managing it effectively is crucial for maintaining tool integrity and product quality.

1.8 FRICTIONAL HEAT MANAGEMENT IN MANUFACTURING

The frictional heat can be managed in two ways: by reducing the frictional force by use of a third body with lower shear strength between the rubbing bodies—referred to as lubricants and/or by use of a third body that is an effective medium of heat transfer—referred to as coolants. Lubricants can be used in solid (e.g., graphite, tantalum, hexagonal boron nitride, polytetrafluoroethylene (PTFE), and glass), semi-solid (e.g., grease and glass paste), liquid (e.g., vegetable oils, animal fats, and mineral oils), and gaseous state (e.g., nitrogen, argon, and carbon dioxide). Liquid and gaseous tlubricants are often used as coolants (Figure 1.13). A generalized term called "metal working fluid" is used for the materials that serve as lubricant and coolant simultaneously. The classification of lubricants can be on the basis of their physical state or on the basis of applications. Through the scriptures, the use of lubricants on wheel hubs and axles can be dated back to early civilization. The use of lubricants in the metal working process and the cutting process is relatively new. Recorded history indicates that the use of tallow on grinding wheels and the use of water for grinding were in common practice in the 16th century. Scriptures indicate that oil was used for the wire drawing process in the 17th century in 1650. During this period, a process called "sull-coating" that permits easy drawing of steel wires was discovered. In this procedure, iron was exposed to urine to develop a soft layer of coating on the surface. This coating worked as a lubricant during the wire drawing process. The coating was also used as a protection for the tools and dies. With the advent of the development of machine tools and the use of electricity as a power source, the use of metal working fluid grew, and the system for automatic application of cutting fluids evolved.

The functional requirement of lubricants or cutting fluids varies depending on the application. For instance, the expected functions of lubricants meant for metal forming are (1) separation—to separate the parts from tools; (2) lubrication—to reduce friction at workpiece and die interface; (3) corrosion control—to prevent corrosion

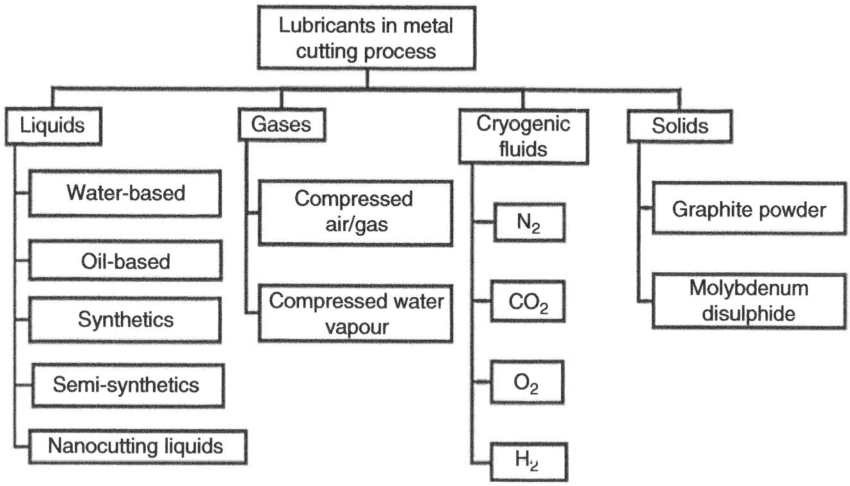

FIGURE 1.13 Various lubricants and cutting fluids utilized in metal cutting operations. (Adapted from Abdelrazek et al., 2020.)

of the tools and workpiece; and (4) easy removal—the lubricants should be easily removable from the workpiece.

The lubricant's primary function in metal cutting is to keep the tool and the workpiece cool. The lubricating roles are the secondary ones. The easy removal, corrosion management, chip flushing, and other tertiary tasks are among them. Based on the functional requirements, the physical properties used to qualify lubricants are viscosity, thermal conductivity, surface tension, etc. Besides the physical properties of metal working fluids needed to meet the functional requirement, the factors influencing the selection of metal cutting fluid involve size of the shop, type of machine used, materials involved, quality of water available (if the lubricant has to be water soluble), type of filtration, operation safety, toxicity, storage and control requirements, recyclability, ease of disposal, availability, performance vs. cost comparison and severity of operations—light duty operations (e.g., polishing), moderate duty operation (e.g., turning and milling), heavy duty operation (centre less grinding and sawing), and extremely heavy duty operations (e.g., extrusion, thread forming, and broaching).

The method of application of lubricants is designed by taking into account the performance of the machine, the quality of the product, and fluid conservation in view. The following are various techniques that are in common practice to deliver the metal working fluid at the tool–workpiece interface.

i. **Mop and sponge**: This is a basic type of lubrication technique, where lubricants are applied manually. This is mostly used for manual operations and for small forgings.
ii. **Drip application**: As the name implies, metal workings are applied through drip applicators. This method is mostly employed for polishing operations.
iii. **Flood lubrication**: In this method, lubricant is continuously poured onto the mating bodies. The tool and workpiece interface always remains submerged

in the lubricant. This technique is mostly employed for turning and sawing operations.

iv. **Roll coaters**: This technique involves using two rollers to apply lubricants to sheet surfaces, commonly used in sheet metal working processes.

v. **Electrodepositing**: In this process, the electrical charge is passed through a fluid, and the opposite charge is applied to the metals. This causes coating of lubricants on the metal surface. It is a costly lubricant deposition technique used for very special applications where the manufacturing process is optimized and automated for mass production.

vi. **Spraying**: In this technique, lubricant is dispersed onto the tool and workpiece interface using a spray device, which may optionally incorporate air.

vii. It is known that there are four main lubrication modes for metal working operations: boundary, mixed, thin film, and thick film lubrication. When the lubricant film between the tool and the workpiece surface is thicker than the total height of their surface imperfections, thick film lubrication takes place. The lubricant layer in thin film lubrication has a thickness equal to the peaks and valleys of the surface. Boundary lubrication produces a discontinuous film since the lubricant is mostly found in the valleys. In a situation known as mixed lubrication, certain surface irregularities are in direct contact with one another, while others are kept apart by a lubricant coating.

1.9 FRICTION ASPECTS IN MANUFACTURING

Casting, shaping, cutting, and metal joining are the four basic categories into which manufacturing processes are typically separated. Friction is present in all these domains, and process improvement requires a deep comprehension of friction. The subsequent sections will explore the specific impact of friction across these different manufacturing domains.

1.9.1 Friction in Casting

1.9.1.1 Continuous Casting

Continuous casting provides a cost-effective and efficient alternative to traditional ingot casting, making it essential for the steel industry. In this process, molten metal is transferred from a ladle or turn dish into a mould, as depicted in Figure 1.14. The mould walls, which are kept at a lower temperature, initiate the solidification of the metal from the outer edges inward.

The solidified outer layer with hot liquid at centre is called a strand. As the process progresses, the strand enters into the spray chamber. In the spray chamber, the strand comes in contact with a series of water-cooled rollers. To increase the solidification rate, the strand is sprayed with water. In high-speed continuous casting, the friction arises from the motion of the liquid metal and the thin shell of the strand in the mould (Nakato et al., 1984).

The consequences of the friction are sticking and tearing of the thin shell of the stand and spillage of molten metal. This can introduce surface defects and can

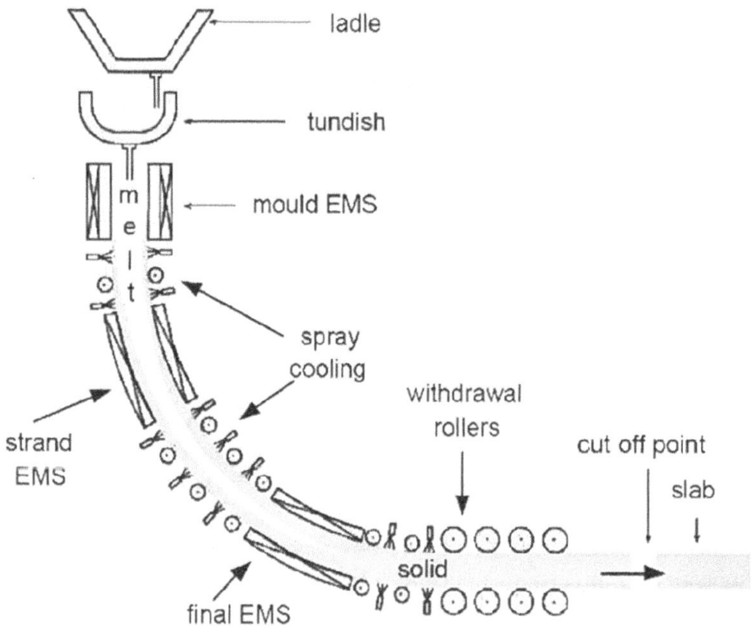

FIGURE 1.14 Various lubricants and cutting fluids utilized in metal cutting operations. (Adapted from Abdelrazek et al., 2020.)

damage the mould wall, causing disruption and delay in process due to clean up and repair. Minimization of friction is, therefore, of paramount interest for the casting industry. Frictions on mould wall can be monitored through mould friction monitoring systems (Wang et al., 2007). To minimize the friction in mould, lubrication is provided through introducing a slag layer between the mould wall and the strand. To facilitate the movement of slag, casting powders with low melting and solidification points are employed. Additionally, synthetic liquid lubricants are applied to the mould walls to ease the flow of metal (Kress et al., 1971). Oscillating the mould wall helps in smooth movement of the strand, but if not done correctly, it can disrupt the even distribution of the lubricant, increasing mould friction (Szekeres, 1996).

1.9.1.2 Injection Moulding

Injection moulding is a widespread technology to manufacture goods of various shapes with high dimensional accuracy. Figure 1.15 presents a basic diagram of an injection moulding machine. The precision of the product's dimensions relies heavily on the mould quality. This technology is used for producing both metals and polymers. For metals, the lost wax process is mostly employed for producing components. For polymers, however, a permanent mould is used. The main function of an injection mould is to shape the processed polymers at the required temperature and, subsequently, cool the product to room temperature, facilitating easy removal of the unmoulded part without any physical deformation/damage. The shape and surface finish of the cavity are, therefore, the most important parameters for injection

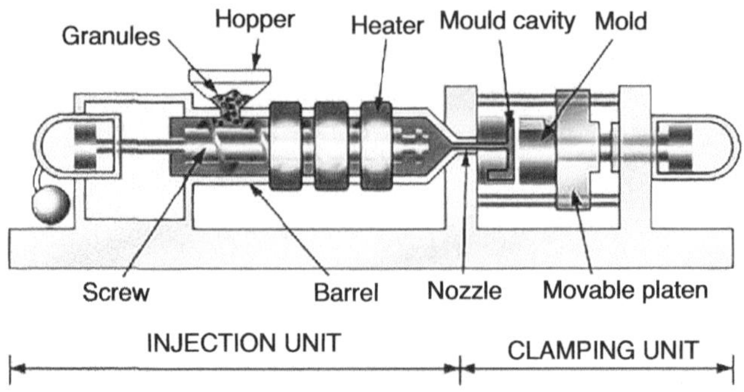

FIGURE 1.15 Basic diagram of an injection moulding machine. (Svečko et al., 2013.)

moulding. Hence, great attention is paid to surface-damaging wear. The intensity of wear is influenced by factors such as mould material, type of polymers used, complexity of the shape, temperature of operation, and injection pressure. Friction and die wear in injection moulding are reduced by the use of appropriate lubricants through a suitable lubrication system. PTFE or perfluoropolyether grease is commonly used for such applications. For certain applications like moulding of elastomer or nanostructured polymers, application of low module lubricants enhances friction. For such applications, microtexturing on the mould is provided using laser technology. Such textures reduce the adhesive friction and wear significantly (Voyer et al., 2017).

1.9.2 Friction in Plastic Deformation Processes

Plastic deformation is the process of changing a material's shape while keeping its volume constant by mass redistribution. Relative movement between the tool and the workpiece is involved in this redistribution, which causes friction at their interface. Friction has different effects on forming processes depending on the shape of the workpiece, the design of the tool, and the kind of motion that occurs between them. Analysing friction in these processes is complex, as the frictional forces do not conform neatly to Coulomb's law of friction. Nevertheless, this fundamental law can still provide a basic approximation of the forces needed for forming operations.

1.9.2.1 Forging

Forging is the oldest known metal forming process. In this process, material is given a shape between two dies with the application of compressive load. Forging is generally categorized into two main types: open die forging and closed die forging. Figure 1.16 portrays the common forging operations in terms of a tribological perspective.

The most basic form of open die forging involves producing discs by compressing cylindrical blanks between two flat dies. During compression, the material expands radially outward. This radial flow is, however, limited by the friction that exists between the tool and the workpiece. Consequently, the material in direct contact with the dies does not move as freely as the material further from the contact surfaces.

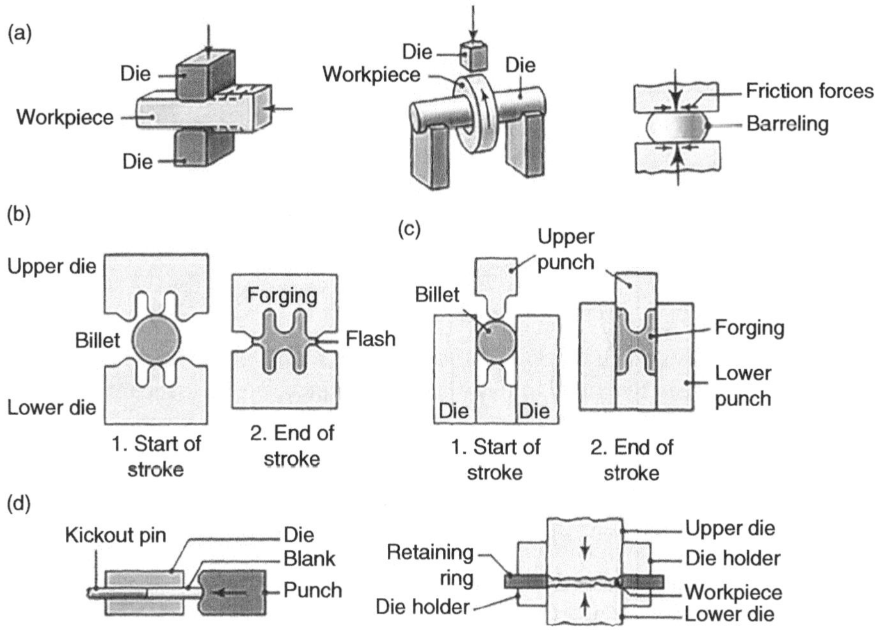

FIGURE 1.16 Common forging processes. (a) Open-die forging processes like ring forging, cogging, and upsetting. (b) Flash or impression die forging with extra material. (c) Forging operations without flash. (d) Extra forging techniques such as coining and heading. (Schmid et al., 2013.)

This frictional resistance generates shear stress at the tool–workpiece interface, resulting in lateral pressure within the material. This pressure is typically lowest at the periphery and highest at the centre.

The pressure at a specific distance from the centre can be quantified as follows:

$$P_a = \sigma_y \exp\left(\frac{2\mu}{h}(r - x)\right) \tag{1.9}$$

In this equation, r stands for the immediate radius of the cylinder, h stands for the instantaneous height of the cylinder, and represents the material's yield stress σ_y. According to the formula, the pressure reaches its maximum at the centre (where $x=0$). Additionally, the peak pressure increases exponentially as the radius of the disc increases. Accordingly, the force to deform the material increases. The sharp rise in frictional forces with an increase in diameter of the workpiece restricts the size of the forgings. In other words, it puts a limitation on the size of the forging machines that can be manufactured, erected, and operated. Lubrication is used to reduce the friction. The lubrication, however, becomes challenging with the complexity of the die. High friction sometimes leads to cracks in the forging. Too-thick lubricating films, on the other hand, lead to surface defects such as orange peel (Wilson et al., 1995). The commonly used lubricants for different forging processes are given in Table 1.4.

TABLE 1.4
Commonly Used Lubricants in Forging Operations

Type of Forging	Lubricants Used	Lubricants
Hot forging	Graphite, tantalum sheet, MoS_2	0.1–0.3
Cold forging	Wax, MOS_2, graphite	0.05–0.1

Source: Davim (2011).

1.9.2.2 Rolling

Rolling is a metal working process that involves the reduction of material thickness by passing it between two rolls. Initially, as the workpiece contacts the rolls, friction helps pull the material into the rolling process. During rolling, the workpiece undergoes significant compressive forces due to the narrow space between the rolls, which is typically less than the initial thickness of the material. The material thickness decreases as a result of this strong compression and the shear stress that results along the surfaces. In the manufacturing sector, rolling is an essential operation that can be done at room temperature (cold rolling) or at a higher temperature (hot rolling). Figure 1.17 illustrates the process of hot rolling, including the aspects of lubrication used during the operation.

Hot rolling is usually employed when a large reduction in thickness is required. It is employed for the initial breakdown of ingots into blooms. Hot rolling is usually carried out by reversing mills to achieve a large reduction—thick slabs are rolled repeatedly by changing the roll rotation direction and clearance after each pass. Cold rolling is employed when close dimension tolerances need to be achieved. Generally, cold rolling is used to create sheets, strips, and foils.

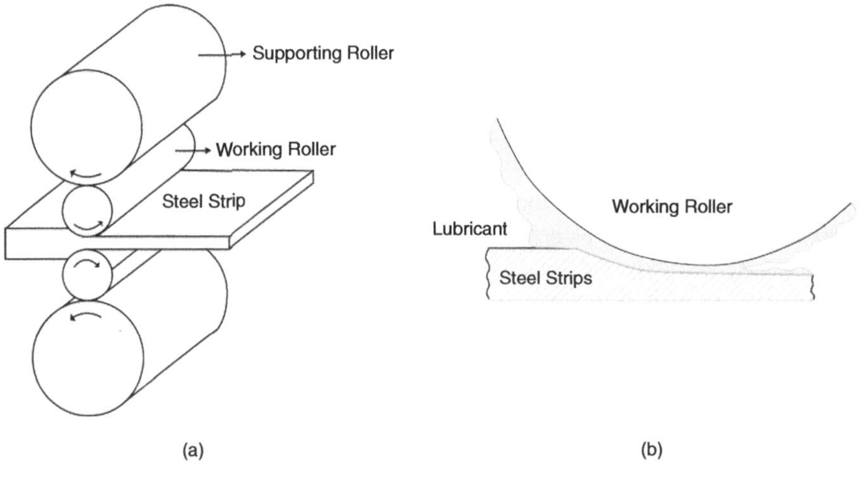

FIGURE 1.17 Sketches of (a) the hot rolling process and (b) lubrication application during hot rolling. (Yanan et al., 2018.)

Friction plays a vital role in the rolling process. The maximum reduction in thickness (Δh_{max}) that is possible in one pass is dependent on the friction coefficient (μ).

$$\Delta h_{max} = \mu^2 R \tag{1.10}$$

where R denotes the rolls' radius. According to the equation, there can be no decrease in the sheet's thickness if friction is zero. A high coefficient of friction between the rolls and the workpiece is beneficial when a significant reduction is required. Grooves or patterns are, hence, deliberately provided on the surfaces of the rolls employed for hot rolling.

Though the high friction coefficient is desirable for a large reduction in thickness, it contributes to the rolling load. The relationship between the friction and the load is expressed as

$$P = \frac{2}{\sqrt{3}} \sigma \left[\frac{h}{\mu L_p} \left(\exp\left(\frac{h}{\mu L_p} \right) - 1 \right) b \sqrt{R \Delta h} \right] \tag{1.11}$$

where $\bar{h}$ is the workpiece's average thickness between the entrance and departure, b denotes the width of the workpiece, L is the anticipated length of the contact arc, and σ is the workpiece's flow stress.

High value of friction also contributes to edge cracking and poor surface finishing. Frictional force reduces with an increase in roll speed or with the application of lubricants. These two factors are essential for cold rolling, where surface finish and dimensional tolerances are important compared to the reduction in thickness. Table 1.5 shows the common lubricants used in various rolling operations and the estimated range of the coefficient friction.

1.9.2.3 Extrusion

Extrusion is a forming process that reduces the cross-section of material by pushing it through a narrow die opening. This process is categorized into two types based on the material flow and force direction: direct (or forward) extrusion and indirect (or backward) extrusion. Extrusions are often performed at elevated temperatures to reduce the force requirements, as higher loads are needed. The force necessary for extrusion varies with factors including the type of material, the ratio of extrusion, the temperature during extrusion, the speed of the ram, and the friction at the interfaces between the material and the die, and the material and the wall of the container.

TABLE 1.5
Commonly Used Lubricants and Coefficient of Friction in Rolling (Davim, 2011)

Type of Process	Coefficient of Friction	Lubricants
Hot rolling	0.3–0.4	Water, emulsion, mineral oil
Cold rolling	0.01–0.03	Vegetable oils, mineral oils, eaters, or soaps

The material flows in the same direction as the movement of the ram during direct extrusion. Friction is produced by this arrangement between the billet and the die as well as between the moving billet and the stationary container wall. On the other hand, with indirect extrusion, the material flows toward the moving die while the billet stays fixed in relation to the container wall, primarily reducing friction at the billet–die contact. Because there is more frictional resistance during direct extrusion, the starting pressure needed is often larger than during indirect extrusion.

Total extrusion pressure $\left(P_{fe}\right)$ in direct extrusion

$$P_{fe} = P_{df} + P_{cf} \tag{1.12}$$

Total extrusion pressure $\left(P_{be}\right)$ in indirect extrusion

$$P_{be} = P_{df} \tag{1.13}$$

where P_{df} pressure experienced due to frictional force at the die–billet interface and P_{cf} is the pressure experienced due to frictional force at the container wall–billet interface.

$$P_{df} = \sigma\left(\frac{1+\mu\cot\alpha}{\mu\cot\alpha}\right)\left(1 - R^{\mu\cot\alpha}\right) \tag{1.14}$$

R denotes the extrusion ration $\left(R = \dfrac{A_0}{A}\right)$ with A_0 represents the billet's initial cross-sectional area, while A denotes the billet's ultimate cross-section. σ denotes the flow stress and α represents the semi-die angle. μ is the friction coefficient between the die and the material. The expression of die friction changes when lubrication is used.

The pressure experienced due to frictional force at the container wall–billet interface $\left(P_{cf}\right)$ is expressed as

$$P_{cf} = \frac{4L\tau_i}{D_c} \tag{1.15}$$

L represents the billet length inside the container, D_c denotes the container liner's inner diameter, and τ_i represents the uniform interface shear stress between the billet and the container liner.

Since the inner wall of the container is subjected to high frictional forces, it is prone to wear. To avoid wear, a container liner is used. The liner is a replaceable part. Lubrication plays a vital role in extrusion. It not only reduces the extrusion pressure but also defines the surface finish and reduction in defects. Lubricants are chosen according to the deformation temperature, which is often decided on the basis of workability of the material. In hot extrusion of steels or super alloys, billets are coated with glass powder before extrusion, and a glass sheet is applied as a pillow in front of the billet. Table 1.6 shows the common lubricants used in extrusion.

TABLE 1.6
Common Lubricants Used in Extrusion

Type of Extrusion	Lubricants	Coefficient of Friction
Hot extrusion	Glass, graphite	0.05–0.2
Cold extrusion	Castor oil	0.01–0.07

Source: Davim (2011).

1.9.2.4 Drawing

In the drawing process, the material's shape is altered by pulling it through a die, where it encounters compressive stress due to the narrowing caused by the die. The material is forced to flow in the direction of the applied tensile force as a result of this compression. Because of their relative movements, friction arises at the contact between the die and the workpiece when drawing solid goods like wire or bars. The force required for drawing is heavily influenced by the coefficient of friction.

Using the basic Coulomb friction, the drawing force (F) can be determined by the following formula:

$$P_d = \sigma A_f \ln\left(\frac{A_i}{A_f}\right)(1+\mu\cot\alpha) \tag{1.16}$$

where μ represents the frictional coefficient between the wall of the die and wire, A_f denotes the final region of cross-section of the wire, and A_i represents the initial region of cross-section of the wire.

In tube drawing processes, plugs or moving mandrels often support the internal surface of the tube, leading to friction at both the die–workpiece and mandrel (or plug)–workpiece interfaces. The tube drawing force P_d, therefore, is a function of the frictional interactions at both these interfaces.

$$P_{dt} = A_f C * \sigma \frac{1+\dfrac{\mu_1-\mu_2}{\tan\alpha-\tan\beta}}{\dfrac{\mu_1-\mu_2}{\tan\alpha-\tan\beta}}\left[1-\left(\frac{A_i}{A_f}\right)^{\left(\mu_1-\mu_2/\tan\alpha-\tan\beta\right)}\right] \tag{1.17}$$

where α represents the semi-die angle, β denotes the semi-cone angle of plug, μ_1 denotes the frictional coefficient between the wall of the die and tube, μ_2 represents the frictional coefficient between the plug and the tube, A_f denotes the final region of cross-section of the tube, andA_i denotes the initial region of cross-section of the tube.

Typically, drawing operations serve as the final step and are performed at room temperature to ensure high surface finish and precise dimensional accuracy in the final products. To achieve this, lubricants are employed, with a list of common lubricants used in different drawing operations provided in Table 1.7.

TABLE 1.7
Common Lubricants Used in Drawing (Davim, 2011)

Type of Process	Lubricants	Coefficient of Friction
Wire drawing	Mineral oil, fatty derivatives, soap, EP additives	0.03–0.1
Tube drawing	Soap, mineral oil	0.05–0.1

1.9.2.5 Forming of Sheet Metal

Sheet metal forming is a crucial technique in the metal working industry, encompassing various operations such as blanking, stamping, coining, bending, and deep drawing, with deep drawing being particularly intricate. The complexity of calculating frictional forces in sheet metal forming escalates as the depth of the drawn product increases. Figure 1.18 illustrates the schematic of the deep drawing process.

For instance, the force required for deep drawing (P_d) is expressed as

$$P_d = \left[\pi D_p t C\sigma \ln\left(\frac{D_b}{D_p}\right) + \mu\left(2P_h\frac{D_p}{D_b}\right)\right]\exp\left(\frac{\pi\mu}{2}\right) + P_b \tag{1.18}$$

P_h denotes the blank holding force, P_b represents the blank bending force, D_p denotes the diameter of the punch, D_b represents the blank diameter, C denotes the constant, σ represents the flow stress, and t denotes the wall thickness.

Various lubrication regimes, including boundary, mixed, thin film, and full film, may coexist at different points across the tool–workpiece interface or may alternate over time at a single point. Table 1.8 lists the typical lubricants used in deep drawing processes.

The quality of sheet metal forming is heavily influenced by the tribology and friction conditions present during the process. These conditions are shaped by the tribological system, which encompasses the sheet material, any coatings, the tooling material, lubrication, and process parameters. Increases in the friction coefficient have been linked to greater thinning and higher forming forces. Research shows

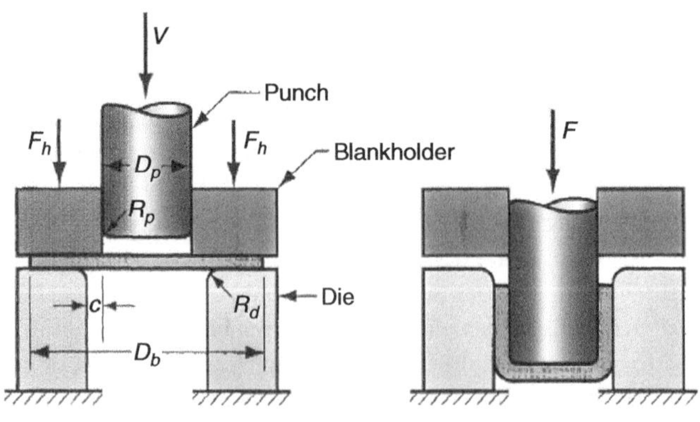

FIGURE 1.18 Deep drawing operation. (Basril et al., 2017.)

TABLE 1.8
Common Lubricants Used in Deep Drawing Process (Davim, 2011)

Type of Sheet Metal Forming Process	Lubricants	Coefficient of Friction
Shearing	Mineral oil, soap solution, EP additives	0.05–0.02
Bending	Soap solution, mineral oil, fat, and emulsion	0.06–0.02
Deep drawing	Wax, soap, tallow, MOS_2, grease	0.07–0.02

that higher friction coefficients lead to increased plastic deformation in both major and minor deformations of the sheet (Tekaut, 2019). The effectiveness of lubrication is assessed by the interplay of two main mechanisms: the roughening of workpiece asperities and the adhesion between contacting surfaces. None of the tested lubricants was able to considerably reduce the friction coefficient in scenarios when a high level of tool surface roughness results in extensive ploughing. The lubricants were able to lower the friction coefficient by roughly 3%–52%, depending on the dies' surface roughness (Trzepiecinski and Lemu, 2020).

1.9.3 Friction in Machining

Machining is a process used to alter the shape and surface texture of a three-dimensional object by removing excess material in the form of chips. Machining can be subdivided into two groups: cutting and finishing. Cutting is more about changing the shape by the removal of a large chunk of material, whereas finishing is focused on the texture of the surface. The rate removal of material from the body is small in magnitude in finishing compared to that in cutting.

1.9.3.1 Cutting Process

Friction plays a complex function in metal cutting, impacting both the quality and financial aspects of the process. It affects tool wear, which directly affects how cost-effective operations are, and it also affects the chip creation mechanism, which affects the product quality. The relative movement of the tool and the workpiece causes material to be sheared off, resulting in the formation of chips. The tool, the workpiece, or both may move in a translational or rotational manner to produce this motion. The types of relative motion seen during different cutting procedures are listed in Table 1.9. Friction is created at the tool–workpiece interface by the relative motion. The consolidation of cutting essentials is depicted in Figure 1.19.

In metal cutting, the interaction between the tool and the workpiece is complex, significantly influencing chip formation, frictional heat, and tool wear. The process begins when the wedge-shaped tool initially contacts the workpiece, leading to a brief period of sliding. This phase is quickly succeeded by a significant increase in the cutting load and the actual contact area due to the tool's or workpiece's feed advancement. The intense normal load during this relative motion causes subsurface plastic deformation of the material. Complete adhesion at the contact zone prompts surface plastic flow, while in some instances, the tool's ploughing action on the

TABLE 1.9
Nature of the Relative Motion Seen in Various Common Cutting Processes

Cutting Process	Motion of Workpiece	Motion of Tool
Turning	Rotary	Translatory
Milling	Translatory	Rotary
Drilling	Fixed	Rotary and translatory
Planning	Translatory	Translatory (intermittently)
Shaping	Translatory	Translatory (intermittently)
Boring	Translatory	Rotary
Hobbing	Rotary	Rotary and translatory
Broaching	Fixed	Translatory

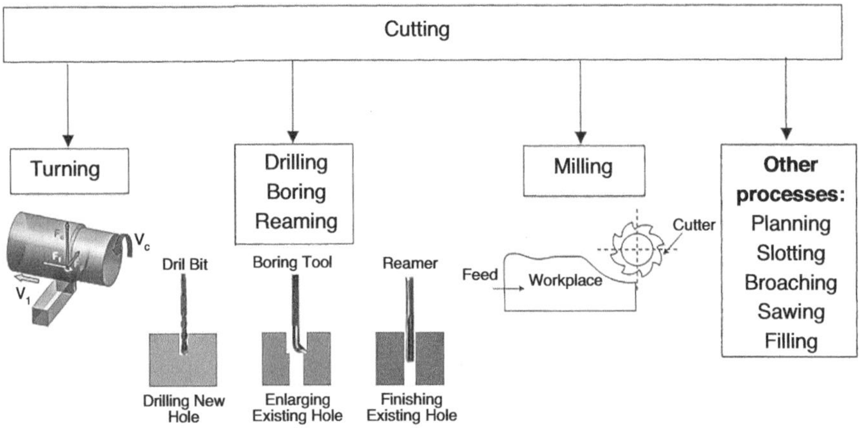

FIGURE 1.19 Cutting fundamentals. (Adapted from Toenshoff, 2014).

plastically deformed material also contributes to this flow. As a result, a layer breaks off from the workpiece surface and moves up against the tool's rake face to create what is called a chip (Figures 1.20 and 1.21).

Three main types of chips are produced during the machining process: (1) continuous, (2) discontinuous, and (3) continuous with a built-up edge on the tool face. The kind of chip that is created depends on a number of variables, including the material's ductility, depth of cut, feed rate, and cutting speed. When ductile materials are machined under mild circumstances, the chips produced are generally continuous, while brittle materials frequently produce discontinuous chips. Slow speeds and high feed rates when machining ductile materials can also lead to discontinuous chips.

In situations involving very ductile materials and low cutting speeds, part of the chip adhering to the rake face may shear off and stick to the tool, forming a built-up edge. This phenomenon significantly impacts the tool's frictional conditions and adversely affects the workpiece's surface finish. Table 1.10 presents the average

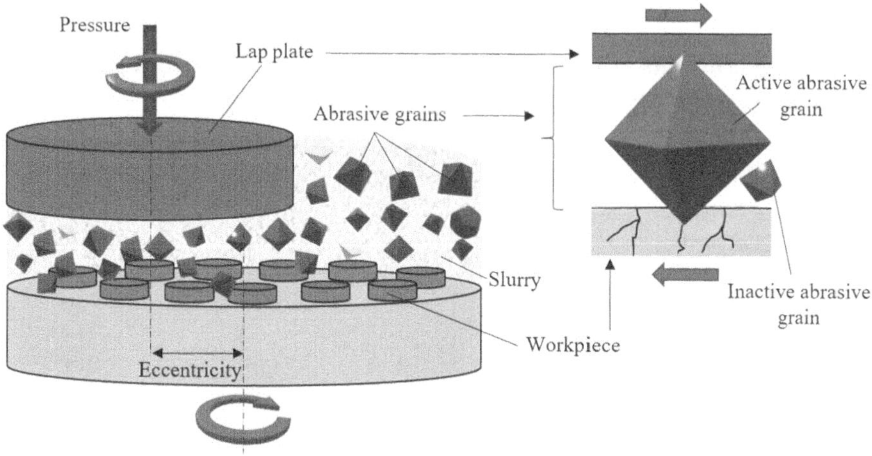

FIGURE 1.20 Diagrams for lapping. (Li et al., 2019.)

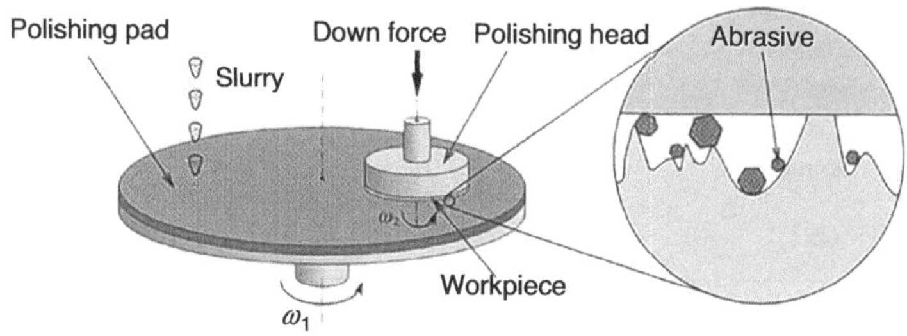

FIGURE 1.21 Schematic of polishing. (Geng et al., 2023.)

coefficients of friction observed on the tool face when cutting various metals, illustrating how the tool and workpiece materials, along with cutting parameters like speed, feed, and depth, affect the friction coefficient during metal cutting (Bailey, 1975).

1.9.3.2 Finishing Processes

Finishing processes can be categorized into two groups as per the use of abrasive (1) processes where bonded abrasive particles are used for the removal of material (e.g., grinding, honing, and super finishing) and (2) processes where free abrasive particles are used for the removal of material (e.g., lapping and polishing).

When the abrasive particles are constrained by the bond on the surface of tool, the contact between the workpiece and abrasive is considered two body types, and the motion is usually considered sliding. In cases where free abrasive particles are used, the contact is considered three body types. Here, the motion could be rotating or sliding as the abrasive grains are free to rotate or slide between the workpiece and the pad.

TABLE 1.10
Average Coefficients of Friction Observed on the Tool Face During the Cutting of Various Metals

Work Material	Tool Material	Depth of Cut (in.)	Tool Rake Angle (deg.)	Cutting Speed (ft/min)	Mean Coefficient of Friction
SAE 1020	Carbide	0.0981	0	384	0.805
SAE 1020	Carbide	0.0981	0	530	0.683
SAE 1020	Carbide	0.0981	0	1010	0.585
SAE 1020	Carbide	0.0981	20	384	0.109
SAE 4135	Carbide	0.0981	0	334	0.640
SAE 4135	Carbide	0.0981	0	540	0.565
SAE 4135	Carbide	0.0981	20	393	0.909
SAE 1112	HSS	0.0981	25	JO	0.577
SAE 1112	HSS	0.0981	40	10	0.624
SAE 1112	Carbide	0.0981	0	84	0.577
SAE 1112	Carbide	0.0981	20	384	0.600
SAE 1112	Carbide	0.0981	15	453	0.687
Al-6061-0	Carbide	0.0981	20	2080	0.500
Al-6061-0	Carbide	0.0981	20	1280	0.485
Al-6061-T6	Carbide	0.0981	20	2070	0.157
α-brass	Carbide	0.0981	20	1720	0.278
Al-2024-T4	Carbide	0.0981	20	2080	0.309

Source: Bailey (1975).

Regardless of the type of contact or movement involved, the main goal of surface finishing is to optimize wear on the workpiece while preserving its surface integrity. Consequently, it is essential to minimize the wear of the abrasive particles and manage frictional heat effectively. In finishing operations, temperatures can briefly spike. As a result, the hardness or strength of the abrasives used should surpass that of the workpiece across a broad temperature range. Table 1.11 lists the strength and

TABLE 1.11
Strength of the Abrasive Materials

Abrasive Type	Hardness (GPa)	Moh's Scale
Aluminium oxide	~21	9.0
Silicon carbide	~24	9.5
Cubic boron nitride	70	9.9
Diamond	100	10

Source: Tabor (1954).

hardness of commonly used abrasive materials. It is important to understand that the effectiveness of an abrasive is not solely dependent on its hardness.

In finishing operations, the interactions between the abrasive particle and the workpiece can be categorized as rubbing, ploughing, and cutting. During rubbing, there is minimal penetration into the workpiece, resulting in a relatively low force exertion. During rubbing, asperities deform and hence surface roughness reduces. Practically, a very small amount of workpiece is removed in rubbing. Rubbing interaction is predominantly seen in methods employed for obtaining mirror finishing. In ploughing, the grains penetrate into the workpiece. However, the penetration level is moderate. Due to penetration, the force experienced is higher in magnitude than that experienced in the rubbing interaction. As a result, material gets displaced from the path of the abrasive grain and scratches are formed. The material removal, however, remains significantly less.

In cutting, abrasive grains penetrate deeper into the workpiece. In comparison to ploughing, in cutting, a higher force is experienced due to deeper penetration. When the force applied reaches a critical threshold, shearing of the material occurs, removing it from the surface in the form of chips. In some finishing operations, all these interactions—rubbing, ploughing, and cutting—occur simultaneously, though to varying degrees.

Fluids play a very important role in surface finishing operations. In processes where bonded abrasive particles are used, the major role of the fluid is to remove the chip and loose abrasive particles from the interface. The secondary role is heat removal, if any heat is generated. In processes where loose abrasive particles are used, the fluid serves as a carrier. The fluid transports the abrasive particle to the interface and helps in the uniform distribution of the abrasive particles across the interface. The processes that employ loose abrasive particles are generally cold processes and operate at a low speed.

1.9.4 Friction in Joining

In joining, heat is used to fuse two products together and form a permanent bond. The heat is supplied by the application of electricity or gas. In the 1980s, a new welding process was developed, where the materials were joined by the use of frictional heat.

1.9.4.1 Friction Welding

In this type of welding process, two material parts are rubbed against each other at high speed and pressure. The rubbing between the two surfaces results in frictional heating. The local heat plasticizes the materials, and the motion between the two promotes the mixing of viscous materials at their contact points, creating a metallurgical bond. Minimal surface preparation is required for friction welding, and no filler metals, flux, or shielding gases are needed. The procedure takes less time to finish and maintains uniform mechanical qualities because it takes place below the melting temperature of the materials involved. Friction welding, first studied in the 1980s by The Welding Institute, the University of Bristol, and Rolls-Royce plc, is currently used to fuse a range of materials together, including plastics, metals, and composites. The aerospace industry finds it particularly useful for building titanium alloy blisks

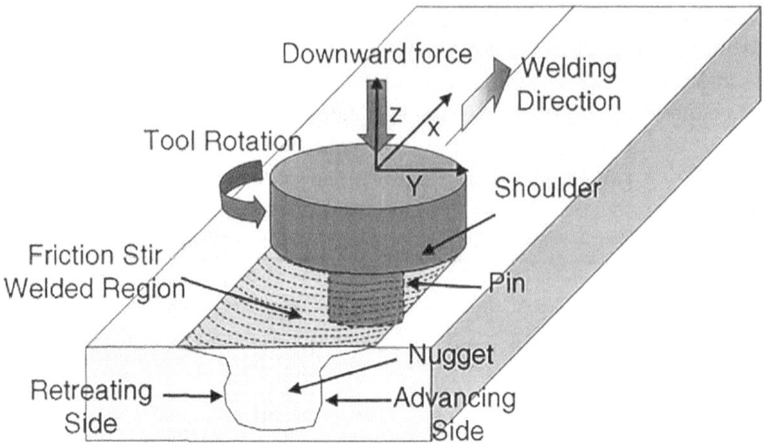

FIGURE 1.22 Diagram of the friction stir welding procedure. (Li et al., 2019.)

in engines because it provides notable weight savings over conventional blade–disc assemblies that employ a "fir-tree" design.

Friction welding encompasses several techniques, each with its unique approach:

- **Rotary friction welding**: This method generates heat through the rotation of one piece against another, combined with a compressive axial force to forge a bond.
- **Linear friction welding**: Heat is produced by rapidly oscillating one component back and forth against another, under significant compressive force.
- **Friction stir welding**: This technique employs a non-consumable rotating tool that traverses along the joint line of the materials to be welded, generating the necessary heat through friction. Notably, the tool does not become part of the final weld. An illustration of this process is depicted in Figure 1.22.
- **Friction stir spot welding**: Similar to friction stir welding, this variant involves rotating a tool at a specific spot on the materials intended for joining, without moving the tool along a seam.

1.10 TOOL WEAR

Tool wear, which .eventually results in tool failure, is the slow degeneration of machine tools brought on by cutting operations. Tools become stressed over time because they are constantly rubbing and undergoing intense friction with workpiece. Metal-to-metal contact, together with extreme pressure and stress, is what causes this stress. Furthermore, it experiences extremely high temperatures. Tool monitoring can be used to investigate and address tool failure due to tool wear, which is a common occurrence. The wear typically develops over time and is a slow breakdown in a cumulative process that shortens the life of the tool. Tool shape, depth, cutting fluid,

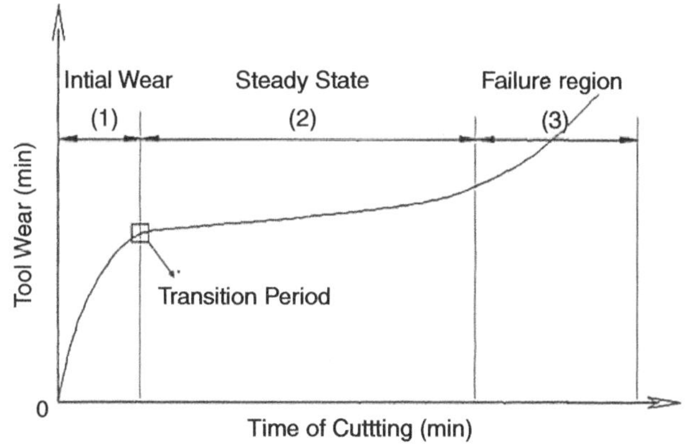

FIGURE 1.23 Tool wear in relationship to cutting time. (Memarianpour et al., 2021.)

and cutting speed all affect how well a tool wears. As a result, several tools lose some of their effectiveness and sharpness (Figure 1.23).

1.10.1 Tool Wear Classification

Tool wear is affected by a number of variables, such as the kind of equipment being used, the hardness of the material being machined, the quantity of operations carried out on the part, and the force utilized during each operation (Lauzier, 2021). The following are typical tool wear patterns:

i. **Abrasive wear**: The purity of the metal feedstock used in workpiece for machining and other industrial processes varies based on the intended purpose. Contaminants have the potential to produce edge fragment accumulation. Over time, these fragments could erode the tool's sharpness and effectiveness. Abrasion is a mechanical type of wear that typically occurs at low spindle speeds.
ii. **Flank wear**: Flank wear, which occurs parallel to the cutting edge, may be the source of the cutting edge failure. As the workpiece undergoes adhesive and abrasive wear, high temperatures shape and affect its performance characteristics. If the flank wear worsens, the cut must be made faster. Figure 1.24 displays the flank wear diagram.
iii. **Crater wear**: Tools damage the rake face by chipping away as they quickly impact the work surface. This flow of face chips leaves the skin with a crater that resembles a scar. Unless it directly impacts the cutting surface, crater wear is a common type of tool degradation that usually has no effect on the tool's performance. This type of wear generally occurs near the cutting edge. Figure 1.25 illustrates the schematic of crater wear.
iv. **Adhesive wear**: Adhesion wear can happen when friction between tools and workpiece raises temperatures. In this situation, chips that are flowing over the tool's face could form a spot-welding-like bond with the tool face. The workpiece's actual dimensional precision may also be affected by this.

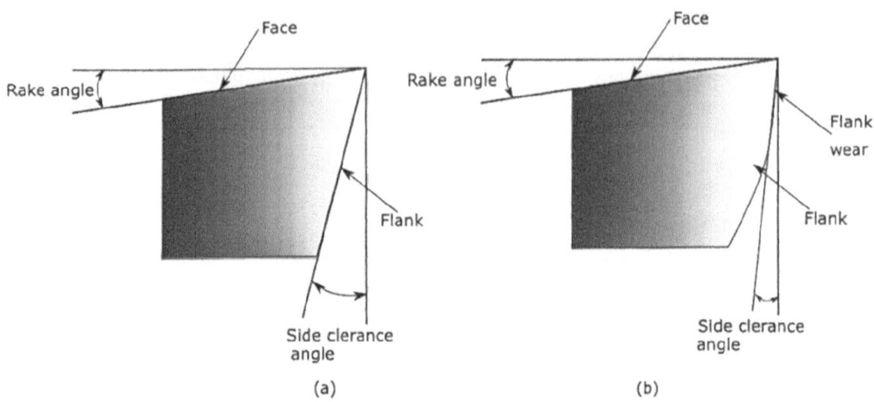

FIGURE 1.24 Schematic of flank wear. (Mouldovan et al., 2017.)

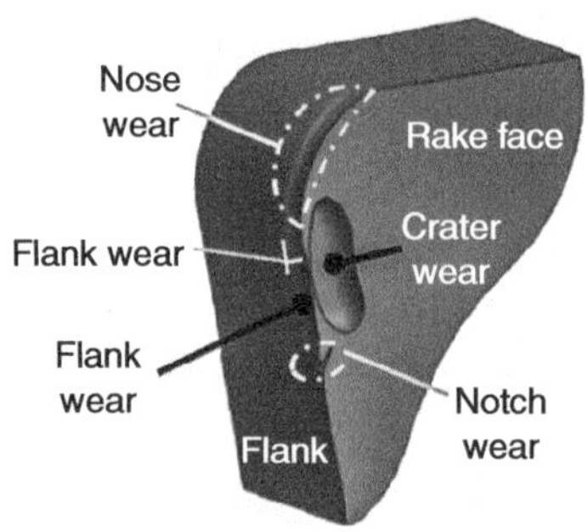

FIGURE 1.25 Schematic of crater wear. (Prado et al., 2020.)

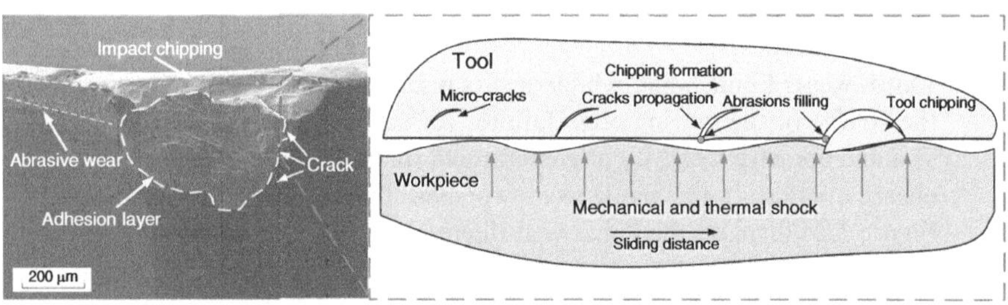

FIGURE 1.26 SEM micrograph of adhesion wear. (Liu et al., 2022.)

Adhesion wear could also happen more frequently if the incorrect fluid or quantity of fluid is insufficient. Figure 1.26 shows the SEM micrograph of adhesion wear.

1.10.2 Tool Wear Effects

The wear on a tool can produce particular wear effects when it reaches the end of its lifecycle or when it has been touched by an event. These outcomes affect the calibre of the components made, the effectiveness of the machinery, or the operator-related downtime. The significance of comprehending tool wear effects is that they can be used to create criteria and algorithms for future tool wear detection and prevention.

i. **Increased cutting force:** Tool wear effect requires additional or increased cutting force. As increased cutting force may be needed to make up for increasing tool wear, both the tool and the workpiece will experience an increase in acceptable tolerance. However, if the wear is substantial, the force can be too great.

   High temperatures are a constant result of metal-to-metal friction during cutting, drilling, and other manufacturing operations. If the cutting fluid used is inappropriate or insufficient, temperatures can rise during the cutting process. Additionally, an increase in feed rate or depth of cut can lead to higher temperatures. Similarly, higher stroke or cutting speeds can further elevate the heat generated.
ii. **Decreased accuracy of produced parts:** The precision of the part declines with wear. Any workpiece will have allowable tolerances. However, a certain kind of tool wear can worsen to the point that the part no longer meets the specifications for that cutting operation, which results in part loss. Multiple wear patterns may coexist, cascading the deformation and affecting the part's precision. As a result, pieces will either need to be scrapped or rebuilt, which can lead to serious quality issues. A key component of efficiently maintaining good quality is managing tool wear and anticipating when failure will occur.
iii. **Decreased tool life:** Reduced tool life can be brought on by many types of wear, the hardness difference between the tooling and the workpiece, and often high and low temperatures. Based on knowledge of the workpiece and tool material, training, and experience of the operator, these factors can be controlled. But smart tool monitoring software from Machine Metrics can also keep an eye on them (Lauzier, 2021). Reduced tool life is a problem that can be managed if the proper automated monitoring is in place.
iv. **Inadequate surface finish:** The impact on the surface finish increases as tool wear becomes more significant. Uneven or jagged cut faces on the workpiece may result from dull tools. Additionally, drilling and cutting could result in surface build-up or an increase in wear land, which would impair the surface finish by causing chipping and cratering. This can lead to serious quality problems, especially in high precision machining, which necessitates the need for rework or scrapping of the affected items.
v. **Cost-effectiveness of cutting operations:** Machining and cutting are costly production techniques. Tool wear can have an effect on the high cost of the machinery, operator training, high-quality tooling, and suitable material selection of the workpiece feedstock. Each time a tool's lifetime is

shortened and a piece of work is trashed, the cost of the run increases and the profit margin decreases. Additionally, outdated tools put the operator and machine in danger and could lead to more expensive maintenance work.

The following six steps have to be taken care for reducing the tool wear: (1) recognize warning signs, (2) recognize types of tool wear, (3) be attentive to temperature effects, (4) use proper coolant, (5) consider the material of the workpiece, and (6) match the tool to the task.

1.11 CONCLUDING OBSERVATIONS AND EMERGING TRENDS IN INDUSTRY

Wear, lubrication, friction, and contact are examples of phenomena that are essential to industrial technology. The chapter covered the fundamentals of tribology and how it affects the main manufacturing processes; yet it also acknowledged how complicated the analysis of friction is in these processes. Numerous factors influence the lubrication and friction conditions in manufacturing. These variables are influenced by the machine, workpiece, tool, lubricant, and process. Accurate estimation friction and wear can help in designing the machines, products, and processes in an improved manner. Understanding of lubrication and lubricants can improve the efficiency of frictional heat management. These goals are achievable with a holistic approach to integrate the knowledge in materials, tribology, heat transfer, and manufacturing.

REFERENCES

Abdelrazek, A., Choudhury, I., Yusoff, N., and Kazi, S.N. 2020. Metal cutting lubricants and cutting tools: A review on the performance improvement and sustainability assessment. *The International Journal of Advanced Manufacturing Technology*, 106. https://doi.org/10.1007/s00170-019-04890-w.

Amontons, G. 1699, "De la Resistance Causéedans les Machines, tant par les Frottemens des Parties Qui les Composent, Quepar la Roideur des CordesQu'On Y Employe, et la Manière de Calculerl'Un et l'Autre." *Mémoires de l'Academie Royale des Sciences*: 206–27.

ASM International Handbook Committee 1992, *ASM Handbook: Friction, Lubrication, and Wear Technology*. Ohio: ASM International.

ASTM, G-99. 2017, Standard Test Method for Wear Testing with a Pin-On-Disk Apparatus. https://doi.org/10.1520/G0099-17.

Bailey, J.A. 1975. Friction in metal machining—Mechanical aspects. *Wear* 31, no. 2: 243–75.

Basril, M.A.M., Teng, H.M., Azuddin, M., and Choudhury, I.A. 2017. The effect of heating temperature and methods towards the formability of deep drawn square metal cup *IOP Conf. Ser.: Mater. Sci. Eng.* 210: 012067, https://doi.org/10.1088/1757-899X/210/1/012067.

Bharat, B. 2020. *Handbook of Micro/Nano Tribology*. Boca Raton, FL: CRC Press.

Bose, A., Parthasarathi, N.L, Arvinth Davinci, M., Borah U., and Jeevanantham, A.K. 2018. Determination of frictional coefficient of 316L (N) stainless steel by ring compression test using simulation. *Journal of Manufacturing Engineering*, 13, no. 2: 113–7.

Chen, Z., Khajeh, A., Martini, A., and Kim, S.H. 2019. Chemical and physical origins of friction on surfaces with atomic steps. *Science Advances*, 5, no. 8. https://doi.org/10.1126/sciadv.aaw0513.

Ciulli, E. (2019). Tribology and industry: From the origins to 4.0. *Frontiers in Mechanical Engineering,* 5. https://doi.org/10.3389/fmech.2019.00055.

Davim, J.P. 2011. *Tribology for Engineers: A Practical Guide*. Sawston Cambridge: Woodhead Publishing.

da Vinci, L. 1493. *Codice Madrid I*. Madrid: BibliotecaNacional de España.

Dudkiewicz, Ł., Hawryluk, M., Ziemba, J., Miżejewski, A., Polak, S., Marzec, J., and Szymańska, T. 2023. Determination of the friction coefficient in the ring test for selected lubricants dedicated to the hot forging process of precision steel products. *Lubricants*, 11: 399. https://doi.org/10.3390/lubricants11090399.

Dowson, D., Taylor, C.M., Godet, M., and Berthe, D. 1979. Thermal effects in tribology. In *Proceedings of the 6th Sympos*, Leeds-Lyon, and France Lyon. September 18–21.

Geng, Z., Huang, N., Castelli, M., and Fang, F. 2023. Polishing approaches at atomic and close-to-atomic scale. *Micromachines*, 14: 343. https://doi.org/10.3390/mi14020343.

Jost, H.P. 1966. *Lubrication (Tribology) Education and Research. A Report on the Present Position and Industry's Needs*. London: Her Majesty's Stationary Office, Department of Education and Science.

Kelemenová, T., Dovica, M., Božck, P., Koláriková, I., Benedik, O., Virgala, I., Prada, E., Miková, Ľ., Kot, T., and Kelemen, M. 2020. Specific problems in measurement of coefficient of friction using variable incidence tribometer. *Symmetry*, 12: 1235. https://doi.org/10.3390/sym12081235.

Krbata, M., Eckert, M., Bartosova, L., Barenyi, I., Majerik, J., Mikuš, P., and Rendkova, P. 2020. Dry sliding friction of tool steels and their comparison of wear in contact with ZrO_2 and X46Cr13. *Materials*, 13: 2359. https://doi.org/10.3390/ma13102359.

Kress, B.H., Hill, L., and Alexander, D.E. 1971. Lubricants for continuous metal casting operations. United States Patents US 3,620,290, Conshohocken, PA: Quaker Chemical Corporation.

Lauzier, J. 2021. MachineMetrics, Tool Monitoring. *MachineMetrics Blog*, October 14, 2021. https://www.machinemetrics.com/blog/tool-wear.

Ledbetter, H.M. 1982. Temperature behavior of Young's moduli of forty engineering alloys. *Cryogenics*, 22: 653–6.

Li, K., Liu, X., and Zhao, Y. 2019. Research status and prospect of friction stir processing technology. *Coatings*, 9: 129. https://doi.org/10.3390/coatings9020129.

Li, Q., Dimaki, A., Popov, M., Psakhie, S.G., and Popov, V.L. 2014. Kinetics of the coefficient of friction of elastomers. *Scientific Reports*, 4, no. 1: 5795. https://doi.org/10.1038/srep05795.

Liu, D., Liu, Z., and Wang, B. 2022. Effect of cutting parameters on tool chipping mechanism and tool wear multi-patterns in face milling Inconel 718. *Lubricants*, 10: 218. https://doi.org/10.3390/lubricants10090218.

Memarianpour, M., Niknam, S.A., Turenne, S., and Balazinski, M. 2021. Study of the effects of initial cutting conditions and transition period on ultimate tool life when machining inconel 718. *Materials*, 14: 592. https://doi.org/10.3390/ma14030592.

Michael, D. 2010. Unification of friction and wear. In: Nikas, G.K. (ed), *Recent Developments in Wear Prevention, Friction and Lubrication*: 159–96. Trivandrum: Research Signpost.

Mouldovan, O.G., Dzitac, S., Moga, I., Vesselenyi, T., and Dzitac, I. 2017. Tool-wear analysis using image processing of the tool flank. *Symmetry*, 9: 296. https://doi.org/10.3390/sym9120296.

Nakato, H., Ōmiya, S., Habu, Y., et al. 1984. Optimizing mould lubrication for high-speed continuous casting of slabs. *JOM*, 36: 44–50. https://doi.org/10.1007/BF03338407.

Parthasarathi, N.L., Borah, U., and Albert Shaju, K. 2013a. Effect of temperature on sliding wear of AISI 316 L(N) stainless steel – Analysis of measured wear and surface roughness of wear tracks. *Materials and Design*, 51: 676–82.

Parthasarathi, N.L., Borah, U., and Albert Shaju, K. 2013b. Correlation between coefficient of friction and surface roughness in dry sliding wear of AISI 316L (N) stainless steel at elevated temperatures. *Computer Modelling and New Technologies*, 17, no. 1: 51–63.

Parthasarathi, N.L., Muthukannan, D., and Utpal, B. 2012. Effect of plasma spraying parameter on wear resistance of NiCrBSiCFe plasma coatings on austenitic stainless steel at elevated temperatures at various loads. *Materials and Design*, 36: 141–51.

Prado, T., Pereira, A., Fenollera, M., and Mathia, T.G. 2020. Simple discriminatory methodology for wear analysis of cutting tools: Impact on work piece surface morphology in case of differently milled kinetics steel H13. *Materials*, 13: 215. https://doi.org/10.3390/ma13010215.

Rigney, D.A. 1981. Fundamentals of friction and wear of materials. In *Papers Presented at the 1980 ASM Materials Science Seminar*, October 4–5, 1980. Pittsburgh: Pennsylvania.

Röder, J., Bishop, A.R., Holian, B.L., Hammerberg, J.E., and Mikulla, R.P. 2000. Dry friction: Modeling and energy flow. *Physica D: Nonlinear Phenomena*, 142, no. 3–4: 306–16.

Schmid, S.R., Cantó, M.Á.S., and Pasang, T. 2013. Tribology of forging. In: Wang, Q.J., Chung, Y.W. (eds), *Encyclopedia of Tribology*. Boston, MA: Springer. https://doi.org/10.1007/978-0-387-92897-5_618.

Svečko, R., Kusić, D., Kek, T., Sarjaš, A., Hančič, A., and Grum, J. 2013. Acoustic emission detection of macro-cracks on engraving tool steel inserts during the injection moulding cycle using PZT sensors. *Sensors*, 13: 6365–79. https://doi.org/10.3390/s130506365.

Szekeres, E. S. 1996. Overview of mould oscillation in continuous casting. *Iron and Steel Engineer,* 73, no. 7: 29–37.

Tekaut, İ. 2019. The effects of friction coefficient on formability behaviour of dual phase steel. *Materials Research*, 22, no. 5: e20180759. https://doi.org/10.1590/1980-5373-MR-2018-0759.

Tabor, D. 1954. Mohs's hardness scale – A physical interpretation. *Proceedings of the Physical Society. Section B*, 67, no. 3: 249–57.

Toenshoff, H.K. 2014. Cutting, fundamentals. In: Laperrière, L., Reinhart, G. (eds), *CIRP Encyclopedia of Production Engineering*. Berlin, Heidelberg: Springer. https://doi.org/10.1007/978-3-642-20617-7_6633.

Trzepiecinski, T., and Lemu, H.G. 2020. Effect of lubrication on friction in bending under tension test-experimental and numerical approach. *Metals,* 10, no. 4: 544. https://doi.org/10.3390/met10040544.

Voyer, J., Ausserer, F., Klien, S., Velkavrh, I., and Diem, A. 2017. Reduction of the adhesive friction of elastomers through laser texturing of injection moulds. *Lubricants*, 5, no. 4: 45. https://doi.org/10.3390/lubricants5040045.

Wang, X.D., Yao, M., Du, B., Fang, D.C., Zhang, L., and Chen, Y.X. 2007. Online measurement and application of mould friction in continuous slab casting. *Iron Making and Steelmaking*, 34, no. 2: 138–44. https://doi.org/10.1179/174328107X155295.

Wilson, W.R.D., Hsu, T.-C., and Huang, X.-B. 1995. A realistic friction model for computer simulation of sheet metal forming processes. *Journal of Engineering for Industry*, 117, no. 2: 202–9. https://doi.org/10.1115/1.2803295.

Yanan, M., Sun, J., Wu, P., Dong, C., and Yan, X. 2018. The role of nano-TiO_2 lubricating fluid on the hot rolled surface and metallographic structure of SS41 Steel. *Nanomaterials*, 8(2): 111. https://doi.org/10.3390/nano8020111.

2 Tribological Aspects in Nano/Micro-Manufacturing

N. Anand Krishnan, K. Vipindas, Deepak G. Dilip, and Subrahmanyam Adabala

2.1 INTRODUCTION

Nowadays, micro/nano features and parts are in high demand in various engineering applications such as biomedical, aerospace, optics, electronics, automobile, and naval fields. Even though several micro/nano-manufacturing techniques exist, the technologies for precision micro/nano features with excellent surface quality and high productivity are very limited. Considering these aspects, this book chapter mainly focused on the tribological aspects of conventional nano/micro-cutting operations such as micromilling, microturning, and microdrilling. The conventional micro/nano-machining processes are considered promising micro/nano-manufacturing operations to make intricate 3D structures with high precision and manufacturing rate (Masuzawa, 2000; Leo Kumar et al., 2014). Conventional micromachining is considered a tiny version of macro machining with features like size effect and minimum uncut chip thickness (MUCT) (Ng et al., 2006; Aramcharoen and Mativenga, 2009; Anand et al., 2021). Size effect in micro/nano machining is observed when uncut chip thickness (UCT) is below or near the tool cutting edge radius (r_e) (Anand et al., 2021). During micro/nano machining, r_e has a high impact on the material removal mechanism. For UCT above r_e, shearing becomes the dominant cutting mechanism and plowing turns out to be the dominant mechanism when UCT is less than MUCT (Vipindas et al., 2018; Venkatesh et al., 2017; De Oliveira et al., 2015; Anand and Mathew, 2020b). Therefore, the size effect has a major role in tribological behavior during the micromachining process. MUCT in micromilling varies between 0.33 and 0.25 times r_e, regardless of machining parameters, workpiece, and tool material (De Oliveira et al., 2015). Consequently, micromachining experiences difficulties such as plowing effect, edge breakage, high tool deflection, and tool wear. Hence, it is important to improve the micro/nano machinability.

The increasing concern over health and the ecological impact of manufacturing activities forces manufacturers to reduce lubricant use. While machining Ti and Ni alloys, the consumption of coolant is very high to reduce the machining temperature and to maintain the desired product quality. The cost part associated with the usage of coolant comprises the acquisition, disposal, and washing of the machined component. All these costs will be around four times the tooling cost used in the machining

DOI: 10.1201/9781003359333-2

process. Hence, all machining industries are moving toward dry machining or reducing the usage of cutting fluid to reduce the cost of the machining as well as to make the process environmentally friendly. To achieve this, it is essential to understand the function of the cutting fluid in machining operations and to develop new materials for tool and coating.

Dry machining has the following advantages over the flood machining process:

i. No pollution of environment;
ii. No residue of the coolant on machined parts. This eliminates the postcleaning process and allied cost/energy associated with this;
iii. No remainder of the lubricant on the chips. This makes chip disposal much easier.

Before discussing dry cutting, it is essential to understand the basic roles of cutting fluid and its problems. In general, the use of cutting fluid improves the life of the tool by virtue of a decrease in cutting force and cutting temperature. The decrease in machining force is due to the improved lubrication provided by the cutting fluid and the cooling effect of cutting is responsible for reducing the machining temperature. In addition to this, the cutting fluid helps in removing the chips from the machining zone. The heat generated in the machining process generally dissipated in the chip, workpiece, and a small amount in the cutting tool. However, the highest amount of cutting temperature is generated at the cutting region. This causes the formation of built-up edge, diffusion wear, crater wear, and degradation of cutting edge. Therefore, it is very important to study the wear mechanism.

This book chapter aims to investigate the tribological aspects of the tool-based micro/nano-machining and forming processes. The tribological aspects such as friction and wear mechanism during different tool-based micro/nano-machining and forming processes were discussed in detail. This chapter is organized into five sections. Section 2.1 gives an overview of nano/micro-manufacturing by focusing on tool-based micro/nano machining and forming. The commonly used tribological parameters were also discussed in this section. Section 2.2 provides the tribological aspects of tool-based dry nano/micro-machining process. Analysis of friction and wear during tool-based dry micro/nano-machining operations was discussed in this section. The importance of the frictional size effect and wear behavior, such as adhesion, abrasion, diffusion, crater wear, flank wear, and edge rounding, was highlighted in this section. The influence of the adhesion mechanism, the changeover from plowing to shearing mechanism in micro/nano cutting, and the significance of critical edge radius on wear and frictional behavior were also explored in this chapter.

Section 2.3 discusses the frictional and wear characteristics of lubrication-assisted micro/nano machining. The tribological aspects of minimum quantity lubrication (MQL) and flood lubrication environments were discussed and compared with dry micro/nano machining. The influence of micro/nano texturing on wear mechanisms was discussed in section 2.4. In this section, different micro/nano textures created by the manufacturing processes and their tribological behavior were discussed. Finally, tribological analysis of micro/nano forming was presented in section five. In this section, the role of friction on material behavior in micro/nano forming is presented, and its comparison with conventional forming is highlighted. In addition, the role of friction during micro/nano forming on the product quality is also discussed in Section 2.5.

2.2 TRIBOLOGICAL ASPECTS IN DRY NANO/ MICRO-MACHINING PROCESSES

This section discusses the wear and friction behavior during dry micro/nano machining.

2.2.1 Dry Machining

By introducing dry machining, a cost saving of up to 17% can be achieved (Dudzinski et al., 2004) through eliminating the coolant supply, its handling cost, and component cleaning (Klocke, 1997; Lahres et al., 1999). The use of MQL is one of the techniques to reduce the usage of cutting fluid (Lahres et al., 1999; Klocke and Eisenblatter, 1997); apart from this, many hard coatings for the cutting tools were developed. The following section will discuss the idea of MQL and hard coating of the tool that will help in achieving dry machining.

2.2.2 Minimum Quantity Lubrication (MQL)

Under a minimum quantity supply of coolant, a small amount of cutting fluid will be applied at the machining zone. Even though the minimum quantity application of the cutting fluid can substitute most of the characteristics of the conventional flood cooling system, some of the features cannot be met with the minimum quantity application. For example, effective flushing of the chip is not possible and the cooling effect is partial (Dudzinski et al., 2004). Popke et al. (1999) claimed that for drilling operations MQL provides better performance compared to the flood-type lubrication condition. Machado and Wallbank (1997) reported that under MQL condition cutting force and machining temperature were significantly reduced and an improvement in tool life and surface finish was observed. Varadarajan et al. (2002) developed a system for supplying coolant under minimum quantity application. Their investigation revealed that 60% coolant-rich lubricant is the ideal solution for minimum quantity application. The optimum parameter for machining AISI 4340 is 2 mL/min of coolant supply rate, 20 MPa pressure, and a pulsing rate of 600 pulses/min. Compared to flood and dry machining conditions, under MQL condition resulted in lower cutting force, minimum machining temperature, improved surface finish, and enhanced tool life.

Klocke and Eisenblatter (1997) investigated the effectiveness of minimum quantity application during drilling operations. It was observed that a good surface finish and a longer tool life were achieved. Lahres et al. (1999) compared the machining performance under dry, flood, and MQL while machining austenitic 22Mn6 steel. In their study, it was found that under MQL condition the chips produced during the machining process are almost dry in nature. This facilitates direct recycling of the chips without any further processing. Also, it was noted that compared to dry machining, minimum quantity application gave better machining performance.

2.2.3 Tribological Aspects During Dry Nano/Micro-Machining Process

When we eliminate the cutting fluid in any machining process, this implies that we are removing the positive effects of cutting fluid in the machining process. The main difficulties to be overcome in dry machining are the removal of heat from the machining zone and the effective removal of chips from the machining zone. In addition to this, the quality of the machined component and stable wear characteristics of the tool also need to be guaranteed. To meet these requirements, it is required to have tools with high hot hardness, high refractivity, and low coefficients of friction (Dudzinski et al., 2004). To an extent, appropriate tool coating can partially fulfill some of the functions of the cutting fluid so that dry machining can be achieved. Hence, tool coatings play a major role in moving toward dry machining (Schulz et al., 2001).

The major functions of coating in dry machining are to reduce the friction at tool–work and tool–chip interfaces and to protect against heat and diffusion to ensure high wear resistance at elevated temperatures. (Ti, Al)N has a low coefficient of thermal conduction and better oxidation stability compared to other hard coatings, such as TiN coating. Oxidation resistance of (Ti, Al)N is approximately 800°C, whereas for TiN it is about 600°C. The diffusion and oxidation resistance of the (Ti, Al)N film can be improved with an Al_2O_3 top layer. Compared to uncoated and TiN-coated carbide tools, the TiAlN-coated carbide tool has better wear behavior. It was reported that an Al/Ti ratio of one gives the best wear behavior while performing dry drilling of tempered steel (Dudzinski et al., 2004).

Tonshoff et al. (1998) suggested that with the addition of oxygen to (Ti, Al)N coating, TiAlON can be formed, which offers high abrasion resistance due to the formation of Al_2O_3 under dry machining conditions. Alumina is thermally stable and provides good resistance to oxidation. In addition, a graded multilayer structure of TiAlON with Al_2O_3 was developed by providing a stable oxide layer on top of the nitride coating (Bouzakis et al., 1999). Schulz et al. (2001) reported that oxide Al_2O and ZrO_2 physical vapour deposition (PVD) coatings reduce the friction at elevated temperature and thereby provide better wear resistance compared to TiN, TiCN, and TiAlN coatings. This will reduce the chances of workpiece material adhesion on the tool rake face and chip flow will also be improved. They have investigated the performance of different oxide PVD coatings such as TiAlN–Al_2O_3, TiAlN–ZrO_2, and TiZrN–ZrO_2, and the results were compared with uncoated and TiAlN-coated tools. Oxide-coated tools are recommended for machining under dry condition compared to uncoated and TiAlN-coated tools, as they possess a low coefficient of friction and hardness even at high-temperature conditions. Apart from this, oxide-coated tools reduce the machining temperature significantly. Within oxide coating, TiAlN–ZrO_2 gives better performance.

Even though TiAlN coating may increase the tool life by resisting certain types of wear, this coating possesses a high coefficient of friction. Thus, cutting fluids are required to get the best performance from these tools. Application of solid lubricant/titanium composite coating can address this issue and promote dry machining. This will significantly reduce the friction coefficient, cutting force, and machining temperature and reduce the chances of formation of built-up edge. MoS_2/titanium composite coating offers a very low friction coefficient, which makes them suitable for high-speed machining. Other low friction coatings suitable for dry machining are WC/C coatings (Renevier et al., 2000; Derflinger et al., 1999; Navinsek et al., 2002; Renevier et al., 2003). Arndt and Kacsich (2003) developed an AlTiN coating suitable for dry, MQL, and high-speed machining applications.

Hence, the use of cutting fluids in machining is nowadays highly discouraged due to its negative impacts on environment and health of the operators. Also, its use raises the cost of the final product. To cut the usage of cutting fluid in machining, new methodologies have been introduced to lessen its use and by developing new coatings that can promote dry machining.

2.3 TRIBOLOGICAL ASPECTS IN LUBRICATION-ASSISTED NANO/MICRO-MACHINING PROCESS

Even though dry cutting is considered an appropriate method to reduce the harmful ecological impact and workers' health issues associated with the cutting fluid, it creates numerous concerns such as severe tool wear, low surface quality, chip removal difficulty, and high machining temperature (Boswell et al., 2017). Hence, lubrication-assisted micro/nano machining is the best alternative to overcome these problems associated with dry machining. Researchers found that the cutting fluid usage leads to a significant drop in tool wear, burr formation, and surface roughness. This is caused by the improved tribological performance of the cutting fluid, which results in a decrease in friction. Due to the high cost and associated health and environmental concerns of cutting fluids, flood lubrication is not recommended as a lubrication method for cutting. Hence, the research related to MQL has become popular. Under MQL conditions, only a minimal amount of lubricant is delivered to the machining zone using high-pressure air. Hence, MQL reduces the undesirable impacts of flood and dry cutting. As a result, MQL-assisted micro/nano machining has become popular. It was found that the MQL-assisted micro/nano machining significantly reduces the tool wear and surface roughness (Anand and Mathew, 2021; Hassanpour et al., 2016; Anand and Mathew, 2020a). Anand and Mathew (2021) observed that the MQL during micromilling significantly enhanced the tool wear and surface quality. Chipping, delamination, and edge breakage were observed to be more significant in dry cutting compared to MQL conditions. This is due to the higher friction and machining temperature. The occurrence of W (tungsten) in the EDS mapping is a hint of chipping or delamination. It is also common that the chip fragments were extremely gathered in dry conditions than in MQL cutting. As a result, the surface quality becomes poor in the case of dry microcutting. Adhesion and abrasion were found to be vital while micromilling on Inconel 718. The application of MQL significantly reduces these adhesions, chipping, and abrasion. Consequently, the reduction in tool diameter becomes slow and this reduces the dimensional errors in micro/nano machining. 25% average reduction in edge radius and 30% reduction in flank wear were observed for MQL conditions compared to dry due to improved lubricant access into the machining region. The minimum flank wear and edge radius were observed at feed/tooth to cutting edge radius ratio (f_t/r_e) near 1 due to the influence of size effect for all conditions (Anand and Mathew, 2021).

After the size effect region, flank wear and edge radius increase with feed/tooth for both situations. Above the size effect region, an increase in flank wear was observed to be slow. Hence, the MQL considerably reduces the edge fracture due to the better lubricant penetration.

Kajaria et al. (2012) investigated the tool wear during micromilling and found that the tool life under MQL condition is almost 100 times that of dry microcutting. For achieving eco-friendly cutting, the lubricant needs to be sustainable. Therefore, few researchers focused on vegetable oil in MQL. The elongated polar fatty acid chain in vegetable oil offers high film strength and reduces friction and wear (Katna et al., 2020). In addition to that, vegetable oil shows bacterial growth resistance and high corrosion resistance. Vazquez et al. (2015) explored MQL while micromilling of Ti64 and observed that lubrication in the feed path provides low tool wear in comparison with reverse-feed directions.

The reduction in tool wear improves the machined surface characteristics. Hassanpour et al. (2016) investigated the impact of MQL in micromilling of Ti alloy and observed that MQL significantly improved the machined surface quality. It was observed that the surface irregularities were significantly reduced by the usage of MQL-supported machining due to the appropriate cooling and lubrication.

2.4 MICRO/NANO-TEXTURING EFFECTS ON FRICTION AND WEAR

Surface texturing has been applied in numerous day-to-day applications to improve performance by manipulating friction and wear on surfaces. Some of them include texturing along cylinder liners for improving lubrication in IC engines, and various textures are created on vehicle tyres to increase grip and reduce wear, mechanical seals created by laser texturing, etc. Even though the applications are numerous, there are a couple of disadvantages that hamper their use on a larger scale. The major drawbacks are the additional cost incurred in making these textures and the adverse effect they can have on the component's overall structural rigidity and heat transfer characteristics. Hence, additional studies have to be done in that regard to ensure its ability to survive the environments for which it is designed. Over the last few decades, remarkable advances in fabrication, characterization, and computation capabilities have occurred. Hence, creating a texture on a surface and studying its various properties can be easily accomplished. However, answers to certain questions have to be provided. What are the mechanisms involved in improving friction and wear characteristics? Can a universally accepted texture be used for all applications? And are there guidelines that can be followed for texture design for a particular application? The scientific community has been trying to find an answer to these questions, and until it is done, more studies have to be conducted (Gachot et al., 2017).

Numerous manufacturing techniques have been used for making textures. The efficiency of these techniques in improving friction and wear characteristics is discussed in this section.

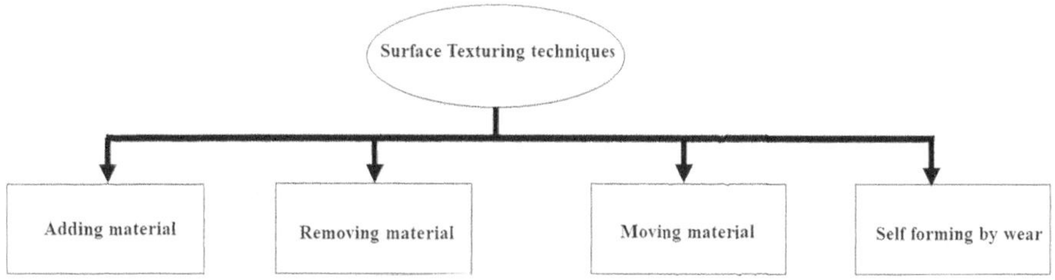

FIGURE 2.1 Classification of various surface texturing techniques.

2.4.1 Fabrication Techniques of Surface Textures

The classification of the various manufacturing processes is given in Figure 2.1.

The quality of surface texturing procedures, including sharpness of the angles, line definition, depth, edge angle control, and pitch accuracy, is an element that will affect tribological characteristics (Costa and Hutchings, 2014).

1. **Addition of material:** In this technique, small features are created on the workpiece surface by adding material. The majority of these methods involve masking the surface with a patterned mask and depositing it on the exposed areas. The different examples include various physical and chemical deposition techniques like electrodeposition (Ying et al., 2019), electrolytic coating (Budinsky, 1988), and additive manufacturing. The main limitation of this technique is the limitation in the thickness of the material that can be added to the surface, as well as its lower mechanical strength in comparison to the substrate. If not controlled, the chances of the textures being sheared off under extreme loads during sliding action are very high.
2. **Removal of material:** The structures are made on the surface by removing the material around them. This is the most commonly used technique for surface texture generation. Here, the material on the surface is removed to create textures like grooves and dimples. Improved strength and load-bearing capacities are generally observed in textures made via material removal compared to material addition. The most commonly used technique is laser texturing. By varying the level of sophistication of the optics used, microfeatures can be easily generated using this method on almost any surface with a very high degree of accuracy. However, the extensive heat energy involved in the ablation process results in microstructural damage, material re-solidification, etc., resulting in the need for additional surface characterization methods to be adopted for laser-textured surfaces (Sorrentino et al., 2020). Another technique that is widely used is chemical or physical etching. This technique is capable of generating surface features with a higher level of accuracy. Masking methods like photolithography and its variants are used in generating the required pattern shapes on the surface, and they have very high flexibility compared to other manufacturing techniques (Mudigonda and Patil, 2015). Also, conventional and non-conventional machining techniques like micromilling (Jain and Bajpai,

2019) and electric discharge texturing (Jithin and Joshi, 2021) are also widely explored as manufacturing techniques for texture generation. The major issue with conventional manufacturing techniques is their lesser flexibility compared with etching and laser-based techniques. Tool and workpiece setup, availability of proper tools, size constraints, etc. are some of the major drawbacks. Hence, they require much more research before being used as a commercial technique for texture manufacturing.

Surface texturing has been extensively researched using material removal techniques, with laser and etching being commercially used. By altering the shape of the textures, factors affecting friction like the area of contact and the lubrication regime of contacting surfaces can be varied, resulting in improved friction and wear characteristics. A study on the effect of lubrication on textured surfaces made by laser shows that textured surfaces and lubricated textured surfaces have better friction characteristics compared to smooth surfaces (Wan et al., 2021) (Figure 2.2).

More studies are now being done on manufacturing different texture shapes for better friction and wear characteristics.

3. **Moving material:** The surface texture generation is attributed to plastic deformation on the surface, which will result in the movement of material from one place to another. This results in texture formation. Short blasting and embossing are major techniques that contribute to texture generation by moving material. Shot-blasting causes surface damage and roughening by impacting sharp, fast-moving particles onto the surface. Due to the numerous process parameters, including size, shape, velocity, and shot hardness, it is challenging to regulate the texture exactly. Furthermore, shot blasting is capable of generating arbitrary textures (Pawelski et al., 1994). Embossing

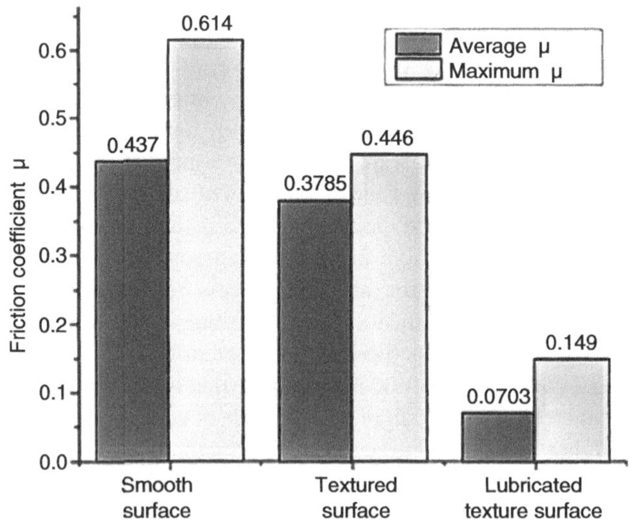

FIGURE 2.2 Comparison of friction coefficients on different surfaces (Wan et al., 2021). (Reproduced with permission from Ref. (Wan et al., 2021). Copyright Taylor & Francis, 2021.)

is a much more sophisticated technique where a textured tool featuring the negative pattern is pressed into the workpiece surface. This method is simple, cheap, and can be used on a very large scale (Szurdak et al., 2014). Since no material removal or addition happens, the structural rigidity of the texture and its ability to withstand shearing forces during sliding are much greater. However, it is very difficult to make textured tools for embossing. So, when we consider the complexity of the textures that can be printed, material removal methods are a better option compared with embossing.

4. **Self-forming by wear:** This is another kind of material removal technique where wear-resistant areas are formed on the surface using processes like diffusion, and during operation, the surfaces with lesser wear resistance are worn off easily with only the wear-resistant areas left standing (Steinhoff et al., 1996).

To conclude, each manufacturing process has its own pros and cons. The selection should be made purely based on the application, ease of operation, cost, and stability of the feature.

2.5 TRIBOLOGICAL ASPECTS IN NANO/MICRO FORMING

Micromanufacturing processes are gaining importance in recent years because of miniaturization, and there is an increase in the use of microcomponents in different applications, i.e., MEMS, electrical/electronics, and biomedical (Geiger et al., 2001). Compared to other micromanufacturing processes, microforming has advantages such as low energy and material consumption and high production rates. Hence, it can be used as a mass production process to produce microcomponents (micropins, cups, springs, screws, etc.). Formally, microforming refers to a forming technique that is used to fabricate micrometallic components, i.e., components with a minimum of two characteristic lengths in the sub-millimeter range (Geiger et al., 2001). Traditionally, microcomponents are formed by scaling down the bulk deformation processes like extrusion, drawing, and stamping. However, due to size effects (Geiger et al., 2001; Wang et al., 2014), the knowledge of bulk deformation processes cannot be directly used in microforming processes. Both material and frictional behavior in microforming processes are different compared to bulk deformation processes as a result of the size effect. Hence, for better predictability and to produce accurate microcomponents, knowledge of size effects in microforming is critical. Size effects in microforming processes can be broadly classified into three categories: effects of grain size, effects of surface grains, and the effect of friction. In the following sections, important experimental observations and conclusions related to frictional size effects are discussed elaborately.

2.5.1 Frictional Size Effect in Microforming Processes

Messner et al. (1994) for the first time conducted ring compression tests (material: brass) on different specimen sizes to understand the frictional size effect. Interfacial friction coefficient is quantified in terms of measured internal and outer diameters of specimen after the test. Their results indicate that there is an increase in the friction coefficient with a decrease in the specimen size. The observations reported by Messner et al. (1994) are further verified by Geiger et al. (2001) by performing scaled-down double-cup extrusion experiments (material: brass) and the ratio of upper and lower cup heights is taken as a measure of friction coefficient. They used four geometrically similar specimens (ratio of diameter to height is 1.0) of different diameters (4, 2, 1, and 0.5 mm). They reported that with a decrease in the specimen size friction increased. Gong et al. (2010) studied the influence of lubrication condition (talc powder, petroleum jelly, soybean oil, and without lubrication) on the frictional size effect during cylindrical compression tests on copper alloy C3602. They used four specimen sizes (6, 4, 2, and 1 mm diameter) with the same height to diameter ratio (1.5). They mentioned that the frictional size effect is noticed only when soybean oil is used, and it is absent for other lubricating conditions. In a continuation work, it was reported that the friction coefficient increased with an increase in punch and roughness and decreased with an increase in punch velocity in the microstrip drawing test (Gong and Guo, 2013).

2.5.2 Models for Frictional Size Effect in Microforming Processes

The observed frictional size effect in forming at microscale is closely associated with the behavior of the lubricant placed between forming and workpiece. Engel (2006) proposed the open and closed lubrication pockets theory to explain the observed frictional size effect in the microforming process. According to this theory, plastic deformation initiates at roughness peaks of lubricated workpieces, and this increases the lubricant pressure present in between the valleys of surface profile. However, the roughness valleys that are close to the free edges of the workpiece surface cannot retain the lubricant and these are called open lubrication pockets (OLPs) (Figure 2.3). With an increase in the forming load (normal pressure), lubricant present in the OLPs is squeezed out, resulting in the transfer of forming load directly to the asperities, hence causing higher friction. Whereas closed lubrication pockets (CLPs) are away from the free edges of the workpiece surface. During forming, the lubricant trapped in CLPs and its pressure increases and transfers the fraction of forming load from tool to workpiece and decreases stress on the asperities in contact, hence causing lower friction. Engel (2006) explained that, under the same forming conditions, the ratio of the number of OLPs to CLPs is high in microforming compared to macro forming processes, hence higher friction is observed.

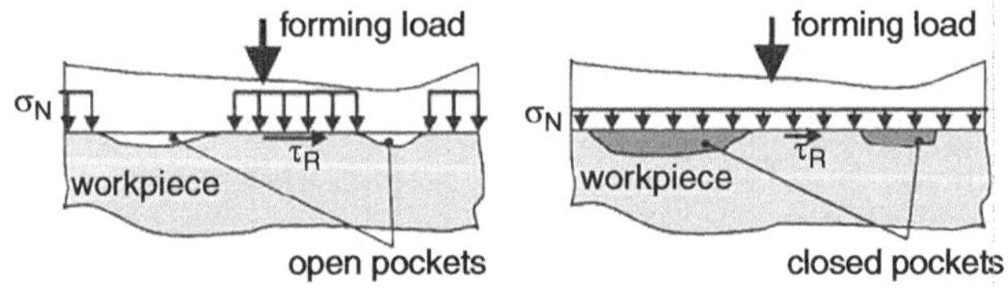

FIGURE 2.3 Pictorial representation of closed and open lubrication pockets (Engel, 2006). (Reproduced with permission from Ref. (Engel, 2006). Copyright Elsevier, 2006.)

Figure 2.3 shows the schematic representation of OLPs and CLPs. Lubricant is placed between tool and workpiece and a normal pressure distribution is shown in the figure.

Krishnan et al. (2007) carried out both experimental and numerical investigations to understand the frictional size effects in microextrusion (material: brass). It was mentioned that the non-uniform friction condition exists during the process and it depends on the size of the micropins extruded. They also reported that the friction reduced in extrusion when dies were coated with diamond-like carbon with silicon (DLC-Si) compared to other coatings (chromium nitride and titanium nitride). Zheng et al. (2013) used a micromechanical model (considering grain effects) to simulate the micro-uniaxial compression under dry friction condition. It was observed that at higher grain size the grain boundary does not significantly affect the tribological behavior during micro-uniaxial compression. Based on the simulation results, they reported that tribological behavior is significantly affected by mechanical properties of individual grains close to and away from the contact area.

2.5.3 Methods of Lubrication in Microforming

Lubrication is essential in reducing the frictional size effects in forming at microscale. Wang et al. (2013a) used a DLC film as a lubricant (because of its excellent tribological properties) in microstrip drawing (material: rolled T2 copper). It was mentioned that the coefficient of friction in DLC film is low compared to lubricants in microstrip drawing. They also reported that polishing and a decrease in the specimen size do not affect the tribological behavior of the DLC film. Brinksmeier et al. (2010) studied the feasibility of using micromilling and grinding for generating microforming tool surfaces and compared their tribological behavior with polished surfaces by using microstrip drawing tests. They reported that machined surfaces (micromilling and grinding) with average surface roughness in the range of 50 and 600 nm resulted in a lower friction coefficient compared to polished surfaces. Based on the above results, they concluded that by using proper cutting conditions, micromilling and grinding can be used to generate surfaces with better tribological properties (for microforming applications) compared to polishing processes. In addition to the lubrication and surface topography of tools, workpiece surface topography also affects the tribological behavior in microforming processes. Wang et al. (2013b) employed ion beam irradiation to modify the surface topography of thin copper sheets to be used in the deep drawing process at microscale. It was ion beam irradiation-induced nanocrystals which caused a decrease in the friction coefficient due to an increase in the storage of lubricant at the tool–sheet contact surface. They reported that with an increase in the irradiation time the punch load decreased and attributed the same to a decrease in the friction coefficient.

2.5.4 Tool Wear in Microforming Processes

Wear of tools used in forming processes at microscale significantly affects the quality of the formed components due to smaller length scales. Hu et al. (2010) studied the wear during microdeep drawing (workpiece material: stainless steel, tool material: tool steel) with a punch diameter of 1 mm. It was mentioned that two failure mechanisms—bottom fracture (due to an increase in frictional force) and cup wall damage (due to non-uniform distribution of wear)—were observed during the micro deep drawing process. They reported that there was no correlation between forming tool wear and maximum punch force during the micro deep drawing process. Shimizu et al. (2013) studied the effect of the scale of the forming process (micro and milli) on progressive adhesion of sheet material (thin-rolled stainless steel foil) during deep drawing under dry friction conditions. Their results indicate that progressive adhesive wear is high in milli-scale compared to that of the microscale deep drawing process and attributed the same to a short sliding distance in the microscale process compared to the milli-scale process. Based on the above results, they concluded that, with miniaturization of the forming process, there is a decrease in the plowing effect of the adhered particles. Flosky and Vollertsen (2014) studied the tool wear behavior during high production rate (200 strokes per minute) one stroke blanking and deep drawing process at microscale (materials: aluminum and copper alloys). They reported that compared to other locations of punch, the edge of the punch wears out at a higher rate and tool positioning significantly affects the tool wear.

2.6 CONCLUSIONS

This chapter mainly aims to discuss tribological behavior during conventional micro/nano-machining and forming processes. The size effect phenomenon, transition from plowing to shearing in microcutting, lubrication behavior, surface texturing, and tribological issues were discussed in detail. The following are major conclusions from this chapter.

- Under minimum quantity application of cutting fluid, a substantial reduction in machining force, temperature, and friction and an improvement in tool life and quality of machined component are observed.
- Chip coming out from the machining zone under minimum quantity application of the cutting is almost dry in nature and hence these chips can be directly taken for recycling without any further processing.
- Oxide PVD coating provides a reduction in friction coefficient at high temperature and offers a higher wear resistance.
- Solid lubricants like MoS_2/titanium composite coating, WC/C coatings, and AlTiN are suitable for dry machining.
- The application of coolants significantly reduces the tool wear, which leads to an improvement in micro/nano machined surface quality.
- MQL, which is considered a sustainable lubrication strategy, helps in the penetration of coolants to the cutting zone and has significantly improved the tool wear and surface quality in micro/nano-machining processes.
- The selection of the type of surface texture required should be made purely based on the application, ease of operation, cost, and stability of the feature required.

- Due to the size effect phenomenon, frictional effects are significant in microforming compared to macro forming processes. In addition, in order to produce quality products using microforming processes, understanding and predicting frictional effects (using models) is necessary.

REFERENCES

Anand K N, and Mathew J. 2020a. "Evaluation of Size Effect and Improvement in Surface Characteristics Using Sunflower Oil-Based MQL for Sustainable Micro-Endmilling of Inconel 718." *Journal of the Brazilian Society of Mechanical Sciences and Engineering* 42(4): 156.

Anand K N, and Mathew J. 2020b. "Studies on Wear Behavior of AlTiN-Coated WC Tool and Machined Surface Quality in Micro Endmilling of Inconel 718." *International Journal of Advanced Manufacturing Technology* 110(1–2): 291–307.

Anand K N, and Mathew J. 2021. "Size Effect and Micro Endmilling Performance while Sustainable Machining on Inconel 718." *Materials and Manufacturing Processes* 36(6): 668–676.

Anand K N, Vipindas K, and Mathew J. 2021. "Microstructure Evolution based Force Model and Surface Characteristic Studies of Inconel 718 during Micro Endmilling." *Machining Science and Technology* 25(6): 875–898.

Aramcharoen A, and Mativenga P T. 2009. "Size Effect and Tool Geometry in Micromilling of Tool Steel." *Precision Engineering* 33(4): 402–407.

Arndt M, and Kacsich T. 2003. "Performance of New AlTiN Coatings in Dry and High Speed Cutting." *Surface and Coatings Technology* 163–164: 674–680.

Boswell B, Islam M N, Davies I J, Ginting Y R, and Ong A K. 2017. "A Review Identifying the Effectiveness of Minimum Quantity Lubrication (MQL) during Conventional Machining." *International Journal of Advanced Manufacturing Technology* 92(1–4): 321–340.

Bouzakis K D, Vidakis N, Michailidis N, Leyendecker T, Erkens G, and Fuss G. 1999. "Quantification of Properties Modification and Cutting Performance of (Ti1-xAlx)N Coatings at Elevated Temperatures." *Surface and Coatings Technology* 120–121: 34–43.

Brinksmeier E, Riemer O, and Twardy S. 2010. "Tribological Behavior of Micro Structured Surfaces for Micro Forming Tools." *International Journal of Machine Tools and Manufacture* 50: 425–430

Budinsky KG. 1988. *Surface Engineering for Wear Resistance*. Prentice Hall: 44–77.

Costa H, and Hutchings I. 2014. "Some Innovative Surface Texturing Techniques for Tribological Purposes." *Proceedings of the Institution of Mechanical Engineers, Part J: Journal of Engineering Tribology* 229(4): 429–448.

De Oliveira F B, Rodrigues A R, Coelho R T, and De Souza A F. 2015. "Size Effect and Minimum Chip Thickness in Micromilling." *International Journal of Machine Tools and Manufacture* 89: 39–54.

Derflinger V, Brandle H, and Zimmerman H. 1999. "New Hard/Lubricant Coating for Dry Machining." *Surface and Coatings Technology* 113: 286–292.

Dudzinski D, Devillez A, Moufki A, Larrouquere D, Zerrouki V, and Vigneau J. 2004. "A Review of Developments towards Dry and High Speed Machining of Inconel 718 Alloy." *International Journal of Machine Tools and Manufacture* 44: 439–456.

Engel U. 2006. "Tribology in Microforming." *Wear* 260: 265–273.

Flosky H, and Vollertsen F. 2014. "Wear Behaviour in a Combined Micro Blanking and Deep Drawing Process." *CIRP Annals - Manufacturing Technology* 63: 281–284.

Gachot C, Rosenkranz A, Hsu S M, and Costa H L. 2017. "A Critical Assessment of Surface Texturing for Friction and Wear Improvement." Wear 372–373: 21–41.

Geiger M, Kleiner M, Eckstein R, Tiesler N, and Engel U. 2001. "Microforming." *CIRP Annals - Manufacturing Technology* 50: 445–462.

Gong F, and Guo B. 2013. "Effects of Influencing Factors on Friction Coefficient in Microsheet Forming." *Materials Research Innovations* 17: 7–11.

Gong F, Guo B, Wang C, and Shan D. 2010. "Size Effect on Friction of C3602 in Cylinder Compression." *Tribology Transactions* 53: 244–248.

Hassanpour H, Sadeghi M H, Rezaei H, and Rasti A. 2016. "Experimental Study of Cutting Force, Microhardness, Surface Roughness, and Burr Size on Micromilling of Ti6Al4V in Minimum Quantity Lubrication." *Materials and Manufacturing Processes* 31(13): 1654–1662.

Hu Z, Huferath-Von Luepke S, Von Kopylow C, and Vollertsen F. 2010. "Characteristic of Wear Behavior of Micro Deep Drawing Tools." *AIP Conference Proceedings* 1315: 335–340.

Jain A, and Bajpai V. 2019. "Mechanical Micro-Texturing and Characterization on Ti6Al4V for the Improvement of Surface Properties." *Surface and Coatings Technology* 380: 125087.

Jithin S, and Joshi S S. 2021. "Surface Topography Generation and Simulation in Electrical Discharge Texturing: A Review." *Journal of Materials Processing Technology* 298: 117297.

Kajaria S, Chittipolu S, Adera S, and Hung W N. 2012. "Micromilling in Minimum Quantity Lubrication." *Machining Science and Technology* 16(4): 524–546.

Katna R, Suhaib M, and Agrawal N. 2020. "Nonedible Vegetable Oil Based Cutting Fluids for Machining Processes – A Review." *Materials and Manufacturing Processes* 35(1): 1–32.

Klocke F. 1997. "Dry Cutting." *CIRP Annals* 46 (2): 519–526.

Klocke F, and Eisenblatter G. 1997. "Machinability Investigation of the Drilling Process using Minimal Cooling Lubrication Techniques." *CIRP Annals* 46 (1): 19–24.

Krishnan N, Cao J, and Dohda K. 2007. "Study of the Size Effects on Friction Conditions in Microextrusion–Part I: Microextrusion Experiments and Analysis." *Journal of Manufacturing Science and Engineering* 129: 669–676.

Lahres M, Doerfel O, and Neumuller R. 1999. "Applicability of Different Hard Coatings in Dry Machining an Austenitic Steel." *Surface and Coatings Technology* 120–121: 687–691.

Leo Kumar S P, Jerald J, Kumanan S, and Prabakaran R A. 2014. "Review on Current Research Aspects in Tool-Based Micromachining Processes." *Materials and Manufacturing Processes* 29(11–12): 1291–1337.

Machado A R, and Wallbank J. 1997. "The Effect of Extremely Low Lubricant Volumes in Machining." *Wear* 219: 76–82.

Masuzawa T. 2000. "State of the Art of Micromachining." CIRP Annals – Manufacturing Technology 49(2): 473–488.

Messner A, Engel U, Kals R, and Vollertsen F. 1994. "Size Effect in the FE-Simulation of Micro-Forming Processes." *Journal of Materials Processing Technology* 45: 371–376.

Mudigonda S, and Patil D H. 2015. "Some Investigations on Surface Texturing on Monel 400 Using Photochemical Machining." *Proceedings of the ASME 2015 International Manufacturing Science and Engineering Conference* 1, V001T02A045. ASME, Charlotte, NC.

Navinsek B, Panjan P, Cekada M, and Quinto D T. 2002. "Interface Characterization of Combination Hard/Solid Lubricant Coatings by Specific Methods." *Surface and Coatings Technology* 154: 194–203.

Ng C K, Melkote S N, Rahman M, and Senthil Kumar A. 2006. "Experimental Study of Micro- and Nano-Scale Cutting of Aluminum 7075-T6." *International Journal of Machine Tools and Manufacture* 46(9): 929–936.

Pawelski O, Rasp W, Zwick W, Nettelbeck H J, and Steinhoff Z. 1994. "The Influence of Different Work-Roll Texturing Systems on the Development of Surface Structure in the Temper Rolling Process of Steel Sheet Used in the Automotive Industry." *Journal of Materials Processing Technology* 45: 215–222.

Popke H, Emmer Th, and Steffenhagen J. 1999. "Environmentally Clean Metal Cutting Processes—Machining on the way to Dry Cutting." *Proceedings of the Institution of Mechanical Engineers, Part B: Journal of Engineering Manufacture* 213: 329–332.

Renevier N M, Lobiondo N, Fox V C, Teer D G, and Hampshire J. 2000. "Performance of MoS2/Metal Composite Coatings Used for Dry Machining and other Industrial Applications." *Surface and Coatings Technology* 123: 84–91.

Renevier N M, Oosterling H, Konig U, Dautzenberg H, Kim B J, Geppert L, Koopmans F G M, and Leopold J. 2003. "Performance and Limitation of Hybrid PECVD (Hard Coating)–PVD Magnetron Sputtering (MoS2/Ti Composite) Coated Inserts Tested for Dry High Speed Milling of Steel and Grey Cast Iron." *Surface and Coatings Technology* 163–164: 659–667.

Schulz H, Dorr J, Rass I J, Schulze M, and Leyendecker T, Erkens G. 2001. "Performance of Oxide PVD-Coatings in Dry Cutting Operations." *Surface and Coatings Technology* 146–147: 480–485.

Shimizu T, Murashige Y, Iwaoka S, Yang M, and Manabe K. 2013. "Scale Dependence of Adhesion Behavior under Dry Friction in Progressive Micro-Deep Drawing." *Journal of Solid Mechanics and Materials Engineering* 7: 251–263.

Sorrentino L, Parodo G, and Turchetta S. 2020. "CFRP Laser Texturing to Increase the Adhesive Bonding: Morphological Analysis of Treated Surfaces." *Journal of Adhesion* 97(14): 1322–1335.

Steinhoff K, Rasp W, and Pawelski O. 1996. "Development of Deterministic-Stochastic Surface Structures to Improve the Tribological Conditions of Sheet Forming Processes." *Journal of Materials Processing Technology* 60: 355–361.

Szurdak A, Rosenkranz A, Gachot C, Hirt G, and Mucklich F. 2014. "Manufacturing and Tribological Investigation of Hot Micro-Coined Lubrication Pockets." *Key Engineering Materials* 611–612: 417–424.

Tonshoff K, Karpuschewski B, Mohfeld A, Leyendecker T, Erkens G, Fuss H G, and Wenke R. 1998. "Performance of Oxygen-Rich TiAlON Coatings in Dry Cuttings Applications." *Surface and Coatings Technology* 108–109: 535–542.

Varadarajan A S, Philip P K, and Ramamoorthy B. 2002. "Investigations on Hard Turning with Minimal Cutting Fluid Application (HTMF) and its Comparison with Dry and Wet Turning." *International Journal of Machine Tools and Manufacture* 42: 193–200.

Vazquez E, Gomar J, Ciurana J, and Rodríguez C A. 2015. "Analyzing Effects of Cooling and Lubrication Conditions in Micromilling of Ti6Al4V." *Journal of Cleaner Production* 87(1): 906–913.

Venkatesh V, Swain N, Srinivas G, Kumar P, and Barshilia H C. 2017. "Review on the Machining Characteristics and Research Prospects of Conventional Microscale Machining Operations." *Materials and Manufacturing Processes* 32(3): 235–262.

Vipindas K, Anand K N, and Mathew J. 2018. "Effect of Cutting Edge Radius on Micro End Milling: Force Analysis, Surface Roughness, and Chip Formation." *International Journal of Advanced Manufacturing Technology* 97(1–4): 711–722.

Wan Q, Gao P, and Zhang Z. 2021. "Friction and Wear Performance of Lubricated Micro-Textured Surface Formed by Laser Processing." *Surface Engineering* 37(12): 1523–1531.

Wang C, Guo B, and Shan D. 2014. "Friction Related Size-Effect in Microforming – A Review." *Manufacturing Review* 1: 23.

Wang C, Guo B, Shan D, and Bai X. 2013a. "Tribological Behaviors of DLC Film Deposited on Female Die Used in Strip Drawing." *Journal of Materials Processing Technology* 213: 323–329.

Wang C, Shan D, Guo B, Xu J, and Zhang H. 2013b. "Effect of Nano-Crystals at Surfaces Induced by Ion Beam Irradiation on the Tribological Behaviour in Microforming." *Vacuum* 89: 267–270.

Ying L, Kai C, Jingjun L, Zhengping Z, Jing J, Feng W, and Zhilin L. 2019. "Electrodeposition of Copper-Doped SnS Thin Films and their Electric Transmission Properties Control for Thermoelectric Enhancement." *Journal of Materials Science: Materials in Electronics* 30: 15880–15888

Zheng W, Wang G, Zhao G, Wei D, and Jiang Z. 2013. "Modeling and Analysis of Dry Friction in Micro-Forming of Metals." *Tribology International* 57: 202–209.

3 Nanocutting Fluids for Machining
An Overview of the Tribological Aspects

B. Depu Kumar Patro, M. Shanmuka Srinivas, P. S. Suvin, and M. Ravi Sankar

3.1 INTRODUCTION

3.1.1 Overview and Need for Machining

Machining refers to a manufacturing process involving the removal of material from a workpiece to shape or refine it into a desired final form. This is an important process that finds extensive application across diverse industries such as automotive, aerospace, construction, and electronics, contributing to the production of a wide range of parts and products. Various machining methods, including turning, milling, drilling, and grinding, are employed, each serving specific purposes and necessitating distinct equipment and techniques. The selection of the appropriate machining process hinges on factors such as the shape and size of the workpiece, required tolerance, material composition, and the desired surface finish. Machining is a crucial part of the manufacturing process because it allows for the precise and accurate production of parts and products. This trade demands advanced skills, necessitating a comprehensive grasp of the machinery and tools employed. It also requires meticulous attention to detail and a proactive attitude toward ongoing learning and adaptation to emerging technologies.

3.1.2 Need for Cutting Fluids

Cutting fluids are used to cool and lubricate the cutting process in metal working and machining operations. They can contribute to enhancing the surface finish of the workpiece, extending tool life, and reducing cutting forces, which can result in increased machining speeds and improved dimensional accuracy. By mitigating the heat generated in cutting, cutting fluids aid in preventing thermal deformation of the workpiece. Moreover, these fluids facilitate the removal of chips and debris from the cutting area, enhancing visibility and minimising the potential for machinery damage. Cutting fluids find applications in various metal working processes, including grinding, turning, milling, drilling, and broaching.

DOI: 10.1201/9781003359333-3

3.1.3 Need for Nanocutting Fluid

Nanocutting is a type of machining that involves cutting materials with extremely small cutting tools, often on the order of nanometres (billionths of a metre) in size. Because the cutting tools are so small, the forces involved in the cutting process are much higher than they would be for larger cutting tools, and the heat and friction generated by the cutting process can be much greater as well.

The utilisation of nanocutting fluid aids in cooling and lubricating the cutting tool and workpiece, thereby decreasing heat and wear on the tool. This, in turn, enhances the precision and quality of the cut. Nanocutting fluids may also contain additives that help reduce friction and wear, and they may be designed to be compatible with specific types of cutting tools or materials. Overall, incorporating nanocutting fluid can enhance the effectiveness and efficiency of nanocutting processes, contributing to an extended lifespan for cutting tools by minimising wear.

Nanocutting fluids are applied in machining operations to enhance the quality and efficiency of cuts, as well as to prolong the longevity of cutting tools. These fluids are suitable for a range of applications, including turning, milling, drilling, and grinding.

There are multiple rationales for employing nanocutting fluids in machining procedures:

- **Lubrication**: Nanocutting fluids have the capability to minimise friction and diminish wear on cutting tools, leading to an extended tool lifespan and enhancing the quality of the cut.
- **Cooling**: Nanocutting fluids aid in dispersing the heat generated during cutting, thereby enhancing the efficacy of cutting tools and lowering the likelihood of thermal damage.
- **Chip evacuation**: Nanocutting fluids can help remove chips and debris from the cutting area, which can improve the visibility of the cut and reduce the risk of tool failure.
- **Corrosion protection**: Nanocutting fluids can provide protection against corrosion for both the cutting tools and the workpiece.
- **To protect the tool**: When machining materials at the nanoscale, the cutting tool is subjected to extremely high temperatures and pressures. This can cause the tool to wear out or break. The application of nanocutting fluids can diminish the temperature and pressure exerted on the tool, safeguarding it from harm and elongating its operational lifespan.
- Nanocutting fluids play a crucial role in enhancing the final product's quality by minimising the heat and pressure applied to the machined material. This leads to a refined, more precise finish, elevating the overall quality of the end product.

In summary, the incorporation of nanocutting fluids proves beneficial in boosting the efficiency and precision of machining processes while also contributing to reduced tooling and maintenance costs.

3.1.4 Types of Nanocutting Fluids

Various categories of nanocutting fluids used in the machining process are as follows:

- **Water-based cutting fluids**: This prevalent category of nanocutting fluids is composed of water and various additives such as emulsifiers, anticorrosion, and biocides. Known for their ease of use and disposal, water-based cutting fluids generally come at a lower cost compared to other fluid types.
- **Oil-based cutting fluids**: Crafted from diverse oils and additives, these cutting fluids serve to lubricate and cool the cutting tool during machining. In contrast, they excel in reducing heat and wear on the tool compared to water-based cutting fluid. Oil-based cutting fluids can be more challenging to dispose of and may involve higher costs.
- **Semisynthetic cutting fluids**: Blending attributes of water-based and oil-based fluids, these cutting fluids aim to provide optimal lubrication and cooling properties. They offer the advantages of oil-based fluids while maintaining ease of use and disposal, which is typically more straightforward than with pure oil-based fluids.
- **Synthetic cutting fluids**: Formulated from synthetic polymers, these cutting fluids are engineered for superior lubrication and cooling properties. Although they come at a higher cost, synthetic cutting fluids boast longer durability and increased resistance to bacteria and contaminants compared to other fluid types.

In conclusion, the selection of the most suitable nanocutting fluid for a particular application hinges on the specific demands of the machining process and the materials being worked on.

3.1.5 Properties of Nanocutting Fluids

Critical properties influencing the performance of nanocutting fluids include:

- **Lubrication**: Nanocutting fluids must possess the capability to diminish friction and wear on the cutting tool, thereby enhancing its efficiency and extending its tool life.
- **Cooling**: Effective nanocutting fluids should disperse the heat produced during the machining process, safeguarding both the cutting tool and the material being machined.
- **Viscosity**: The flow of the cutting fluid is influenced by its viscosity and adhesion to the cutting tool and machined material. A fluid with lower viscosity flows more easily, although it might offer less lubrication compared to a high viscosity fluid.
- **Chemical stability**: Nanocutting fluids should be stable and resistant to chemical breakdown, even when subjected to high temperatures and pressures.

- **Compatibility**: Nanocutting fluids should be compatible with the materials being machined and the cutting tool, and they should not react with other fluids or chemicals that may be present in the machining process.
- **Environmental impact**: Nanocutting fluids need to be environmentally sustainable, causing no pollution or adverse effects on human health.

In summary, the ideal nanocutting fluid should strike a balance among various properties, ensuring effective lubrication and cooling while also maintaining stability, compatibility, and environmental friendliness. Nanocutting fluids display diverse effectiveness across different machining processes and materials. Distinct tool–workpiece interactions and cutting conditions involved in different machining operations like turning, milling, and drilling contribute to varying outcomes. Additionally, materials made of ferrous alloys and composites possess unique properties that influence their behaviour during machining. The effectiveness of nanocutting fluids is not uniform and can differ depending on the particular machining operation and the type of material involved. For instance, certain nanoparticles may excel in reducing tool wear and improving surface finish in one material but might yield different results in another due to variations in hardness, thermal conductivity, or chemical reactivity. Furthermore, different machining parameters, such as cutting speeds, feed rates, and depth of cut, affect the performance of nanocutting fluids differently across materials and operations. Understanding these variations is crucial for tailoring the selection of nanoparticles and fluid compositions to optimise performance in a range of machining scenarios. Rigorous experimental studies and tribological analyses are essential to establish guidelines for the effective application of nanocutting fluids across diverse processes and materials. The goal is to develop adaptable formulations capable of consistently enhancing machining efficiency, tool life, and surface quality across various industrial applications and materials. This variability underscores the necessity for ongoing research to refine and expand the scope of nanocutting fluid applications.

3.1.6 Nanocutting Fluids in Turning

In turning operations, nanocutting fluids play a crucial role in enhancing various machining aspects, including cutting forces, temperature, surface roughness, and chip morphology.

- **Cutting force**: Nanocutting fluids contribute, resulting in a decrease in cutting forces needed for machining a workpiece by minimising the friction between the cutting tool and the workpiece. This not only extends the cutting tool's lifespan but also enhances the overall efficiency of the machining process.
- **Temperature**: Nanocutting fluids aid in dissipating heat produced during the machining process, thereby lowering the temperature of both the cutting tool and the workpiece. This serves to prevent thermal damage to the workpiece and enhances the accuracy and quality of the finished product.

- **Surface roughness**: By decreasing frictional interaction between the cutting tool and the workpiece, nanocutting fluids improve the surface finish of the workpiece, resulting in a smoother and more accurate finish.
- **Chip morphology**: Nanocutting fluids have the potential to influence the size and form of chips generated in the machining process. Specifically, they contribute to the production of smaller chips that are easier to extract from the machining area.

In turning operations, the application of nanocutting fluids possesses the potential to enhance both the efficiency and precision of the machining process, leading to a superior-quality finished product.

3.2 TURNING

The inclusion of 0.5% $n\text{MoS}_2$ in coconut oil exhibits superior performance when compared to alternatives such as sesame and canola vegetable-based oils. This leads to a notable decrease in cutting force by approximately 37%, cutting temperature by about 21%, and a reduction in tool wear by 44%. Additionally, there is a 39% decrease in surface roughness in comparison to the given conditions of dry machining (Padmini et al., 2016).

In another study using $n\text{MoS}_2$, the optimised results were evaluated by Taguchi-based Grey Relation Analysis and found coconut oil (CC)+0.5% nano-molybdenum disulphide ($n\text{MoS}_2$) with a cutting speed of 40 m/min and a feed rate of 0.14 mm/rev (Rapeti et al., 2016). Another study on $n\text{MoS}_2$ showed that the lamellar structure of $n\text{MoS}_2$ produces a nanolayered lattice structure and, when the sulphur atoms react with the machining surface results in a persistent lubricant film, which separates the tool–work interface, leads to a decrease in frictional machining force and cutting temperature as a result of decreasing coefficient of friction (COF) and tool wear.

In another study (Sharma et al., 2017), Al-MoS_2 hybrid nanofluid is used for turning experimentation. The hybrid nanofluid of Al-MoS_2 has demonstrated notable decreases of 18.08%, 7.35%, 5.73%, and 2.38% in feed force (Fx), cutting force (Fz), thrust force (Fy), and surface roughness (Ra), respectively, in comparison to the nanofluid mixed with Al_2O_3. Using atomisation cutting fluid spray systems (like minimum quantity lubrication (MQL)) for titanium alloy results in about 75% increased tool life against wet machining. The contact area of the machining surface with coolant increases, which will help in raising heat transfer from the machining area, and it was found to be 38.6° at 0.25 vol% of hybrid lubricant as compared with the base fluid, that is, 54.6°, and was higher in all concentrations of lubricants (Figures 3.1 and 3.2).

Al_2O_3 nanofluid (a colloidal suspension) of 23% with Multi-walled Carbon Nanotubes (MWCNTS) nanoparticles with a 90:10 volumetric ratio mixed in the base fluid (A stabilised base fluid is created by incorporating 5% vegetable oil into distilled water along with 0.5 vol% emulsifier.) is used for turning experimentation (Sharma et al., 2017). The outcomes indicated that an optimised concentration of approximately 1.08 vol% of Al-MWCNT nanoparticles in hybrid nanofluids yielded

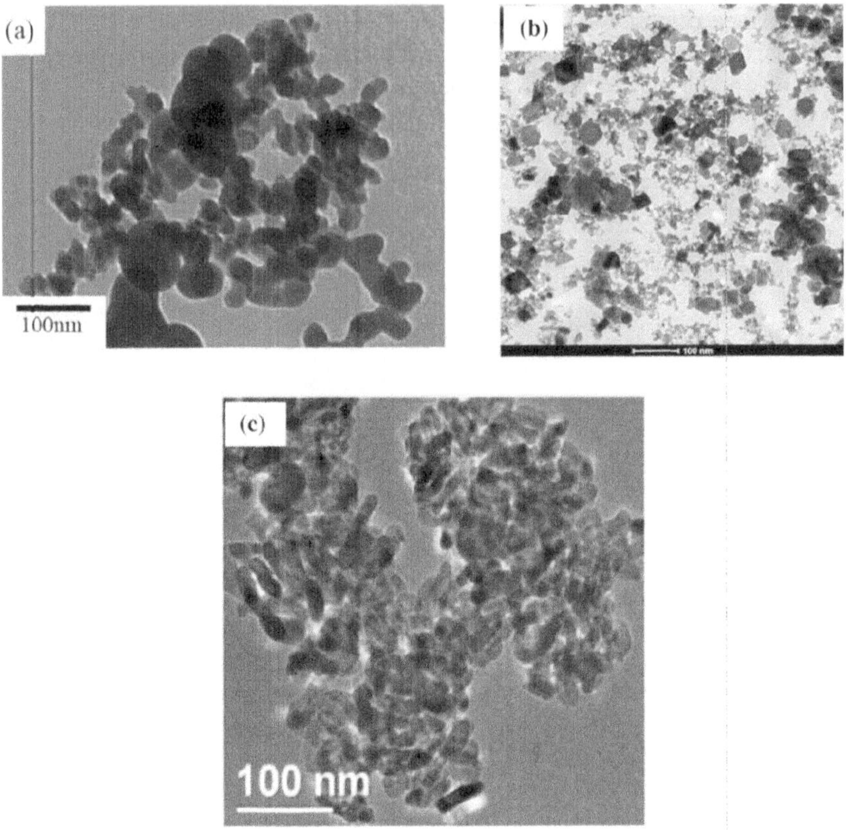

FIGURE 3.1 Surface morphological characterisation of (a) molybdenum disulphide, (b) alumina nanofluid, and (c) alumina–molybdenum disulphide hybrid nanofluid. (Reproduced with permission from Ref. (Sharma et al. 2017). Copyright Elsevier, 2017.)

the minimum tool flank wear during the turning process of AISI 304 stainless steel. Furthermore, the nodal temperature experienced a decrease of 29.06%, dropping from 85.53°C to 60.67°C when compared to the alumina nanofluid. Additionally, the average tool wear exhibited a reduction from 65.39 to 58.16 µm, representing an 11% decrease compared to the alumina nanofluid with the same concentration. This reduction in wear was attributed to the facile exfoliation of the weak MWCNT structure, facilitated by the shearing effect of the chip on the tool surface.

In another study of the MQL technique, the nanolubricant was forced to supply into the machining zone, which assists the nanoparticles in getting embedded between the workpiece and tool during machining, called the mending effect (Gajrani et al., 2019). In this work, a 0.3% MoS_2-based green nanofluid has a lower cutting force compared to a CaF_2-based nanofluid because of its superior specific heat, thermal conductivity, wettability, and viscosity that leads to better lubrication and cooling during machining (17% cutting force and 28% feed force with only mineral oil; 11% reduction in COF and 37% improved surface finish) (Figures 3.3 and 3.4).

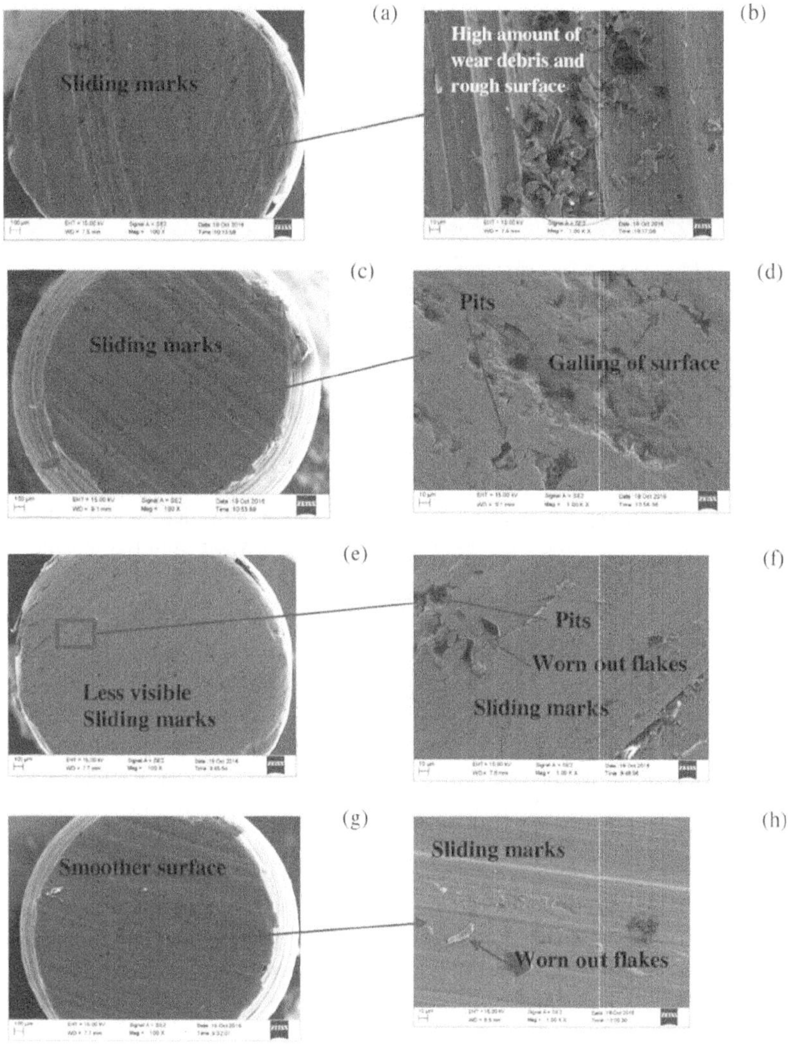

FIGURE 3.2 Surface morphological analysis of pin wear is conducted in the pin-on-disc test under different conditions: (a, b) dry condition, (c, d) base fluid, (e, f) alumina, and (g, h) alumina-molybdenum disulphide nanofluids. (Reproduced with permission from Ref. (Sharma et al. 2017). Copyright Elsevier, 2017.)

The dispersion and stabilisation of nanoparticles within nanocutting fluids is fundamental to their functionality and effectiveness during machining processes. Achieving and maintaining a uniform dispersion of nanoparticles in the fluid matrix is critical to harness their tribological benefits. Nanoparticles have a tendency to agglomerate due to their high surface energy and attractive forces between particles. This agglomeration can lead to inconsistent performance and hinder their desired effects. To address this, various dispersion methods are used to ensure the stable dispersion of nanoparticles in the cutting fluid. Sonication, mechanical stirring, and

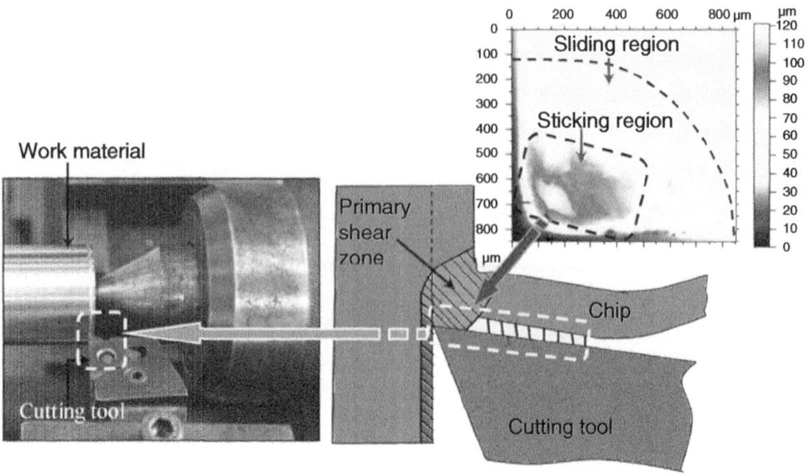

FIGURE 3.3 Illustrating the machining area, highlighting both sticking and sliding zones on the rake face of the cutting tool. (Reproduced with permission from Ref. (Gajrani et al., 2019). Copyright Elsevier, 2019.)

chemical surfactants are commonly used methods to break down particle agglomerates and achieve a homogeneous dispersion. Proper dispersion prevents the settling of nanoparticles, ensuring their availability at the cutting zone throughout the machining process. Stabilisation mechanisms are also employed to maintain the dispersed state of nanoparticles in the fluid. Surface modification of nanoparticles with suitable coatings or surfactants helps prevent agglomeration by creating repulsive forces between particles. This helps maintain the dispersion and preserves the intended properties of the nanocutting fluid. Moreover, the stability of the nanocutting fluid over its operational lifetime is crucial. Long-term stability ensures consistent performance and maintains the fluid's tribological properties over extended machining periods. Continuous monitoring and optimisation of the fluid composition and dispersion techniques are necessary to uphold stability and maximise the benefits of nanoparticles in cutting applications (Figure 3.5).

Despite their potential advantages, nanocutting fluids encounter challenges related to stability and long-term performance. One of the primary concerns revolves around the stability of nanoparticle dispersion within the cutting fluid over extended periods of use. Over time, nanoparticles have a tendency to clump together or settle, causing alterations in the tribological properties of the fluid. This instability can reduce the efficacy of the fluid, affecting its lubrication, cooling, and wear-reducing capabilities during machining operations. Addressing stability issues requires continuous research into innovative stabilisation techniques. Surface modification of nanoparticles, development of stabilising agents, and optimisation of dispersion methods are avenues explored to maintain the uniform distribution of nanoparticles within the cutting fluid. Furthermore, advancements in nanotechnology and fluid engineering aim to create stable formulations capable of withstanding prolonged machining processes. Another challenge is ensuring consistent performance across varying

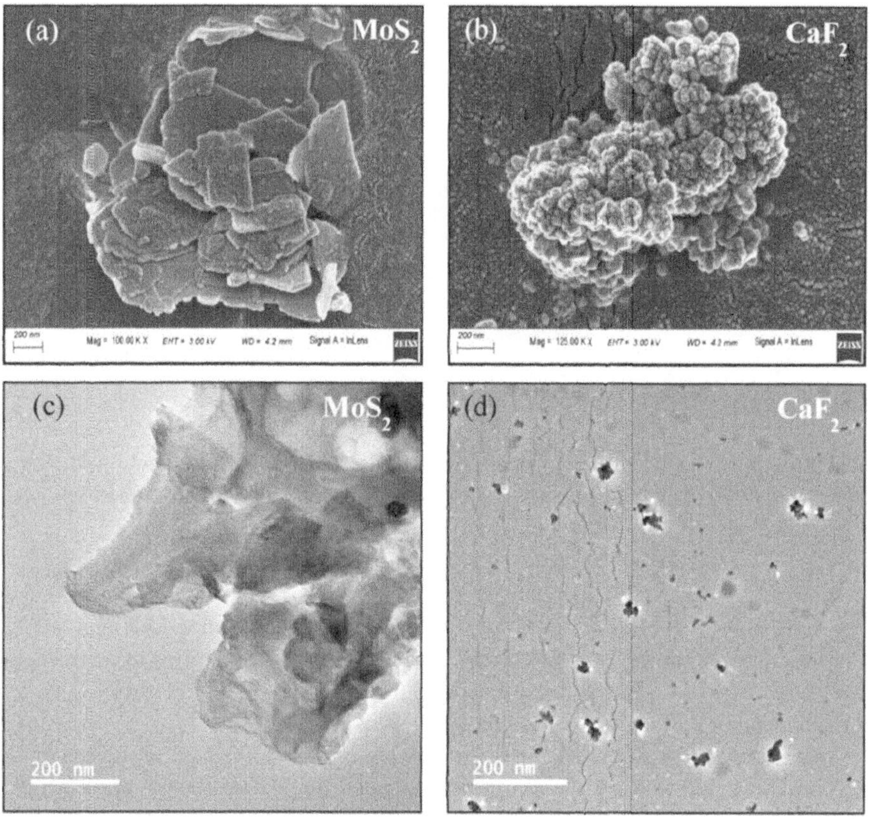

FIGURE 3.4 Surface micrographs of (a) molybdenum disulphide and (b) calcium disulphide and transmission electron micrograph of (c) molybdenum disulphide and (d) calcium disulphide. (Reproduced with permission from Ref. (Gajrani et al., 2019). Copyright Elsevier, 2019.)

operational conditions. Nanocutting fluids might exhibit different behaviours under diverse cutting parameters, material types, or machining environments. Achieving a balance between stability and adaptability is essential to guarantee reliable and effective performance in real-world machining scenarios. Additionally, concerns about health, safety, and environmental impacts associated with nanoparticle exposure require attention. Proper handling, disposal practices, and thorough risk assessment studies are crucial to mitigate potential hazards associated with nanoparticle usage in cutting fluids (Figures 3.6–3.8).

In another work, water-based cutting fluid with nanoparticles of silicon oxide and copper oxide (optimum 0.4–0.7 wt%) uses a surfactant of sodium dodecyl sulphate of 10 wt% used for better dispersibility and stability of nanolubricant (Musavi et al., 2019). The optimum parameters (nanofluid flow rate, injection pressure, nozzle angle, and gap between nozzles and tooltip) set for MQL cutting condition are 150 cm^3/h, 600 kPa, 60°, and 3 cm, respectively. *n*CuO particles perform better than $nSiO_2$ particles in terms of surface quality, and surface roughness improved for *n*CuO-based

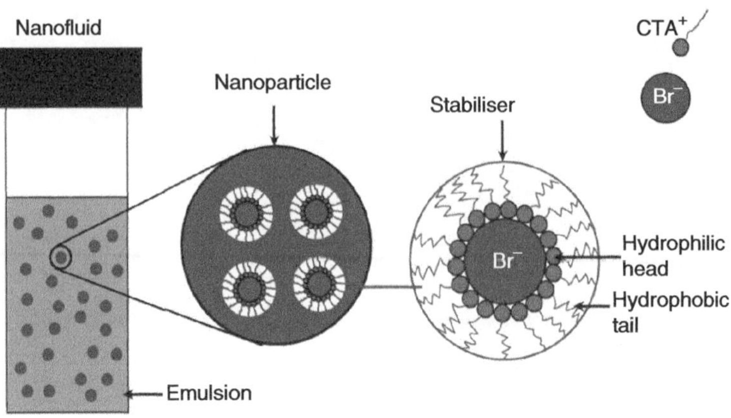

FIGURE 3.5 Schematic representation of nanofluid showing the hydrophilic and hydrophobic tails with a suitable stabiliser. (Reproduced with permission from Ref. (Gajrani et al., 2019). Copyright Elsevier, 2019.)

nanofluid by about 26.5% and 17.8% compared with flood and MQL mode of operation. The surface quality of nanofluid lubricant with surfactant and without surfactant used for machining operations saw around 14% and 4% improvement. It shows how the presence of a surfactant influences the dispersion of nanoparticles, preventing the formation of aggregates (Figures 3.9–3.12).

Two different types of nanoparticles, namely copper and silver, are dispersed in coconut oil (Khan et al., 2019). It was noted that adding 0.25% copper nanoparticles with coconut oil gives the optimum concentration, resulting in the lowest COF compared with the same concentration of Ag nanoparticles blended with coconut oil; this is because the nanoparticles present in between the surface contact facilitate rolling rather than sliding, leading to reduced COF. The mending effect of dispersed nanoparticles is evident in the creation of a tribo-chemical layer between the interacting surfaces, which is particularly beneficial for soft metals like silver, copper, and lead. This layer aids in mitigating direct contact, consequently reducing the shear strength of the tribological pair. The optimal concentration for achieving minimal cutting force and cutting temperature is observed at 0.25% when using copper nanoparticles as compared to silver nanoparticles. There is the same trend followed by cutting force as COF, i.e. beyond 0.5%, it increases and then decreases at a concentration of 1%. Therefore, the decreasing COF can be correlated with the reduction in the cutting force. The optimised MQL flow rate of nanolubricant for alloy steel turning operation was 10 mL/min. The enhanced concentration of the nanoparticles results in an increase in the viscosity of the nanofluid.

In another work, Jojoba is used as a base fluid with $n\mathrm{MoS_2}$ (0.1% optimum) as a nano additive (Gaurav et al., 2020). The usage of the base fluid with nano additive reduced cutting force by 34.39% and 30.21% with only base fluid (Jajoba fluid); the observed phenomenon could be attributed to the laminar crystalline structure of MoS_2 nanoparticles, which imparts low friction resistance. Additionally, in nano-MoS_2 particles, the presence of S-Mo-S structures held together by weak

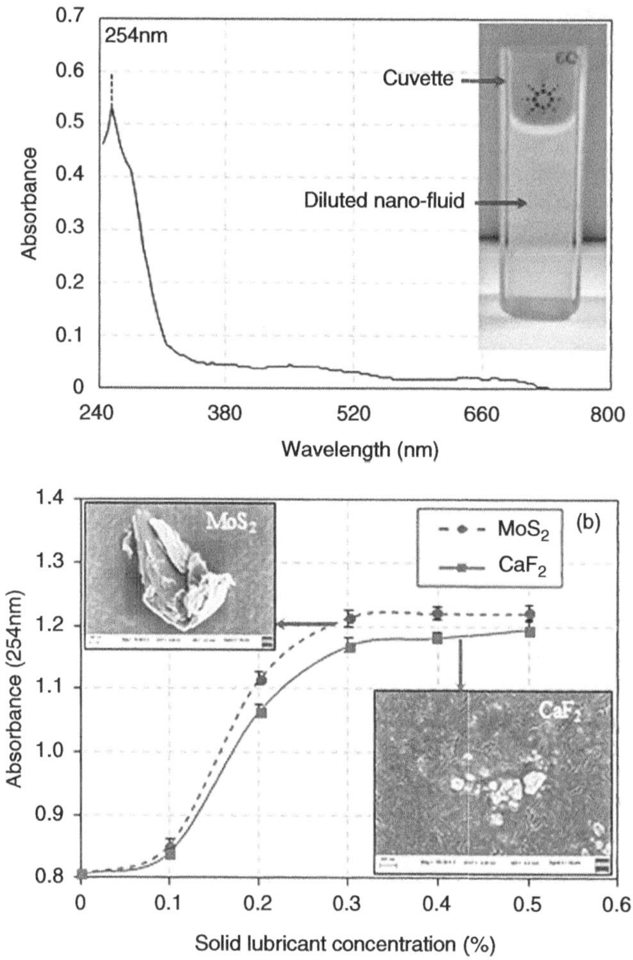

FIGURE 3.6 Changes in absorbance concerning (a) wavelength and (b) concentration of solid lubricant during dispersion stability testing. (Reproduced with permission from Ref. (Gajrani et al., 2019). Copyright Elsevier, 2019.)

van der Waals forces in a hexagonally packed arrangement facilitates the sliding of sulphur atoms against each other, contributing to the attainment of low frictional resistance. Further, as jojoba oil has the capability to absorb more sulphur and at high temperatures, sulphurisation of jojoba oil takes place, leading to improved stability and lubricant properties of cutting fluid. A rise in the weight percentage of nanoparticles results in the agglomeration of particles within the base fluid. This agglomeration, when entrapped in the asperities of the machining surface, restricts the movement of surrounding particles. Consequently, this leads to elevated stress concentration and abrasion, manifesting as wear marks on the machine's surface. $n\text{MoS}_2$ (0.1%) + jojoba base fluid reduces surface roughness by about 40.67% against dry cutting and 10.3% against pure jojoba oil in MQL cutting conditions; it happens

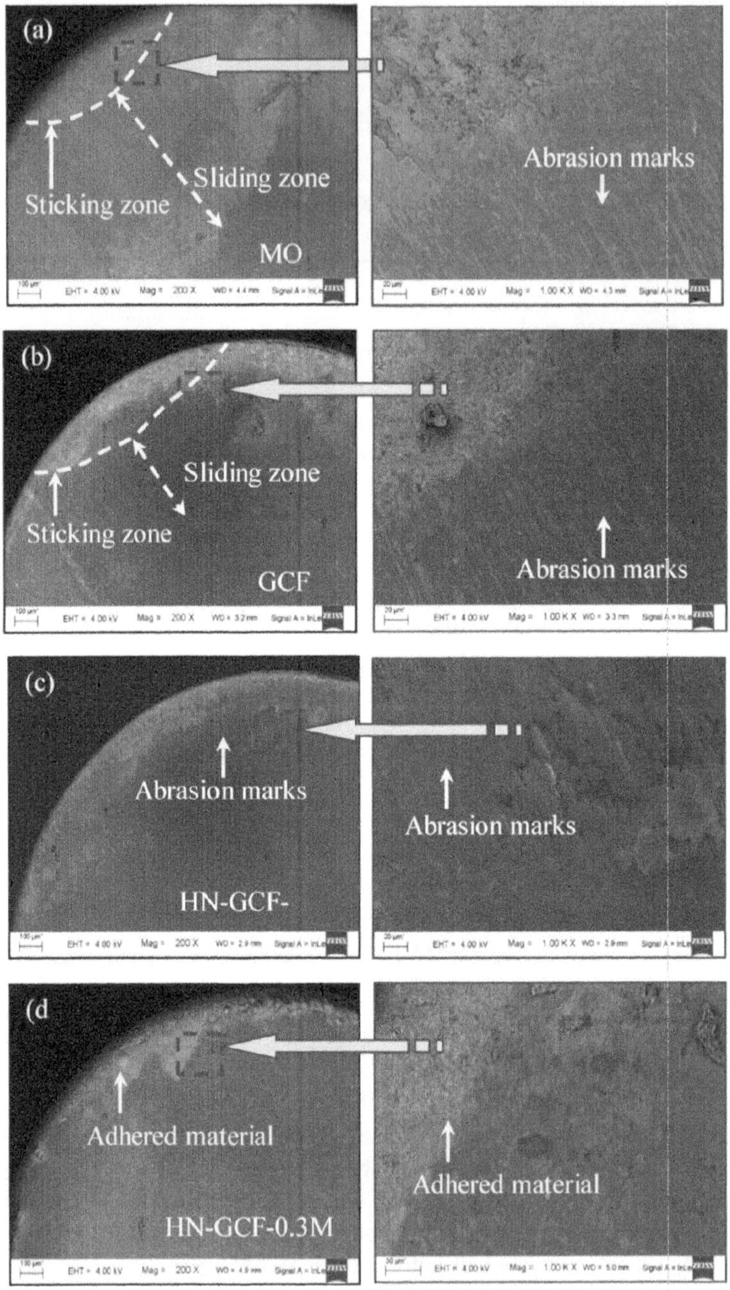

FIGURE 3.7 The surface structure of the cutting tool's rake face after machining with MQCF is observed using (a) mineral oil, (b) eco-friendly cutting fluid, (c) a hybrid nanogreen cutting fluid (HN-GCF-0.3C) with a 0.3% CaF_2 concentration, and (d) a hybrid nanogreen cutting fluid (HN-GCF-0.3M) with a 0.3% MoS_2 concentration. (Reproduced with permission from Ref. (Gajrani et al., 2019). Copyright Elsevier, 2019.)

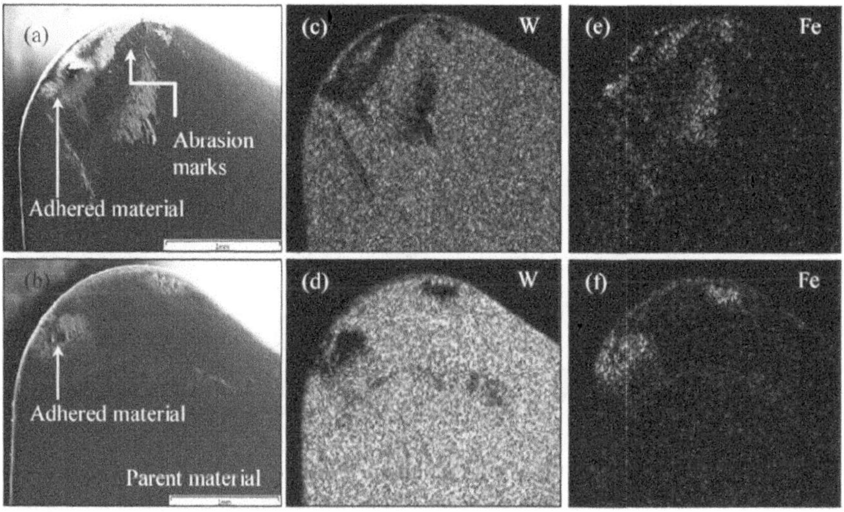

FIGURE 3.8 The surface characteristics of the cutting tool's rake face after machining with MQCF were examined under different conditions: (a) Mineral oil (MO). (b) A hybrid nanogreen cutting fluid (HN-GCF-0.3M) with a MoS_2 concentration of 0.3%. (c) Elemental mapping of Tungsten (W) for the examined cutting tool's rake face using mineral oil (MO). (d) Elemental mapping of tungsten (W) for the examined cutting tool's rake face using HN-GCF-0.3M. (e) Elemental mapping of iron (Fe) for the examined cutting tool's rake face using mineral oil (MO). (f) Elemental mapping of iron (Fe) for the examined cutting tool's rake face using HN-GCF-0.3M. (Reproduced with permission from Ref. (Gajrani et al., 2019). Copyright Elsevier, 2019.)

as the nanoparticles have good heat-carrying capacity and lubricating properties at elevated temperature and MoS_2 nanoparticles also act as a ball in a ball bearing helps reduce COF ultimately the cutting force and surface finish reduced. Tool wear with 0.1% MoS_2 + Jajoba base oil decreased up to 47.37% and 31.58% for only base oil when compared with dry cutting. The primary reason for this phenomenon is the creation of a protective film formed by the long-chain monoester and high-viscosity base oil at elevated temperatures. This protective film serves to restrict friction and thermal damage, effectively safeguarding tools from experiencing a high wear rate. But when an increase in wt% of nanoparticles produces a thick layer on the tool surface, it shows a built-up edge (BUE), which is detrimental for other parameters while getting an advantage in case of tool wear. It was also proven economical and showed about 27% cost reduction compared with commercial mineral oil.

3.3 DRILLING

The choice of nanoparticle material and its concentration in nanocutting fluids profoundly influences their tribological properties and overall efficacy in machining operations. Various nanoparticles, such as graphene oxide, carbon nanotubes (CNTs), and metal nanoparticles (like TiO_2, SiO_2, or MoS_2), exhibit distinct tribological characteristics when dispersed in cutting fluids. Each nanoparticle type offers unique advantages.

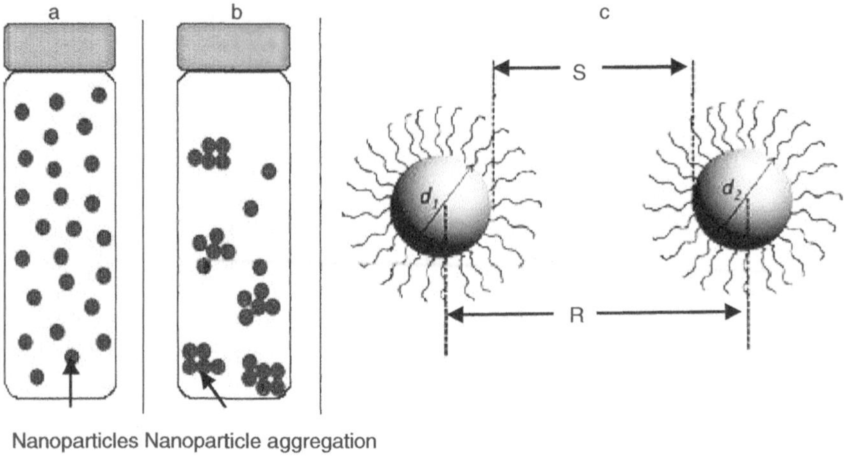

FIGURE 3.9 Illustration depicting the deposition mechanisms of nanoparticles: (a) nanofluid containing the surfactant and (b) nanofluid without surfactant. (c) Positioning of two nanoparticles. (Reproduced with permission from Ref. (Musavi et al., 2019). Copyright Elsevier, 2019.)

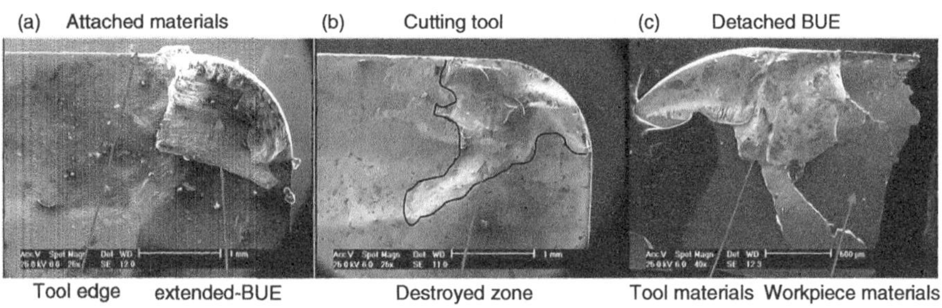

FIGURE 3.10 SEM image of disengaged particles of BUE on the rack face of the cutting tool. (Reproduced with permission from Ref. (Musavi et al., 2019). Copyright Elsevier, 2019.)

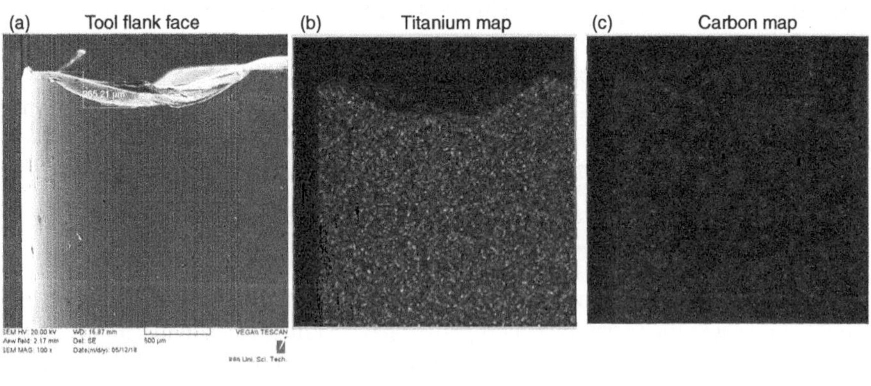

FIGURE 3.11 Surface morphological characterisation and elemental distribution of the tool flank face. (Reproduced with permission from Ref. (Musavi et al., 2019). Copyright Elsevier, 2019.)

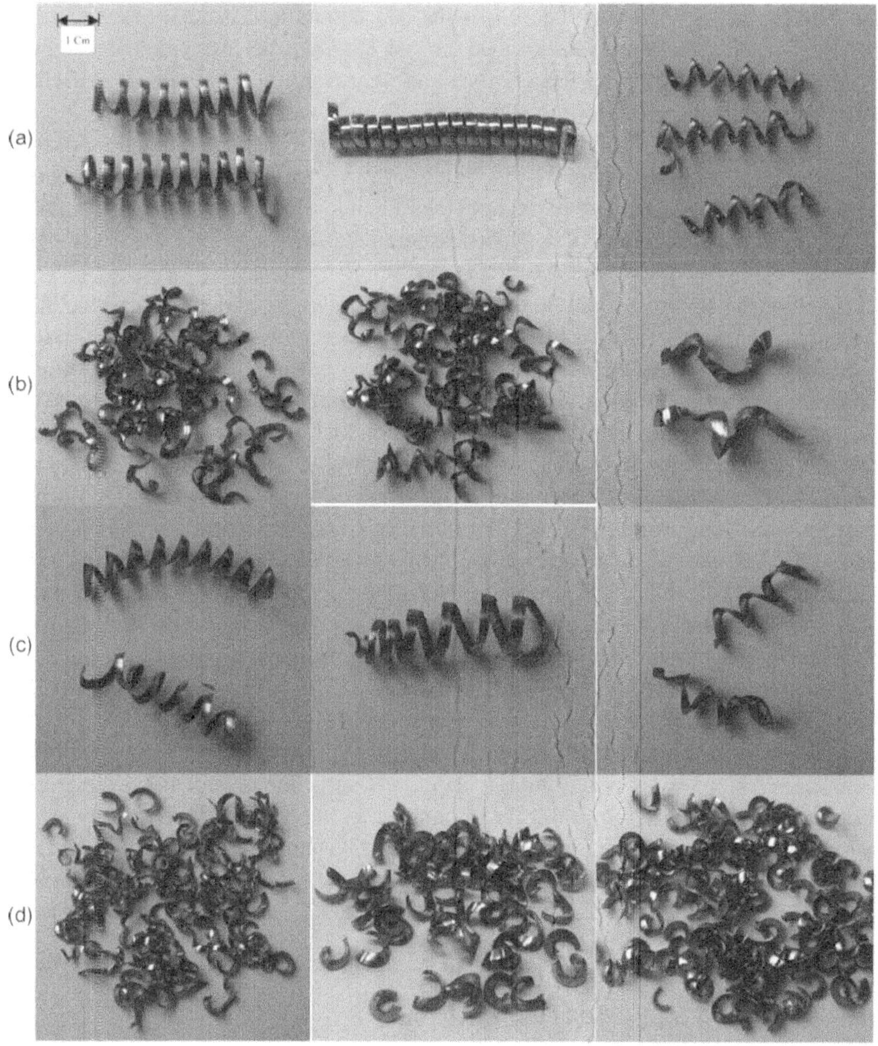

FIGURE 3.12 Morphologies of chip formation under various conditions, specifically at a cutting speed of 70 m/min, a feed rate of 0.22 mm/rev, and a depth of cut of 1 mm. The condition includes (a) conventional fluid (semicontinuous), (b) nanofluid without surfactant after 5 minutes (discontinuous), (c) nanofluid without surfactant after 2 hours (semicontinuous), and (d) nanofluid with surfactant after 2 hours (discontinuous). (Reproduced with permission from Ref. (Musavi et al., 2019). Copyright Elsevier, 2019.)

For instance, graphene-based nanoparticles possess exceptional mechanical strength, high thermal conductivity, and low friction properties. Metal oxide nanoparticles contribute to improved lubrication and wear resistance while also enhancing cooling effects due to their high thermal conductivity. The choice of nanoparticle material is often tailored to specific machining requirements, considering factors such as material compatibility, cost, and environmental impact. Moreover, the tribological behaviour of the cutting fluid is notably influenced by the concentration of nanoparticles. Optimal nanoparticle concentration is determined through careful experimentation and analysis to achieve the desired lubrication, cooling, and wear-resistant properties. A concentration that is too low might not provide significant improvements, while an excessively high concentration can lead to agglomeration or adversely affect fluid flow characteristics, necessitating a balance for optimal performance. Balancing material choice and concentration is crucial to harnessing the maximum benefits of nanocutting fluids. Careful selection and control of nanoparticle parameters play a pivotal role in achieving desired tribological effects and optimising machining processes.

To prepare nanofluid, titanium dioxide nanoparticles are added into soluble oil as the base oil for good stability and gum Arabic surfactant in 1:1 weight concentration (Salimi-Yasar et al., 2017). The roughness and scratch depth of the drill bit exhibited improvement when compared to other machining conditions. This improvement can be ascribed to the high thermal, anti-friction, conductivity, and anti-erosion properties of the nanocutting fluid. Employing nanofluid in the machining zone leads to a notable reduction in temperature, primarily attributed to the enhancement of the heat transfer coefficient through an increase in nanoparticle concentration. Additionally, an increase in nanoparticle concentration was observed to enhance viscosity, while the viscosity of the nanofluid decreased with rising temperature. This temperature–viscosity relationship is attributed to the weakening of intermolecular forces. Also, the density of nanofluid decreases while temperature increases; this is attributed to thermal expansion. At 1 wt% of TiO_2 in nanofluid at temperatures 20°C–60°C, the density decreases by 1.46% and 1.56%, respectively (Figures 3.13 and 3.14).

In another work, graphene oxide powder is used as the nanoparticles (Yi et al., 2017). Graphene oxide powder of a 1–5-layer nanosheet suspended nanolubricant increases thermal conductivity by up to 47.81% in comparison to conventional coolants, which helps improve heat transfer results and decrease cutting temperature. In this experiment, the COF decreased by 37.14% due to the formation of a lubrication film, which also helped to reduce abrasion. The formation of microcracks at the machined surface is

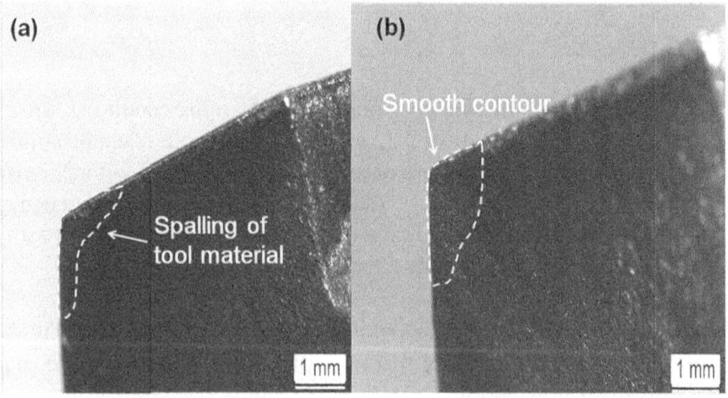

FIGURE 3.13 Variations in flank wear when using different coolants: (a) Conventional coolant. (b) Fluid containing suspended graphene oxide. (Reproduced with permission from Ref. (Yi et al., 2017). Copyright Elsevier, 2017.)

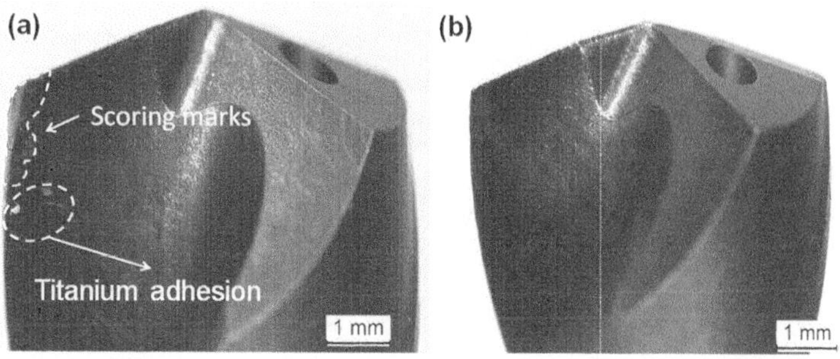

FIGURE 3.14 Comparison of tool wear: (a) worn-out tool showing the scoring marks and titanium adhesion. (b) Nascent tool. (Reproduced with permission from Ref. (Yi et al., 2017). Copyright Elsevier, 2017.)

caused by plastic deformation, and this crack propagation causes mechanical and thermal instability as a result of fluctuation in cutting temperature in the machining zone. Due to the use of GnP in base fluid, it is easy for GnP to penetrate into machining zones, resulting in better lubrication and leading to a decreased COF compared to conventional lubricants. Under spindle speeds less than 1600 rpm and a feed rate of 0.12 mm/rev, the utilisation of GnP-based nanolubricant and conventional lubricant resulted in the formation of continuous spiral chips (referred to as excellent chips) and squeezed string chips (characterised as discontinuous chips). When cutting speed increases, the friction force also increases, and it was seen that the friction force decreases on average by 6.67% compared with conventional nanofluid (Figures 3.15 and 3.16).

When MQL was used for lubricating purposes, a boundary film formation between the interface of tool and work and the nanoparticles in lubricants leads to a polishing or rolling effect at the interface that helps to obtain lesser COF between the surface of the drill and NiTi workpiece and thrust force (Rosnan et al., 2019). With the same cutting condition as MQL nanofluid, when comparing flood lubricants, the magnitude of thrust force is about 2–4 times higher. It also believes that flood coolant was not able to decrease high heat generation at the cutting zone, which is the

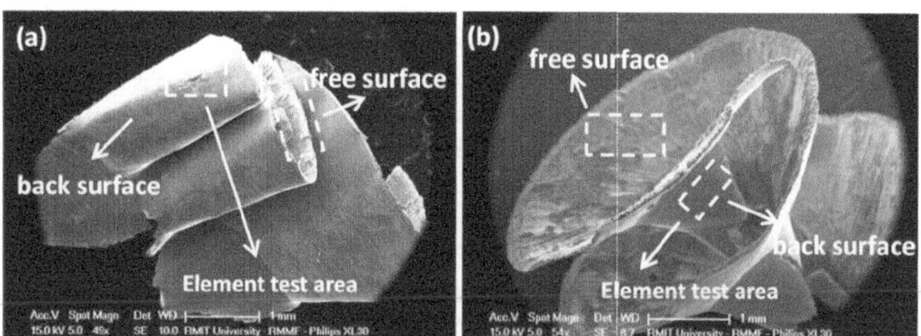

FIGURE 3.15 Surface morphological characterisation of formed chips with (a) conventional coolant and (b) graphene oxide suspended fluid. (Reproduced with permission from Ref. (Yi et al., 2017). Copyright Elsevier, 2017.)

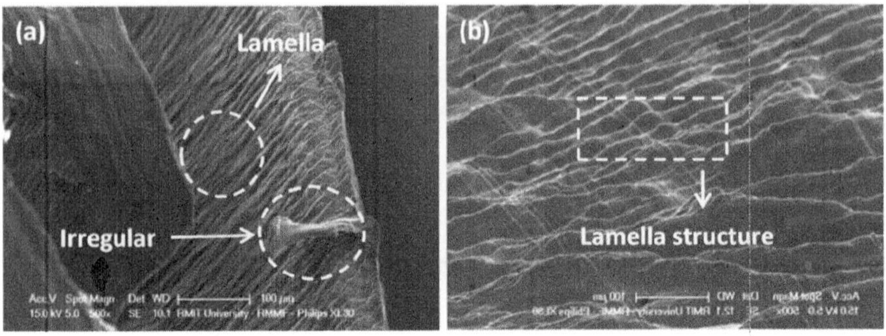

FIGURE 3.16 Morphology of chip-free surface with the application of (a) conventional coolant and (b) graphene oxide suspended fluid. (Reproduced with permission from Ref. (Yi et al., 2017). Copyright Elsevier, 2017.)

main cause of increased thrust force. The optimum cutting speed for nanolubrication in the MQL environment was 10–20 m/min to achieve the lowest tool wear. Surface roughness values ranging from 0.7 to 1.2 μm have maintained uniform and consistent results regardless of the drilling speed, cooling condition, and types of coating on the drill. However, surface roughness values are slightly higher under MQL nanolubricant compared to flood conditions (Figure 3.17).

In another work, the MQL technique is utilised for machining with conditions such as 120 mL/h, 6 bar compressed air with 1.5% graphene nanoparticles (GnPs) presence in vegetable-based base oil (Pal et al., 2020). The outcomes indicated that utilising MQL with nanofluid resulted in significant reductions in torque (64.9%), COF (51.7%), surface roughness (33.8%), and thrust force (27.4%) when compared to drilling conditions with pure MQL. During pure MQL drilling, minute droplets of base oil are deposited on both the tool and machined surface, resulting in the formation of a less stable protective tribofilm due to poor heat dissipation. However, when utilising GnP-based nanolubricant, the cooling performance improves significantly owing to its high thermal conductivity. During the primary drilling operation, the predominant wear mechanism observed is adhesion under various drilling conditions. The inability to disperse the generated heat in the machining area results in the fusion of work material and chips, forming a BUE and causing elevated adhesion wear. However, when employing nanofluid with a nanoparticle concentration of 1.5 wt%, adhesion wear is significantly reduced, potentially attributed to the increased viscosity with higher nanoparticle concentration. Additionally, the presence of GnPs acts as a roller between the contacting surfaces, contributing to a reduction in the COF (Figures 3.18 and 3.19).

In another study (Şirin et al., 2021), it was found that the addition of surfactants in nanofluid results in a slightly decreased thermal conductivity. The increase in wettability, indicated by a decrease in contact angle, demonstrates a stronger affinity of liquid particles for solid surfaces, leading to a broader spread of liquid droplets. Nanolubricants formulated with SDS surfactants exhibit superior wettability compared to GA. The incorporation of nanoparticles and surfactants into the base fluid, despite causing an increase in surface tension, requires careful optimisation of their concentration to achieve desirable outcomes. Elevated cutting force and torque

FIGURE 3.17 The experimental setup depicts the minimum quantity of lubrication nozzle, dynamometer, workpiece, and jig. (Reproduced with permission from Ref. (Rosnan et al., 2019). Copyright Elsevier, 2019.)

contribute to increased power consumption during machining. Employing hBN/GNP-SDS nanofluid at cutting speeds of 30 m/min and feed rates of 0.04 mm/rev results in a substantial 23.51% reduction in thrust force compared to dry machining cutting conditions. This decrease can be ascribed to the improved lubricating properties facilitated by the minimal cutting angle, a feature advantageous for hybrid nanofluids. Importantly, the introduction of hBN-GnP-SDS nanofluid significantly mitigates changes in torque (by 46.99%) and thrust value (by 38.79%) in comparison to both dry cutting and the base fluid, thereby effectively reducing thrust force. Among the monodisperse nanofluids, the combination of hBN-SDS demonstrates significant improvements, reducing torque by 41.9% and 32.10% compared to dry cutting conditions and base cutting fluid conditions, respectively, because of the spherical shape of hBN nanoparticles. The use of hybrid hBN/GnP nanofluid results in a temperature reduction of approximately 27.37% during machining compared to dry cutting. The decrease in surface roughness is credited to several mechanisms, such as mending, rolling, cushioning, polishing, and interlayer sliding effects. These effects are a result of the spherical structure of hBN, and the lamellar structure of

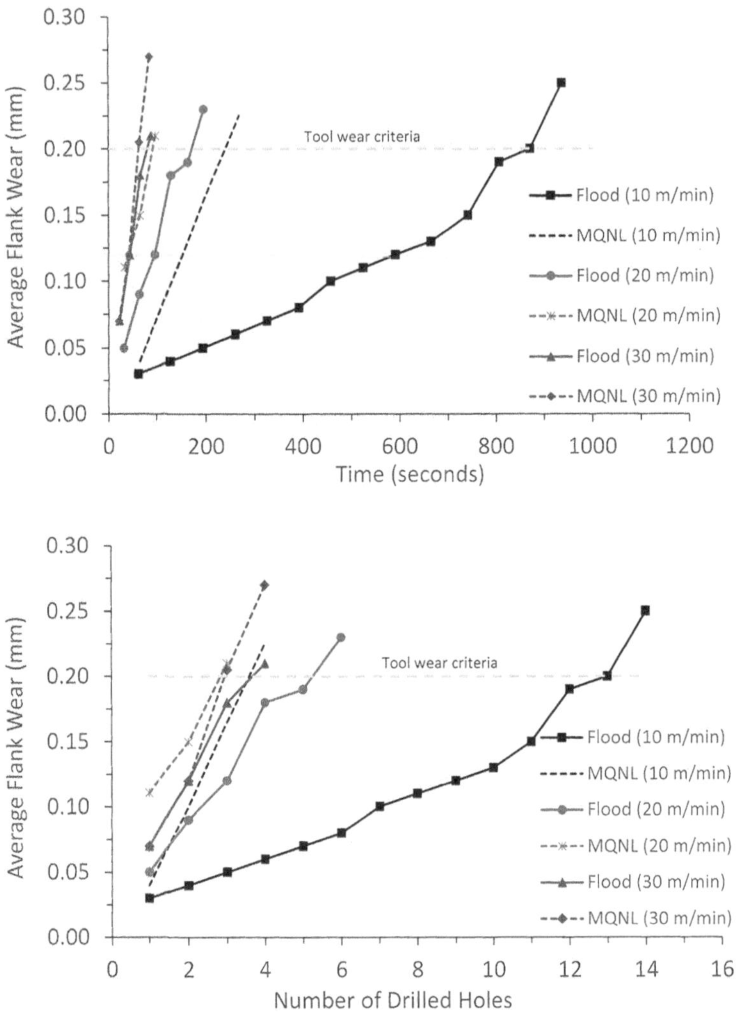

FIGURE 3.18 Variation of the average of flank wear with the change in (a) time (seconds) and increase in the number of holes drilled. (Reproduced with permission from Ref. (Rosnan et al., 2019). Copyright Elsevier, 2019.)

GnP particles present within the hybrid nanofluid. In the presence of hBN/GnP-SDS lubricant, there is a reduction in surface roughness by 46.83% compared to dry cutting and 45.56% compared to cutting with the base fluid. This reduction can be linked to the lower contact angle, facilitating improved nanofluid distribution over the machining surface and resulting in reduced surface roughness. Moreover, tool wear demonstrates a 51.96% improvement under hBN/GnP-SDS nanolubricant conditions compared to dry cutting conditions. Both hybrid and monodisperse nanofluids with SDS surfactants demonstrate good performance in minimising tool wear in contrast to those containing GA surfactants. While various parameters indicate the effectiveness of hBN/GnP-SDS hybrid nanofluid in reducing thrust force and torque,

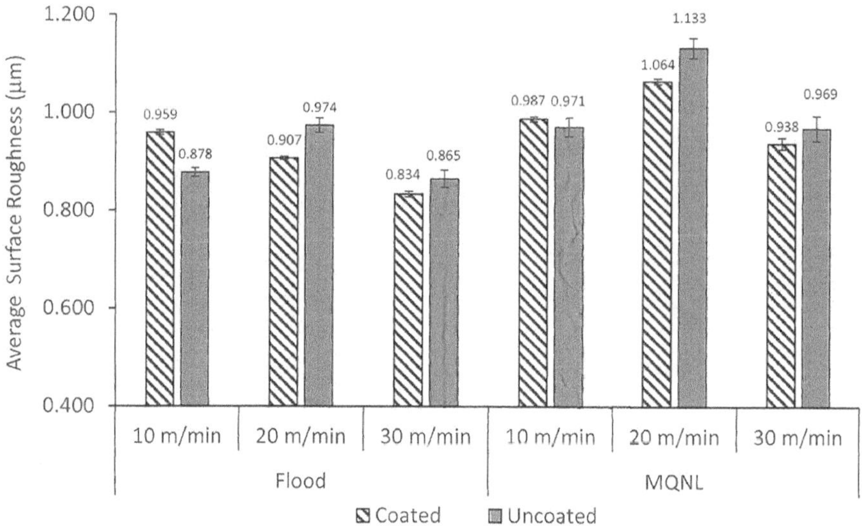

FIGURE 3.19 Variation of surface roughness for coated and uncoated tools for the feed rates of 10, 20, and 30 m/min. (Reproduced with permission from Ref. (Rosnan et al., 2019). Copyright Elsevier, 2019.)

improving hole quality, and minimising tool wear, hBN/GnP cutting stands out as particularly effective in lowering cutting temperatures (Figures 3.20–3.23).

In the work using CuO nanoparticles (Ni et al., 2021), with increasing nanoparticle concentration, the lubrication effect increases with a reduction in cutting force and at 1.1 wt%. It shows negative results because the formation of agglomeration in nanoparticles and optimising CuO nanoparticles of concentration 0.54 wt% with sesame base oil (30 g) gives load reduction of about 62.7% and 68.6% compared to pure base oil. In general, carboxylic groups found in fatty acids and ester groups within triglycerides are inherently polar. These groups exhibit strong affinity and can be physically adsorbed onto the metal surface through van der Waals forces, forming a lubricant film that reduces the COF and the corresponding load. Unsaturated fatty acids, characterised by C=C double bonds, are susceptible to oxidation, which can lead to the degradation of the base oil. This degradation results in a reduction in the antioxidant and thermal stability of the base oil, ultimately causing the failure of the physically adsorbed film. On the contrary, saturated fatty acids possess a linear carbon chain, enabling a strong bonding of film molecules, especially under high-pressure conditions within the workpiece–tool–chip interface. From the experiment, it was proved that about 10–15 μm CuO nanoparticles of 0.54 wt% show the best combination apart from other concentrations and particle sizes (Figures 3.24–3.26).

3.4 MILLING

Nanoparticles such as MoS_2, graphite, and hBN are employed for dispersion in cutting fluids as nanolubricants (Şirin and Kıvak, 2019). The optimal nanoparticle concentration in base liquids, including water, synthetic oil, and vegetable oil, was obtained to

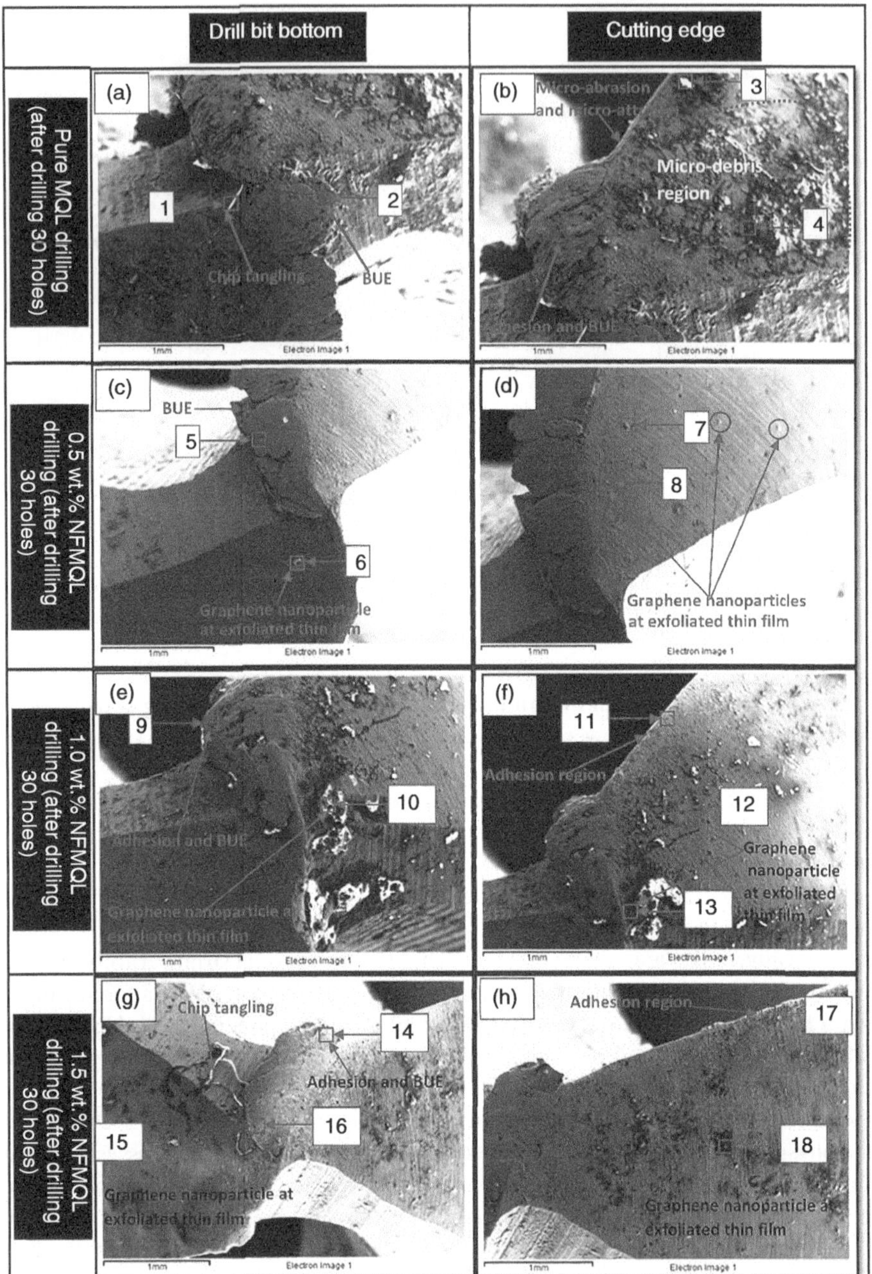

FIGURE 3.20 Surface morphological characterisation of wear on the drill bits under different MQL conditions. (Reproduced with permission from Ref. (Pal et al., 2020). Copyright Elsevier, 2020.)

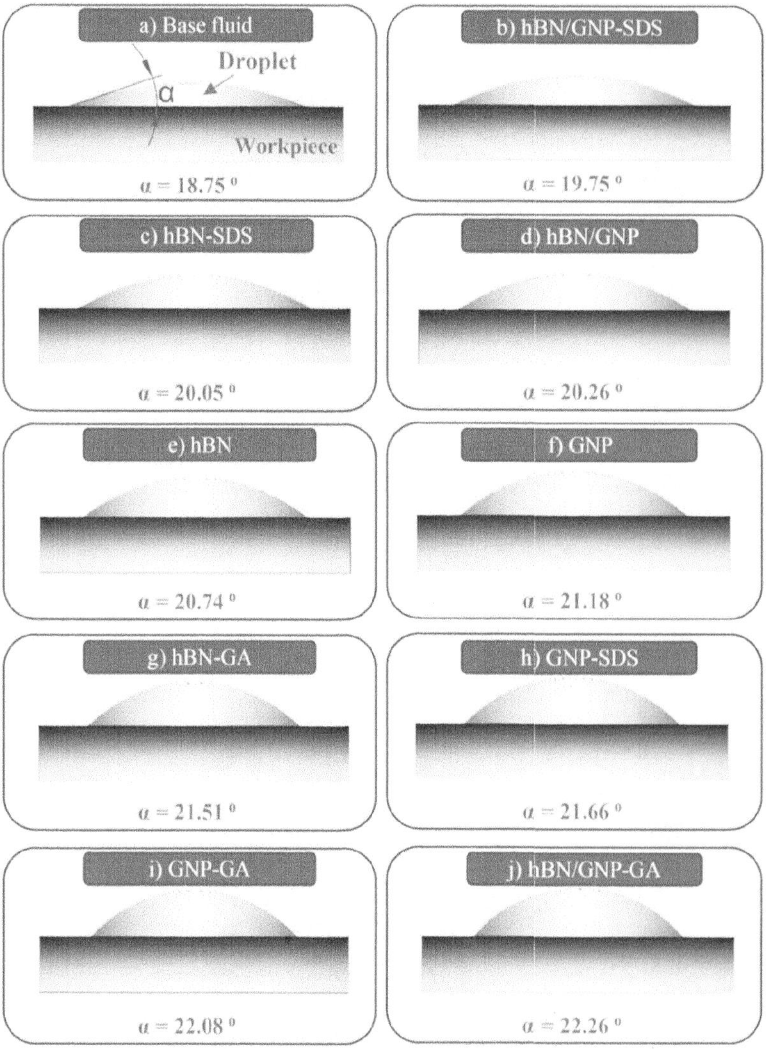

FIGURE 3.21 Contact angle results for different nanofluids. (Reproduced with permission from Ref. (Şirin et al., 2021). Copyright Elsevier, 2021.)

be 0.50 vol%, with hBN nanoparticles exhibiting superior performance compared to MoS_2 and graphite nanoparticles. The hBN nanofluid of concentration 0.50% provided a decreased Ra value, while the comparison with dry and base fluid was 54.26% and 45.42%. Ra values in the case of MoS_2 nanoparticles (45.45% & 34.92%) and with graphite nanoparticles (41.76% & 30.51%) decrease compared with dry and base fluid cutting environments. The minimum Ra value, as compared to other nanoparticles, was observed in hBN due to its lower viscosity. This property aids in the penetration of the lubricant into the cutting area under high pressure through MQL, resulting in a rolling effect of the dispersed nanoparticles and the formation of a tribo film. Also,

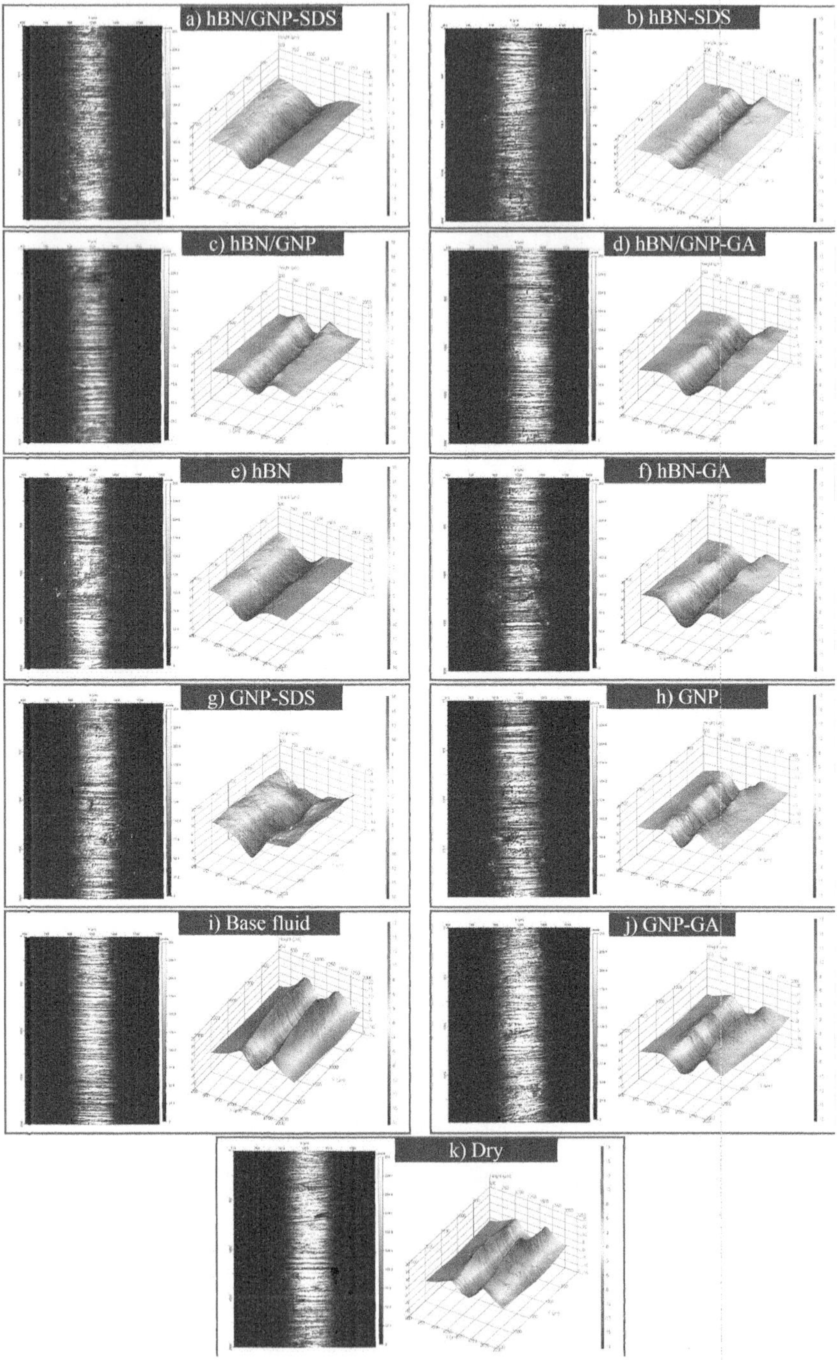

FIGURE 3.22 The surface characteristics and appearance of the hole vary with distinct cutting conditions. (Reproduced with permission from Ref. (Şirin et al., 2021). Copyright Elsevier, 2021.)

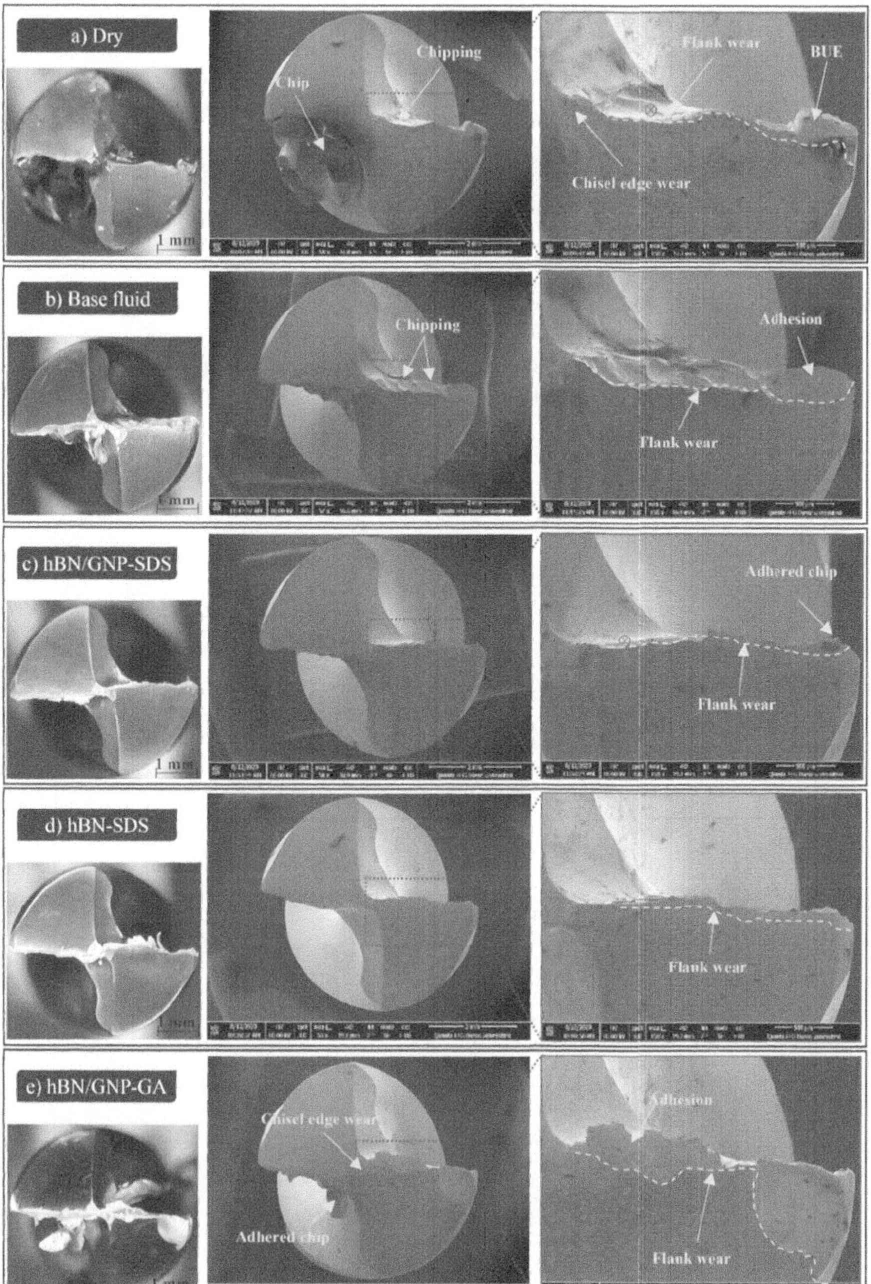

FIGURE 3.23 Morphological analysis of the worn tool surface under different cutting conditions. (Reproduced with permission from Ref. (Şirin et al., 2021). Copyright Elsevier, 2021.)

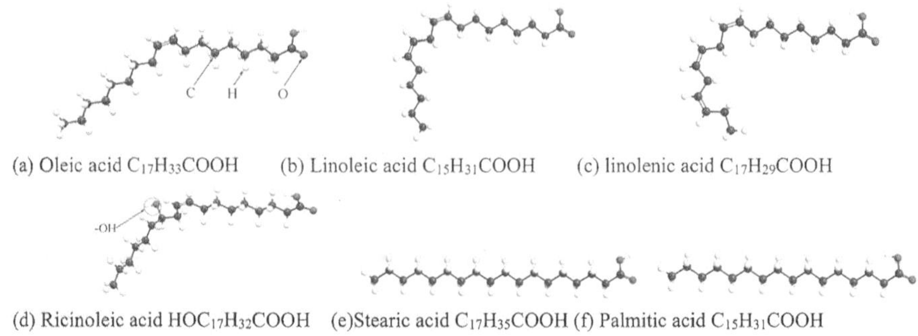

FIGURE 3.24 Molecular structures of different fatty acids. (Reproduced with permission from Ref. (Ni et al., 2021). Copyright Elsevier, 2021.)

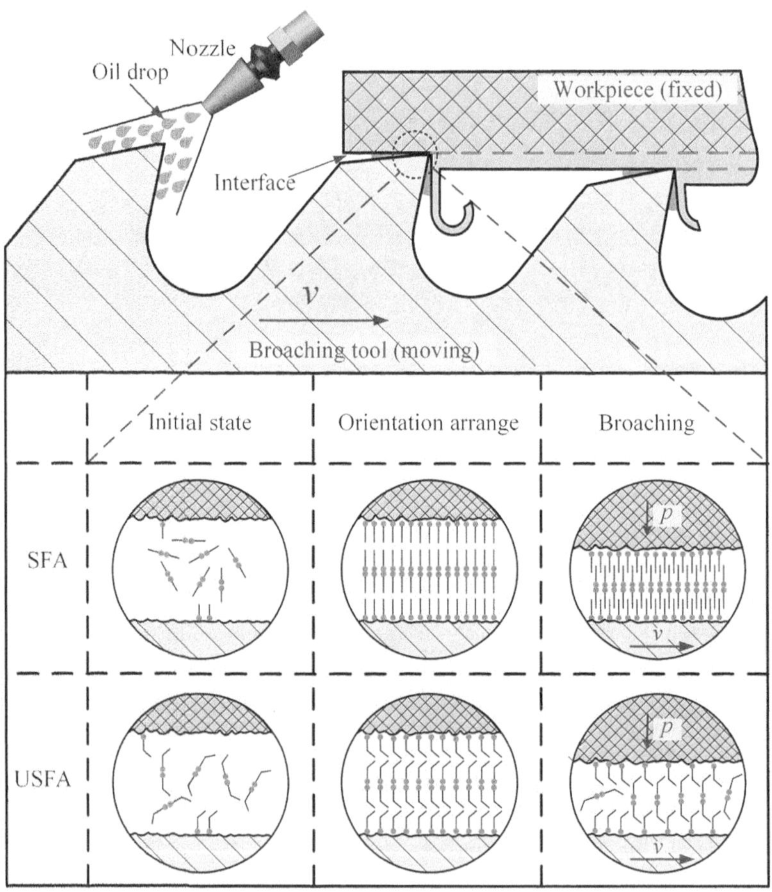

FIGURE 3.25 Schematic diagram of typical fatty acid lubrication. (Reproduced with permission from Ref. (Ni et al., 2021). Copyright Elsevier, 2021.)

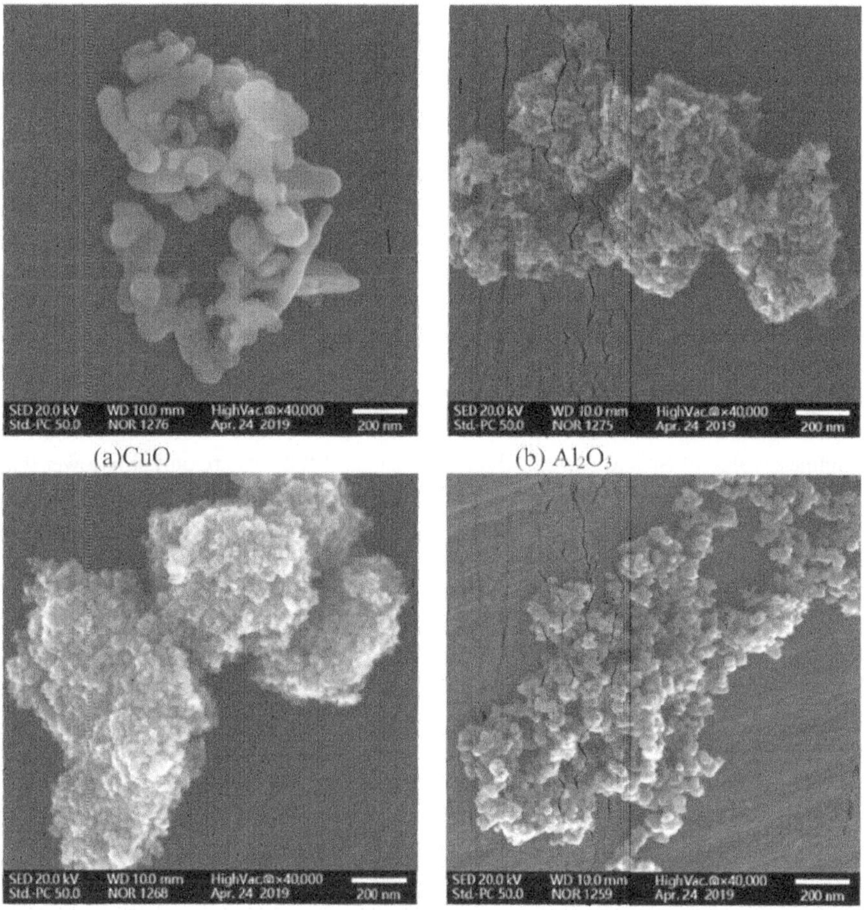

FIGURE 3.26 Surface morphological characterisation of different nanoparticles. (Reproduced with permission from Ref. (Ni et al., 2021). Copyright Elsevier, 2021.)

the high thermal stability of hBN nanofluid results from its hexagonal crystal structure with strongly bonded atoms in the same layer, while between each layer, the van der Waals force is weak. These layers stick to the surface in contact and slide against one another, effectively minimising friction. The nanofluid's cutting temperature is observed to be the lowest in MoS_2 nanofluid, followed by graphite and then hBN nanofluid. This is mainly attributed to the higher thermal conductivity of hBN compared to others, as well as the optimal concentration of the nanofluid. Nanofluid increased tool life by 48.7%, 33.3%, and 20.5% in the case of hBN, MoS_2 and graphite nanoparticles (0.50 vol%) compared with base fluid cutting conditions.

Electrostatic MQL (EMQL) is an advanced version of the MQL technique (Lv et al., 2021). The SiO_2 water PEG-based nanolubricant at 0.5 wt% remained stable for a duration of 30 days without experiencing any precipitation. The findings indicated that with an increase in flow rate, the concentration of PM10 and PM2.5 also rose. Notably, when comparing charged and uncharged cutting fluids at identical flow rates, the charged cutting fluid exhibited a lower oil mist concentration. Consequently, it was chosen for further experiments at a flow rate of 20 mL/h. Modifying the oil–air pressure led to a notable rise in concentrations of PM10 and PM2.5, presenting potential health risks. To optimise conditions and prevent excessive particle concentration, 0.3 MPa was utilised for this experiment. Furthermore, elevated spindle speed and increased depth of cut resulted in heightened concentrations of PM10 and PM2.5. Remarkably, under identical conditions, the oil mist concentration from EMQL was lower in comparison to that from MQL, with the most effective values determined at a spindle speed of 1000 rpm and a depth of cut of 1 mm. Adhesive wear observed on the tool flank face emerged as the primary failure mode due to elevated heat at the contact zone, causing chip melting and adhesion. Comparative analysis demonstrated that, although tool wear and cutting force were higher in the SiO_2 water-based nanolubricant of the EMQL method compared to vegetable oil-based EMQL, they still stood lower than those observed in oil-based MQL, highlighting the effectiveness of EMQL as a technique. Moreover, the surface roughness value of SiO_2 water-based nanolubricant EMQL was remarkably lower than that of the other lubricants examined in the study. This indicates its potential as a feasible substitute for conventional oil-based MQL, presenting an environmentally friendly approach by utilising SiO_2 water-based nanolubricant EMQL (Figures 3.27–3.29).

In comparison to the base fluid, the nano-hBN cutting fluid in the milling operation of nickel alloy X-750 in terms of thermal conductivity and viscosity increased by 12.78% and 42%, respectively, at 0.5 vol% of hBN in nanoparticles in base fluid (Şirin et al., 2021). The surface roughness value of hBN nanofluid compared with dry machining conditions was reduced by 47.2%. The cutting force decreases by 0.3% and 6% in base fluid and nanolubricant when compared with dry conditions. In the crystal structure of hBN, the connection between boron and nitrogen atoms forms a robust covalent bond. This structure plays a crucial role in preventing direct contact between the machining surface and effectively bears the load applied on the surface. The hexagonal rings stack extensively atop each other, forming a lattice structure where individual plates are bound together by relatively weak van der Waals forces. Consequently, when a force acts parallel to the plane, these planes exhibit easy movement, shearing against each other due to the inherent weakness of the van der Waals bond. This results in a reduction in friction and wear, which leads to reduced cutting force. The mending and rolling effect also positively impacts the reduction of cutting force. The highest temperatures of the base fluid and hBN mixed nanofluid were reduced by 23.1% and 27.8% when compared with the dry

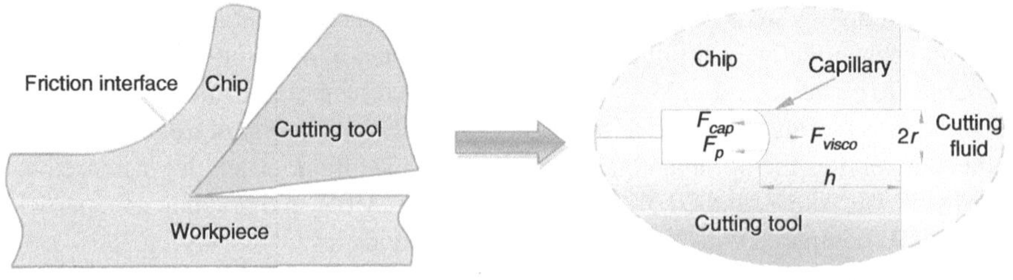

FIGURE 3.27 Penetration model of cutting fluid with insight showing the cutting tool, chip and capillary action of cutting fluid. (Reproduced with permission from Ref. (Lv et al., 2021). Copyright Elsevier, 2021.)

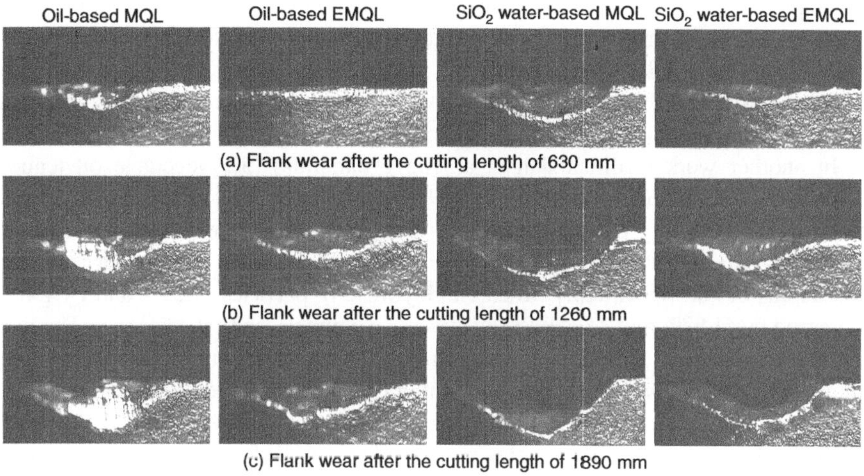

FIGURE 3.28 An optical micrograph illustrates the wear on the tool flank after distinct cutting lengths with diverse lubrication conditions. (Reproduced with permission from Ref. (Lv et al., 2021). Copyright Elsevier, 2021.)

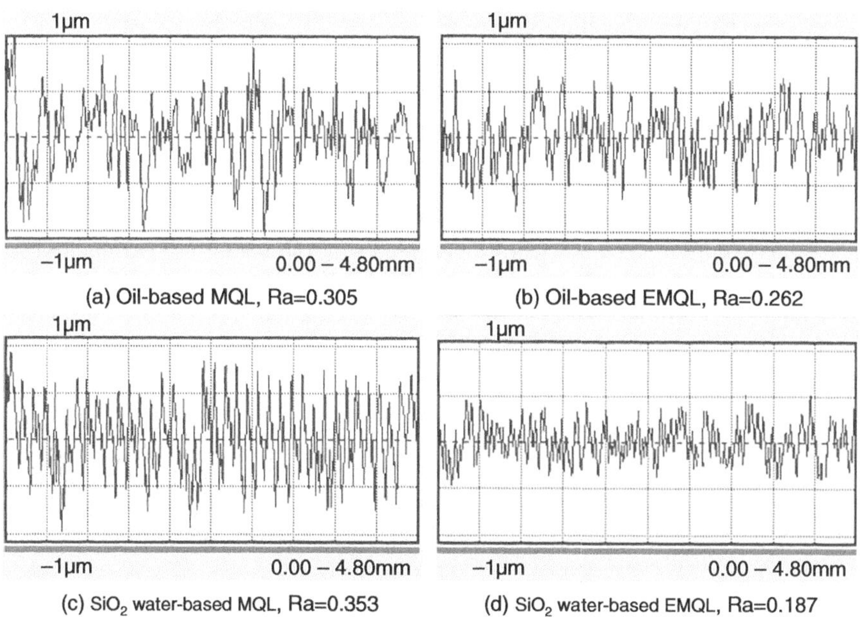

FIGURE 3.29 2D surface roughness measurements of the workpiece under various machining conditions. (Reproduced with permission from Ref. (Lv et al., 2021). Copyright Elsevier, 2021.)

milling case, respectively. Average flank wear of base fluid and nanolubricant increased by 39.8% and 34.4%, in contrast with dry cutting operation. The effect of tool wear in case of nanoMQL and MQL cutting conditions is more than dry cutting, but surface roughness still shows better performance in case of nanofluid because there is no chip sticking on the surface of material milled by nanofluid cutting conditions (Figures 3.30–3.33).

In another work (Kilincarslan et al.,2021), nanofluid (5% vegetable oil (canola based) + 5% *n*Ag solution with 90% deionised water) is used. The machining experiments showed that the Fz cutting force was negligible compared with the Fx and Fy cutting forces. The experiment revealed that cutting forces escalate with higher feed rates and depths of cut. Specifically, when employing *n*Ag particles, the Fx and Fy forces increased by 51.87% and 25.48%, respectively, in comparison to the base oil. The high thermal conductivity of nanoparticles contributes to a reduction in heat transfer, subsequently mitigating the softening effect on metallic materials. However, this reduction in heat transfer results in increased cutting force. Nonetheless, an advantage arises in the form of improved surface roughness and reduced BUE formation when compared to the base oil. Significantly, the application of nanolubricant leads to a notable increase in surface roughness, amounting to 79.71%, as compared to dry conditions. The pivotal factor in curbing cutting temperature is the nanofluid's heat transfer capability, amplified by the high thermal conductivity of the nanomaterial. This enhanced thermal conductivity enables nanolubricants to augment heat transfer within the cutting zone. The stable oil film formation on chip surfaces, facilitated by nanoparticles, contributes to a decrease in cutting temperature, consequently preventing adhesion issues.

Upon optimisation, it became apparent that the nanocutting fluid demonstrated superior performance in attaining minimal surface roughness and cutting temperature. However, concerning minimum cutting force, the base oil showcased favourable outcomes compared to the nanolubricant. Nanocutting fluids present a promising avenue for sustainable machining practices due to their potential to reduce resource consumption, waste generation, and environmental impact. The inclusion of nanoparticles in cutting fluids aims to improve their efficiency, thereby reducing the amount of fluid required during machining operations. This reduction in fluid volume not only minimises resource consumption but also decreases waste generation, contributing to a more sustainable manufacturing process. Moreover, the enhanced performance of nanocutting fluids,

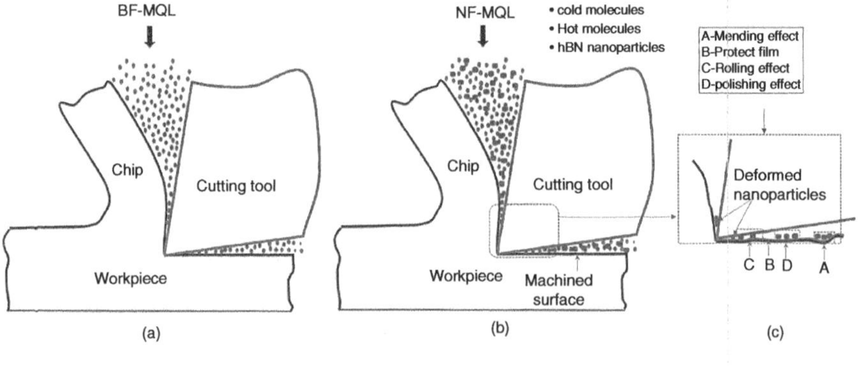

FIGURE 3.30 Schematic representation of lubrication mechanism for different nanofluids. (Reproduced with permission from Ref. (Sirin et al., 2021). Copyright Elsevier, 2021.)

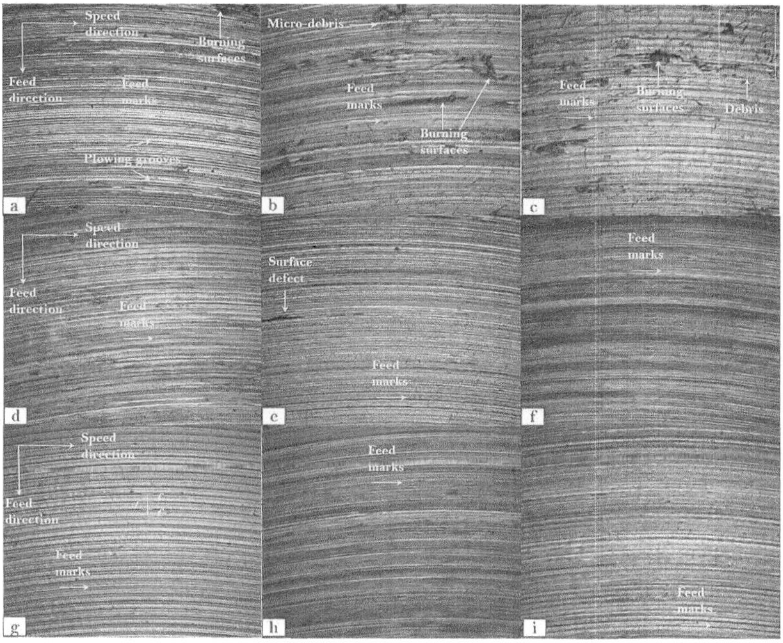

FIGURE 3.31 Optical surface morphology post milling for various machining conditions showing the direction of speed and feed and the surface defects like burning surface, micro debris, surface defects, and feed marks. (Reproduced with permission from Ref. (Sirin et al., 2021). Copyright Elsevier, 2021.)

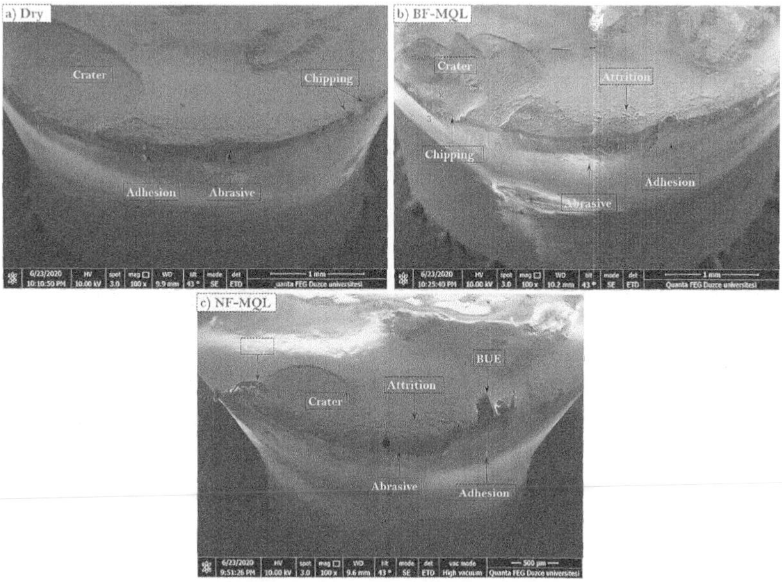

FIGURE 3.32 Different tool wear types on the milling cutter for different machining conditions. (Reproduced with permission from Ref. (Sirin et al., 2021). Copyright Elsevier, 2021.)

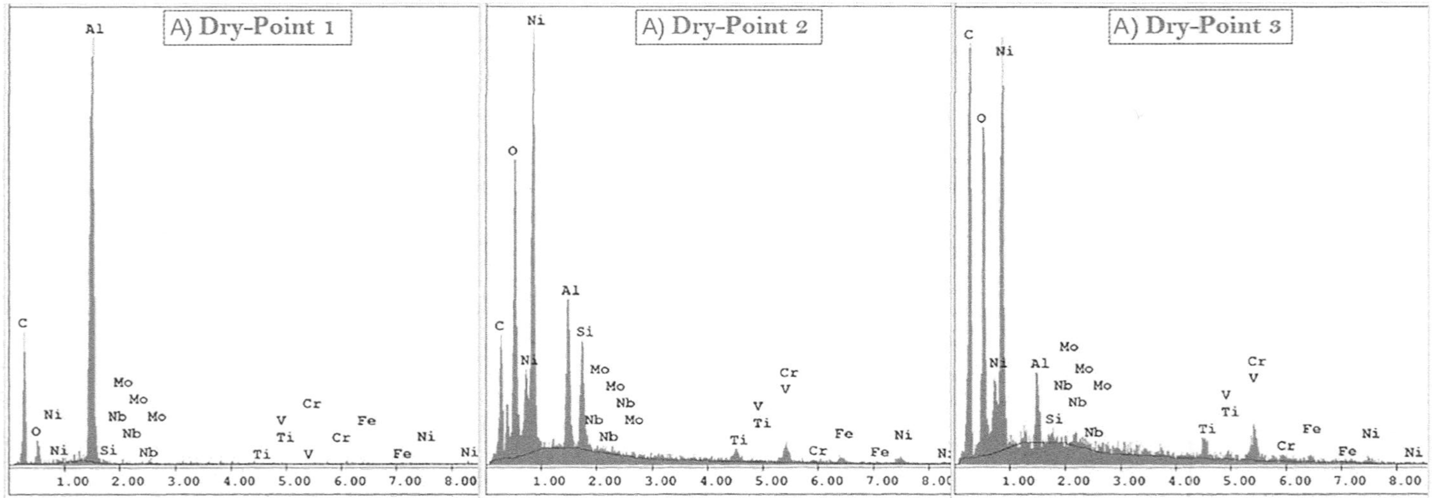

FIGURE 3.33A Energy dispersive spectroscopy on the worn tools for different machining conditions. (Reproduced with permission from Ref. (Sirin et al., 2021). Copyright Elsevier, 2021.)

(Continued)

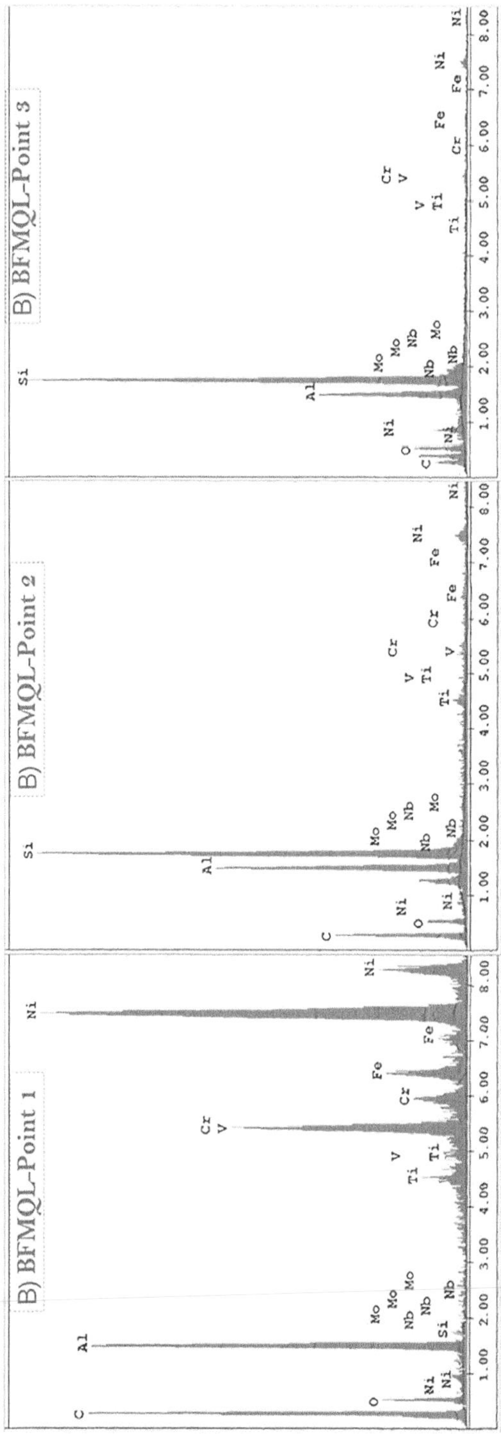

FIGURE 3.33B Energy dispersive spectroscopy on the worn tools for different machining conditions. (Reproduced with permission from Ref. (Sirin et al., 2021). Copyright Elsevier, 2021.)

(Continued)

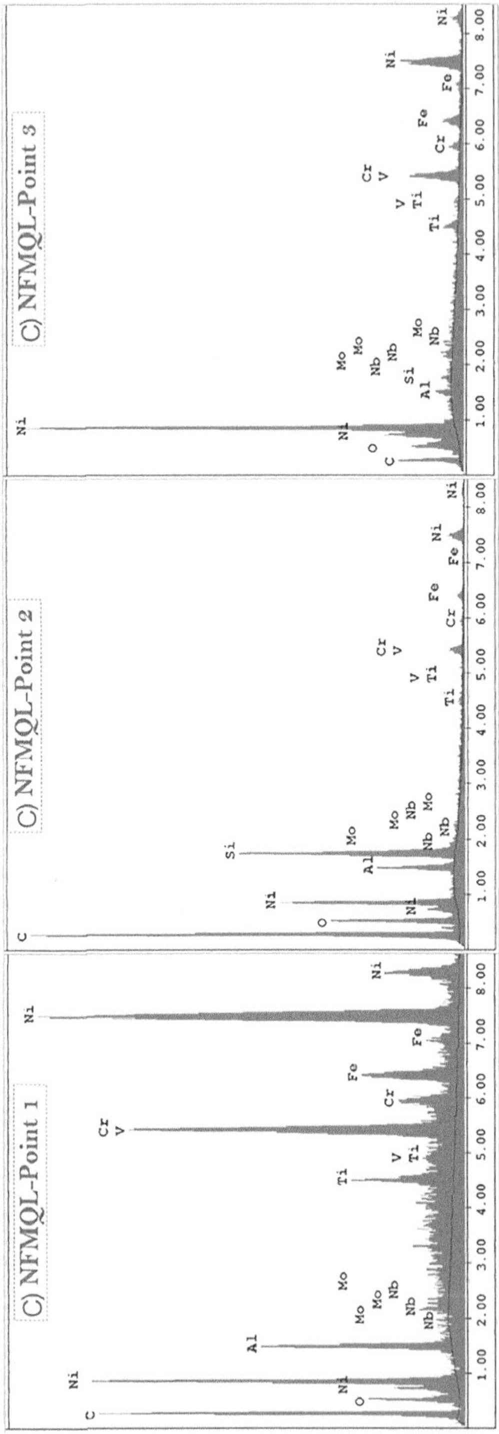

FIGURE 3.33C *(Continued)* Energy dispersive spectroscopy on the worn tools for different machining conditions. (Reproduced with permission from Ref. (Sirin et al., 2021). Copyright Elsevier, 2021.)

such as extended tool life and improved machining efficiency, leads to reduced energy consumption and lower carbon emissions. These fluids allow for higher cutting speeds and feed rates while maintaining or improving machining quality, resulting in more productive and energy-efficient manufacturing processes. Additionally, the potential for reusing or recycling nanocutting fluids further supports sustainability in machining. With proper filtration and purification systems, these fluids can be reclaimed and reused, reducing the environmental impact associated with their disposal. Some nanoparticles used in these fluids are biodegradable or environmentally benign, further aligning with sustainable machining practices. However, it is crucial to consider the environmental and health implications of nanoparticles. Their potential release into the environment and their effects on human health require thorough assessment and regulation to ensure safe handling and disposal practices (Figures 3.34 and 3.35).

The efficacy of nanocutting fluids is determined by the intricate interplay of nanoparticle characteristics, cutting parameters, and material interactions. Various factors, including nanoparticle size, shape, surface chemistry, and dispersion, engage with cutting parameters, including cutting speed, feed rate, and depth of cut. For instance, smaller nanoparticles generally exhibit better dispersion and can penetrate smaller gaps within the tool–workpiece interface, leading to improved lubrication and wear resistance. However, larger nanoparticles might offer superior load-bearing capabilities due to their increased contact area. Nanocutting fluid efficiency and effectiveness are directly influenced by cutting parameters, such as cutting speed and feed rate. Higher cutting speeds might necessitate better cooling capabilities while

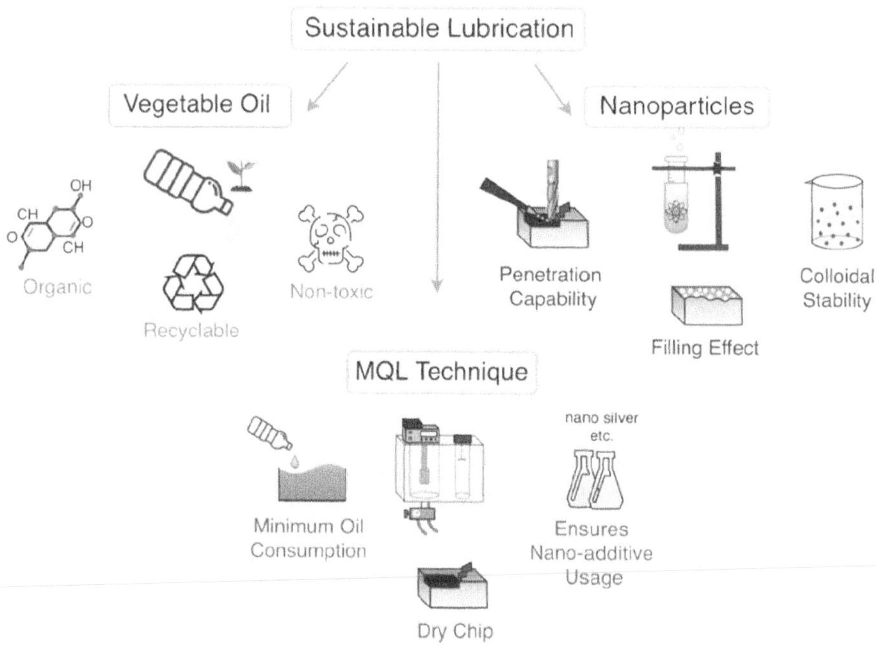

FIGURE 3.34 Different types of sustainable lubrication techniques. (Reproduced with permission from Ref. (Kilincarslan et al., 2021). Copyright Elsevier, 2021.)

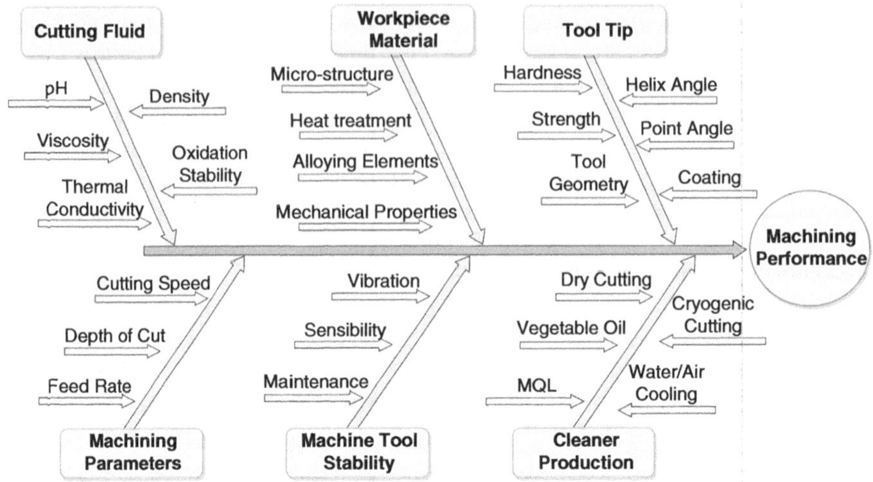

FIGURE 3.35 The cause-and-effect diagram depicts different factors that affect machining performance. (Reproduced with permission from Ref. (Kilincarslan et al., 2021) Copyright Elsevier, 2021.)

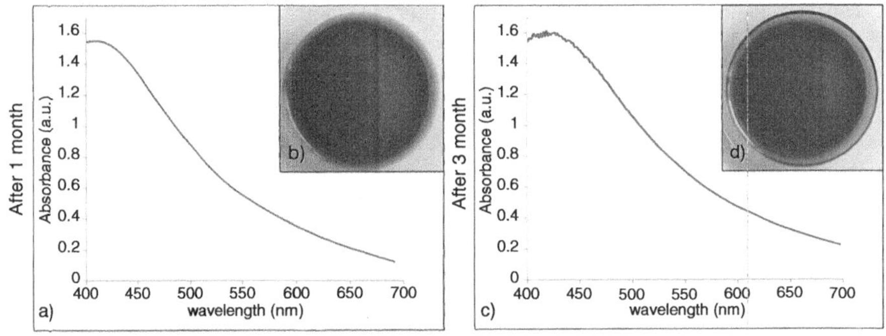

FIGURE 3.36 Nanosilver particle stability assessment: (a) UV spectrum of the sample stored for 1 month. (b) Visual representation of the sample after 1 month. (c) UV spectrum of the sample preserved for 3 months. (d) Visual representation of the sample after 3 months. (Reproduced with permission from Ref. (Kilincarslan et al., 2021). Copyright Elsevier, 2021.)

varying feed rates could affect the fluid's capability to sustain a stable film at the cutting interface. Furthermore, material interactions between the nanoparticles, cutting tool, and workpiece material play a pivotal role. The compatibility of nanoparticles with the materials being machined influences the overall tribological performance. Understanding these intricate interactions is crucial for optimising nanocutting fluids for specific machining applications, ensuring enhanced performance and productivity while maintaining machining quality. Continuous research and experimentation are essential to unravel and comprehend the complexities of these interactions, facilitating the development of guidelines for optimal usage of nanocutting fluids across diverse machining scenarios (Figures 3.36–3.38).

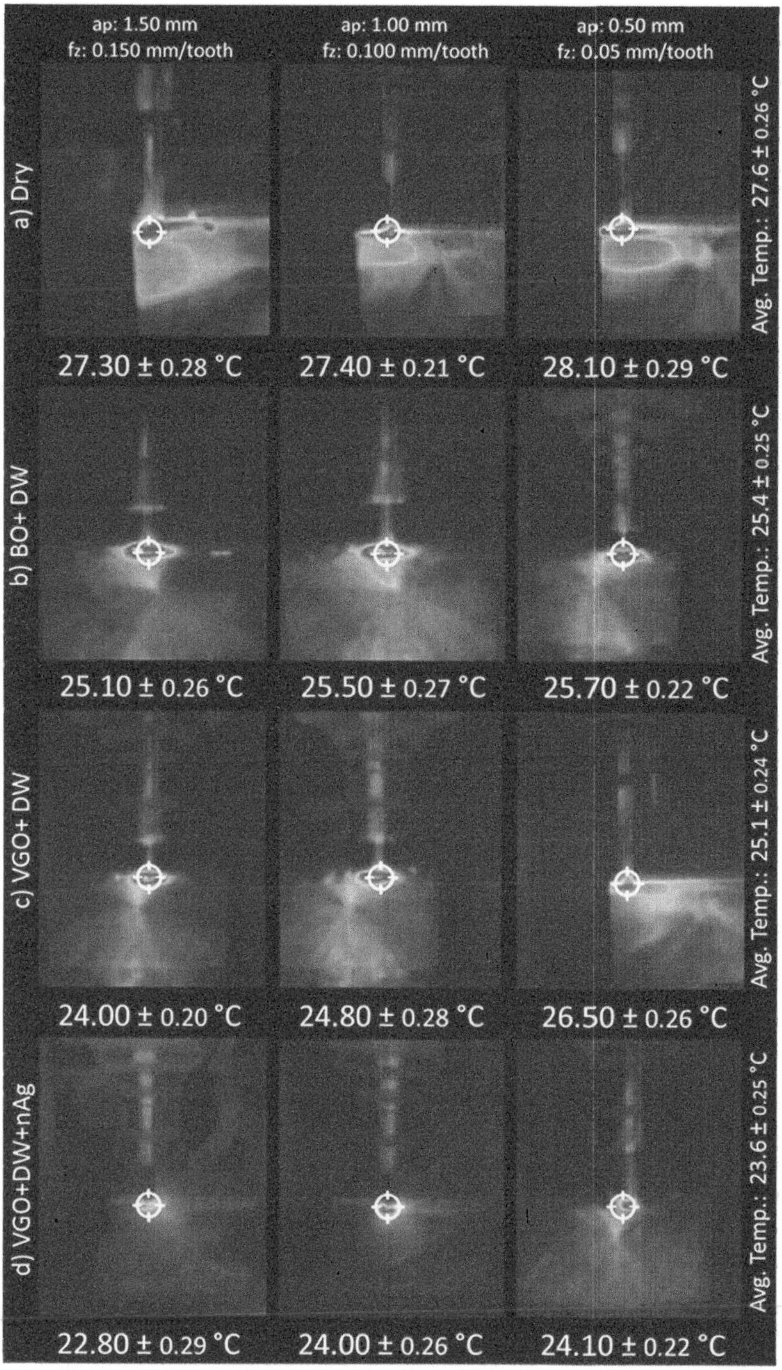

FIGURE 3.37 Machining temperatures for different conditions. (Reproduced with permission from Ref. (Kilincarslan et al., 2021). Copyright Elsevier, 2021.)

FIGURE 3.38 Surface micrographs depicting the tool wear for different machining conditions. (Reproduced with permission from Ref. (Kilincarslan et al., 2021). Copyright Elsevier, 2021.)

3.5 GRINDING

The addition of nanoparticles improves the grinding performance. Zhang et al. (2017) performed the grinding experimentation on Inconel 718 with the application of a 2:1 volume fraction of Al_2O_3/SiC nanoparticles of size (30, 50, and 70 nm) with additives in the base oil using the MQL technique. The mixture of 70:30 nm nanoparticles ratio of Al_2O_3/SiC gives the minimum specific normal grinding force, i.e. 2.645 N/mm, and specific tangential grinding force among all combinations of nanoparticles. In general, the removal rate reflects the efficiency of machining, and the mix 70:30 gives the maximum removal rate or the highest machining efficiency, i.e. 189.05 mm^3/s N, and the experiment shows that the material removal rate is inversely proportional to SiC nanoparticle size. The result showed that Ra, mix(50:30) gives the smallest value: 0.298 μm, and RSm, mix(70:30) shows the lowest value: 0.0381, representing the high workpiece surface quality or the smallest furrows. From the scanning electron microscopy (SEM) image, it can be noticed that the grinding debris is smoother in the 30:70 and 70:30 nanomixtures compared to other nanomixtures, which indicates a superior surface quality (Figure 3.39).

A small contact angle is desirable because of the large coverage area of the droplet, and the reverse will result when using a large contact angle, which affects the smooth lubricant. They proposed that the major mechanisms behind the reduction of the COF are the ball mechanism and the filling mechanism. These mechanisms help reduce the COF during the surface grinding process (Figures 3.40–3.42).

In the work conducted by Sinha et al. (2017), different nanoparticles, such as silver and zinc oxide, were used in the base fluid for grinding Inconel 718. The results suggest that an increase in the wt% of nanoparticles in base oils helps raise the thermal conductivity of the nanolubricant. This uptick can be attributed to the collective impact of Brownian motion, liquid layering, and effective nanoparticle aggregation within the mixture. Furthermore, an increase in concentration correlates with a rise in the nanolubricant's viscosity, possibly as a result of heightened density. Regarding machining outcomes, a lower tangential force is evident at a moderate nanoparticle concentration, coupled with a volume flow rate of 150 mL/h. Additionally, the rolling

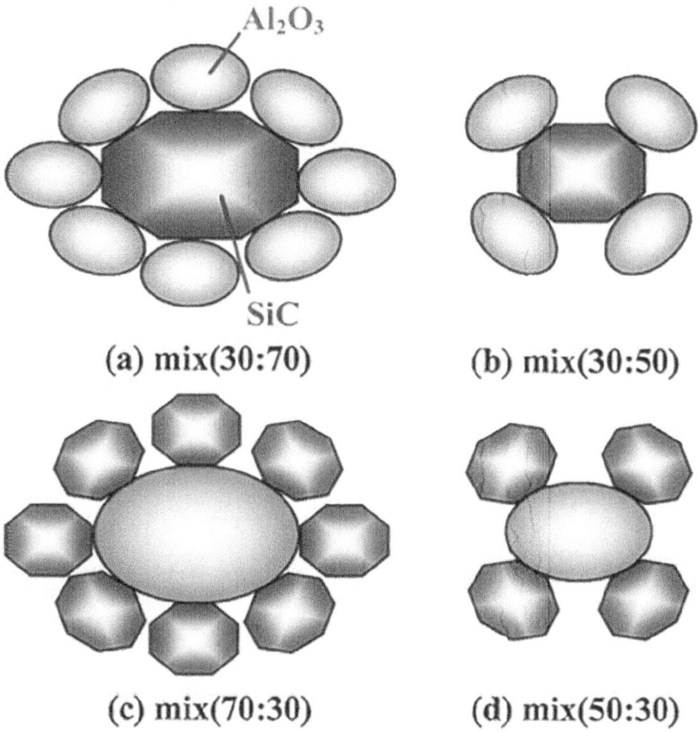

FIGURE 3.39 "Physical coating" phenomenon of Al_2O_3/SiC mixed nanoparticles with varying particle sizes. (Reproduced with permission from Ref. (Zhang et al., 2017). Copyright Elsevier, 2017.)

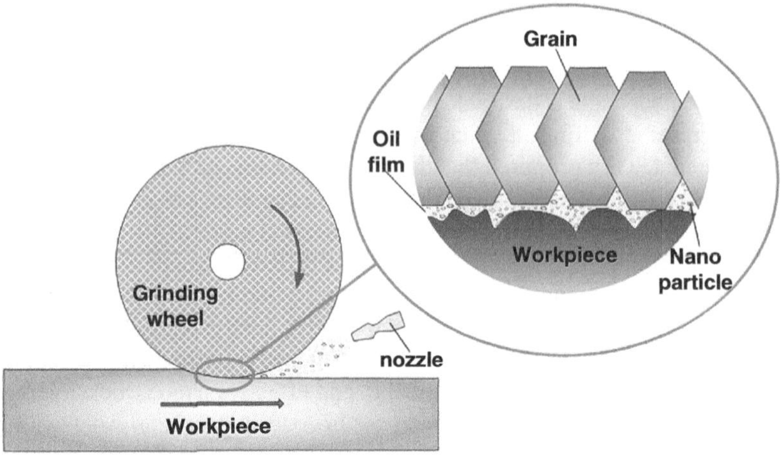

FIGURE 3.40 Schematic representation of the mechanism of oil film interaction at the grinding zone. (Reproduced with permission from Ref. (Zhang et al., 2017). Copyright Elsevier, 2017.)

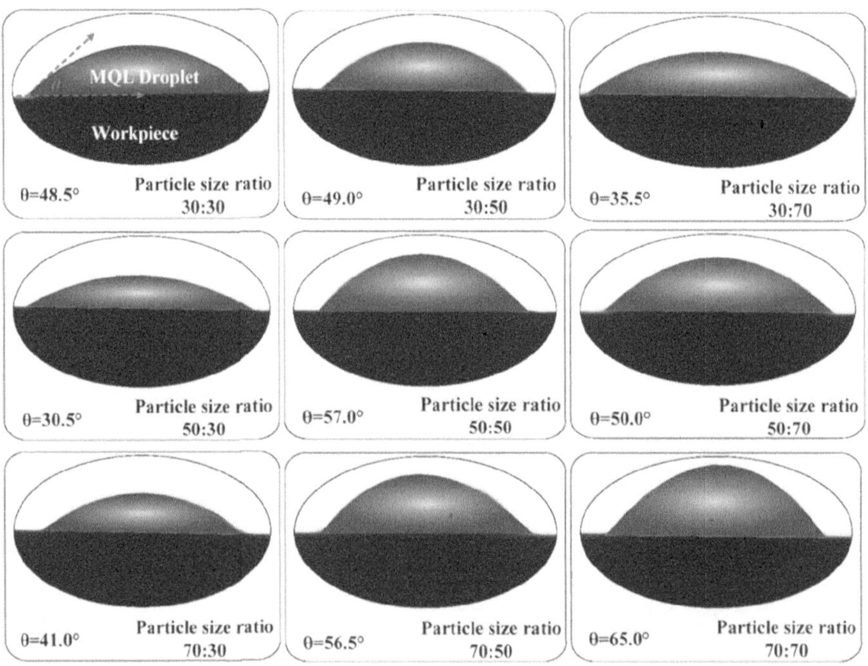

FIGURE 3.41 Contact angle between droplet from the jet and workpiece for varying particle size ratios. (Reproduced with permission from Ref. (Zhang et al., 2017). Copyright Elsevier, 2017.)

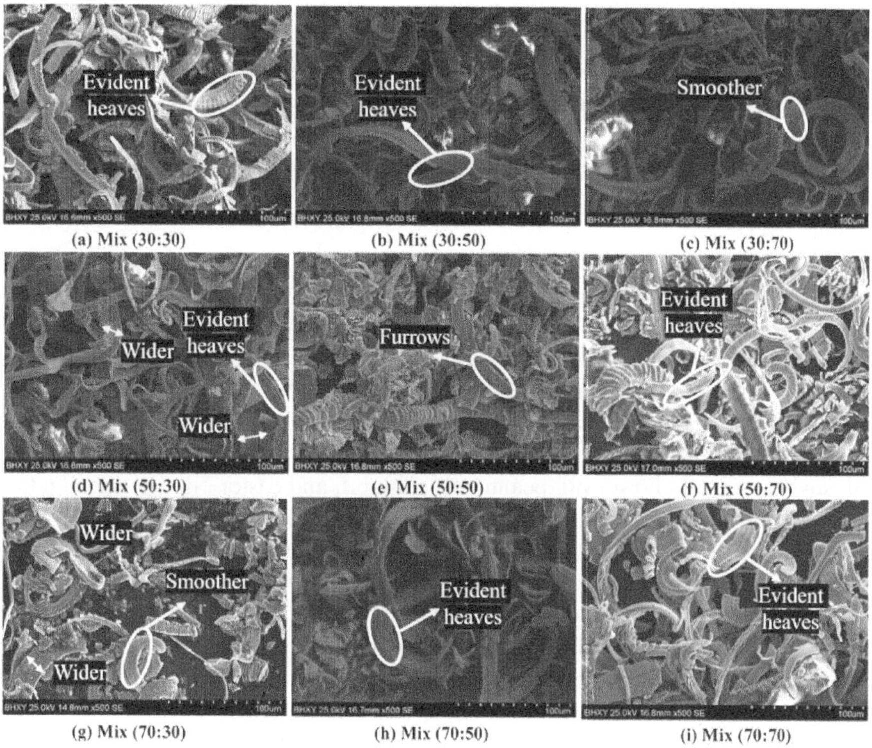

FIGURE 3.42 Surface morphological characterisation of the debris formed post-grinding for different grinding conditions. (Reproduced with permission from Ref. (Zhang et al., 2017). Copyright Elsevier, 2017.)

and sliding of nanoparticles over the work surface due to grinding pressure contribute to a further reduction in grinding force. The surface roughness is higher for the nanolubricant case in comparison to the dry condition due to the formation of microgrooves, which act as microchannels; storing lubricants help reduce contact temperature at friction surface to overcome the weld between the asperities with grinding debris due to decreased localised adhesion, which helps reduce COF and subsequently the grinding force. In the case of dry conditions, the % oxygen level found was 31.2%, while in nanolubricants, it was in the range of 1.2%–1.5%, which proves that the use of nanolubricants has less surface oxidation.

Surface characterisation techniques such as SEM and atomic force microscopy (AFM) offer crucial insights into the modifications taking place at the interface between the tool and workpiece under the influence of nanocutting fluids. SEM offers high-resolution imaging capabilities, allowing for detailed examination of the tool surface morphology post-machining. It enables the observation of wear patterns, tool coatings, and the deposition of nanoparticles on the tool surface. Additionally, AFM facilitates precise topographical analysis at the nanoscale, offering information on surface roughness, adhesion, and material properties. By scanning the workpiece surface, AFM reveals the impact of nanocutting fluids on the finished product's surface integrity, aiding in understanding the effects of lubrication and wear. These techniques not only provide visual data but also allow for quantitative analysis. Measurements of wear rates, surface roughness, and elemental composition analysis using techniques like energy-dispersive X-ray spectroscopy provide comprehensive information about the alterations induced by nanocutting fluids. Insights gleaned from these analyses aid in optimising machining parameters and refining the composition of nanocutting fluids for enhanced performance and surface quality (Figure 3.43).

The nickel-based alloy GH4169 is used for grinding with optimum lubricants such as 8 wt% MoS_2 nanoparticles with T115a & T202 (1:1 proportion) antioxidants 2 vol% + Tween-20 and span-20 dispersing agent (1:1 proportion) with 1.5 vol% + mixed vegetable oil as base oil (soybean/castor with mixture ratio: 1:1) (Jia et al., 2017). Nanoparticles incorporated into the nanofluid contribute to enhancing both the bearing and heat transfer capacities of lubricants. In terms of their lubricating properties, they were ranked in ascending order as follows: maize oil, rapeseed oil, soybean oil, sunflower oil, peanut oil, palm oil, and castor oil. The specific tangential grinding force was minimised under optimum lubrication conditions, measuring at 1.62 N/mm, outperforming other lubrication setups. The force ratio (Ft/Fn) stands as a pivotal parameter in assessing the lubricating conditions within the grinding area. An optimal depiction of lubrication conditions during grinding is indicated by a lower force ratio. For the optimum lubricant, this force ratio was minimised at 0.379 compared to other lubrication conditions, indicating an improved lubricating state in the grinding area. If specific grinding energy (energy/unit vol.) is low, it shows greener as a safe energy source, i.e. conserve energy, and the minimum SGE appear in optimum lubricant, i.e. 83.27 J/mm^3. The minimum surface roughness of the optimum lubricant (soybean/castor) obtained Ra and RSM values, respectively, i.e. 0.325 microns and 0.042 mm.

When the MoS_2 concentration increases, the heat transfer capacity of nanolubricant increases and at an optimal concentration of 8 wt% of nanoparticles, the highest heat transfer from the machining area was achieved, and the peak temperature was reduced in optimal condition by 46.23% while compared with pure oil. Under ideal conditions, specifically with 8 wt% nanoparticles, the surface exhibits smoothness characterised by shallow and narrow furrows, and the phenomena of adherence

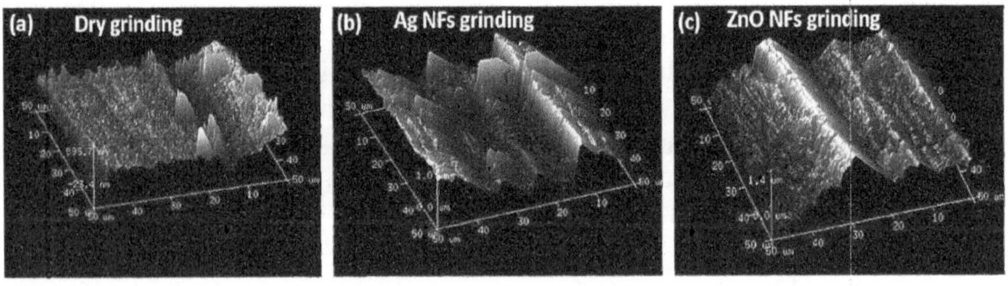

FIGURE 3.43 Atomic force microscopic images of the surface under different grinding conditions. (Reproduced with permission from Ref. (Sinha et al., 2017). Copyright Elsevier, 2017.)

and plastic stacking of materials essentially vanish. If grinding debris is generally bamboo structure and if it was not removed properly, then it will make bulk grinding debris melt due to high temperature at the machining area, causing damage to the lubricating film, but the nanofluid contains MoS_2, which increased the cooling effect, leading to the formation of smooth long and thin grinding debris. The lamellar structure of MoS_2 nanoparticles is characterised by weak van der Waals forces between layers, facilitating smooth movement and, as a result, reducing friction. These nanoparticles exhibit a propensity to be readily absorbed by surface imperfections, aiding in repairing the contact surface and resulting in reduced COF. The elevated surface energy of these nanoparticles promotes irregular diffusion, Brownian motion, and heat diffusion within base oils. This encourages micro-convection, intensifying the transfer of energy between nanoparticles and base oils. However, if the nanoparticle concentration surpasses 10 wt%, agglomeration occurs, impeding the ability to fill micro bulges and potentially damaging the formation of the lubricating film. This phenomenon affects Brownian movement, compromising heat transfer efficiency and ultimately increasing friction and surface roughness while reducing heat transfer capability.

The GH4169 is a nickel-based alloy that comes under the category of difficult to machine. The six nanofluid individuals form among CNT-based nanofluid + Polysorbate 80 as a dispersant (0.3%) + palm oil as base oil is used for the surface grinding process (Li et al., 2017). The Grinding force ratio (GR), representing the ratio of tangential force to the normal force and indicating the interaction between abrasive particles and the workpiece, typically falls within the range of 0.2–0.7. In the case of nanofluids involving SiO_2 and MoS_2 nanoparticles, a low GR value is observed, specifically 0.365 and 0.367. The highest grinding temperature is recorded for ZrO_2 nanofluid, i.e. 125.7°C, and the lowest achieved for CNT-based nanolubricant, i.e. 110.7°C, CNT shows good heat transfer properties like the highest thermal conductivity among all six nanoparticles and heat transfer coefficient. The nanofluid containing CNTs exhibits a notably low energy ratio coefficient of 40.1%. This coefficient signifies the percentage of heat energy effectively transferred to the workpiece in relation to the overall energy generated during the grinding process. A smaller contact angle corresponds to a smoother workpiece surface, indicating effective cooling and lubrication by the nanofluid. This is due to a smaller contact angle signifying a larger infiltration area, facilitating improved cooling and lubrication. The observed contact angle relations in descending order are as follows: CNT, MoS_2, Al_2O_3, PCD, and ZrO_2.

In another experimental endeavour, the unidirectional carbon fibre-reinforced ceramic composite (Cf/SiC) was subjected to grinding with carbon nanoparticles (CNPs) dispersed in deionised water as a base fluid (Qu et al., 2020). The CNPs are strongly hydrophobic, so the oxidation modification process, surfactant, & ultrasonic vibration are used for full dispersal in deionised water. From this experimental condition, the optimum parameters were obtained, such as a nanoparticle concentration of 5 g/L, the fluid flow rate of 80 mL/h, an air pressure of 7 bar, and a nozzle distance of 60 mm.

Grinding these materials under dry conditions results in small and irregularly shaped grinding debris. The rotation of the grinding wheel, coupled with air agitation, poses potential health hazards for workers as inhaling these particles may lead to respiratory issues. Moreover, the fine debris could infiltrate precision machinery, causing internal damage and compromising the integrity of the equipment. Hence, those points of view show that using an MQL-based nanofluid gives an advantage against dry lubrication conditions.

In another study, the 1–5 layer of GnPs (0.1 wt% optimal) are dispersed in Accu Lube LB2000 as the base fluid for the machining of TC4 alloy (Li et al., 2020). The results show that the 0.1 wt% of nanoparticles, the normal force, tangential force and COF decreased by 35.211%, 67.375%, & 49.644% relative to the dry machining. This is because oil film formation during grinding helps reduce friction and wear. The wt% GnPs increase load-carrying capacity. The maximum temperature obtained during dry conditions was 278.932°C, and the minimum temperature was obtained, i.e. 183.612°C while using 0.1 wt% of GnP in nanofluid, which was 34.173% less in comparison to dry conditions. It can be attributed to the oil film generation helping to reduce COF and also to GnPs, which help thermal transmission in liquid and may be the cause of thermal conductivity and heat transfer coefficient.

The maximum specific grinding energy under dry conditions was 598.384 J/mm^3, and the minimum specific grinding energy under GnP nanofluid was 195.222 J/mm^3, which was reduced by 67.375% compared with dry conditions. In dry conditions, due to poor lubrication, COF is large, but when using nanofluid, it gives better lubrication properties due to the formation of a lubricating film. The surface roughness Ra, i.e. 0.295 in the case of 0.1 wt% of GnPs in nanofluid, was reduced by 51.797% while comparing dry conditions. Surface micro-hardness, i.e. 349.37 HV for 0.1% of GnP in nanofluid, which was 19.039% less than the dry condition, i.e. 431.53 HV.

The utilisation of the optimal nanoparticle combination, specifically the MoS_2-WS_2-based nanolubricant, showcased remarkable outcomes (Kumar et al., 2019). A comparison with the base fluid revealed significant reductions in normal grinding force by approximately 27% and the specific grinding energy by 39%. Employing the same nanofluid resulted in a substantial decline in surface roughness, nearly 41%, and a chipping layer depth of 86% of the workpiece when compared to flood grinding conditions. Furthermore, the MoS_2-WS_2 hybrid nanofluid demonstrated a noteworthy decrease of 39% in tangential force and 27% in normal force compared to using only DI water (Figures 3.44–3.47).

The measurement of the dynamic contact angle by the sessile drop method in which a two microlitre droplet with a flow rate of 0.5 μL/s shows the lowest contact angle of 47.03 ± 0.75 degrees in the case of Al_2O_3 nanoparticle in deionised water, but in the case of MoS_2-WS_2-based nanofluid, it shows 53.67 ± 0.73 degrees. The hybrid MoS_2-WS_2 nanoparticles exhibit the lowest grinding force compared to WS_2-hBN and MoS_2-hBN nanofluids. This is attributed to the synergy between MoS_2 nanoparticles, which reduce the unsaturated bonds of WS_2, thereby increasing the oxidation rate. Consequently, this aids in forming a thermally stable phase of WS_2 nanoparticles, specifically WO_3, confirmed by XRD analysis. The presence of WO_3 and MoO_3 is ascertained through these techniques. Specific energy is markedly influenced by tangential force. In both the mono MoS_2 and hybrid MoS_2-WS_2 nanofluid states, the presence of nanoparticles leads to the development of an oxide layer on the workpiece surface, consequently decreasing the contact area between surfaces.

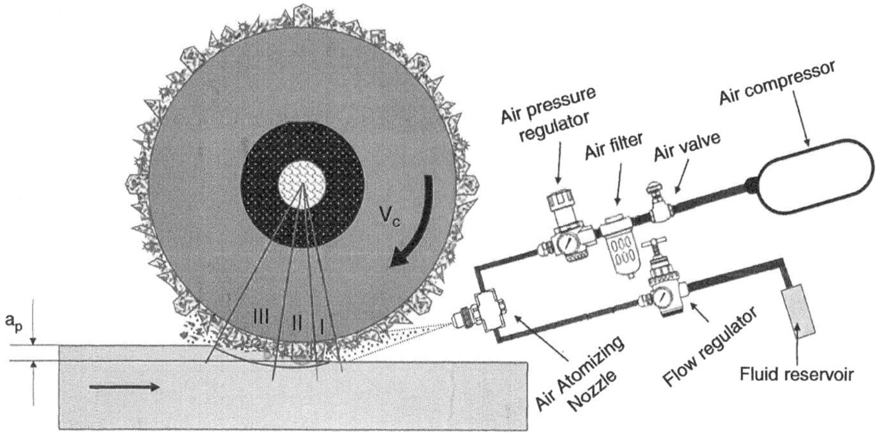

FIGURE 3.44 Schematic diagram of MQL setup. (Reproduced with permission from Ref. (Kumar et al., 2019). Copyright Elsevier, 2019.)

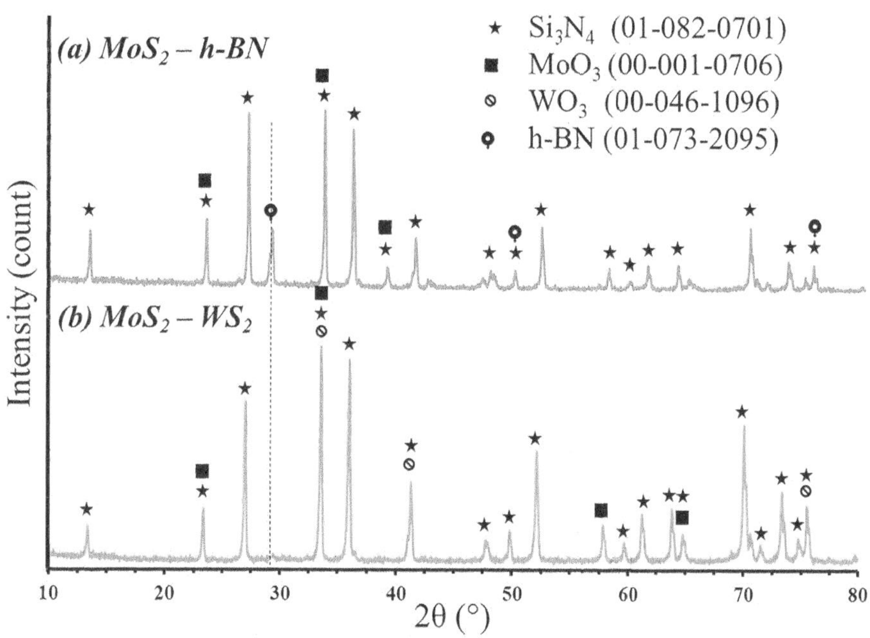

FIGURE 3.45 X-ray diffraction patterns of different nanoparticles. (Reproduced with permission from Ref. (Kumar et al., 2019). Copyright Elsevier, 2019.)

This reduction leads to a decrease in tangential force and undeformed chip thickness during grinding.

The lamellar structure characterising MoS_2, WS_2, and h-BN results in lower hardness due to the weak van der Waals bonding between intermolecular layers.

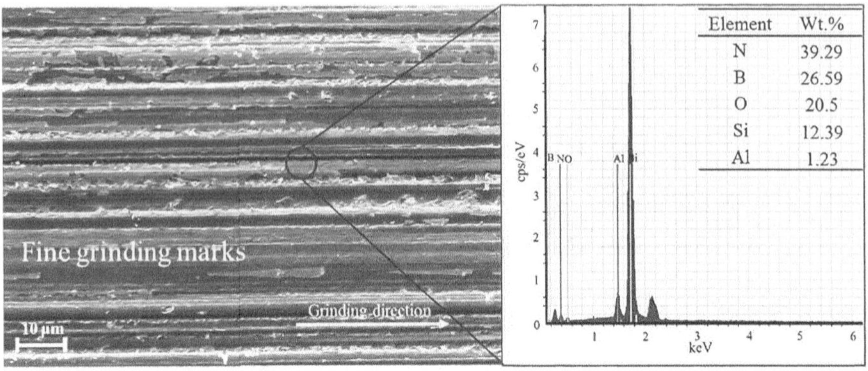

FIGURE 3.46 Surface morphology of the machined surface shows the fine grinding marks and the elemental mapping of the marked surface. (Reproduced with permission from Ref. (Kumar et al., 2019). Copyright Elsevier, 2019.)

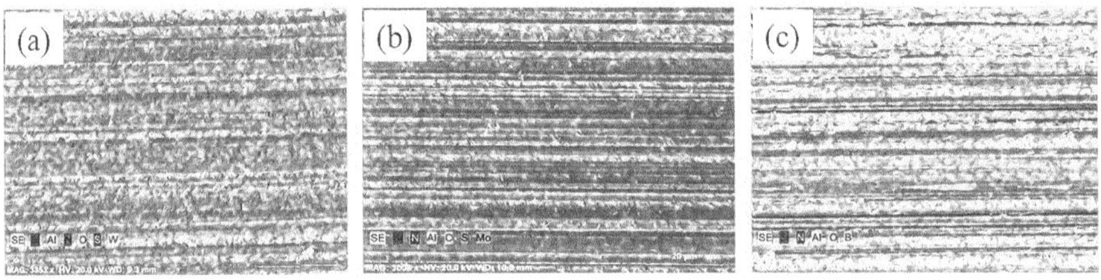

FIGURE 3.47 Elemental distribution of the workpiece machined under different machining conditions. (Reproduced with permission from Ref. (Kumar et al., 2019). Copyright Elsevier, 2019.)

Consequently, this structure generates lower shearing forces during grinding, contributing to decreased wear and grinding force. While hard nanoparticles like Al_2O_3 function similarly to ball bearings between the workpiece and tool, lamellar structure-based nanoparticles operate differently. The GnPs in palm oil as base fluid are used as nanoparticles for the machining of Ti-6Al–4V alloy grade 5 (Ibrahim et al., 2020). It is observed that the long-time use of sonication causes damage to GnP, and also, a short time will not create shear stress and cavitation, which are the main properties that cause dispersion stability. So there is an optimised sonicated parameter, i.e. 40 KHz for 4 hours at 35°C temperature. For the above pour point of palm oil, avoid any agglomeration and sedimentation of nanoparticles and maintain this temperature to observe any kind of sedimentation. At optimised 0.1 wt% GnP in palm oil, it gives the minimum tangential force of 6.68 N in comparison with other modes of lubrication. The optimised 0.1 wt% of GnP in palm oil helps decrease the specific energy to a minimum of 15.739 J/mm^3, and the specific energy is significantly reduced by 91.87% and 80.25% compared to flood and commercial lubricant.

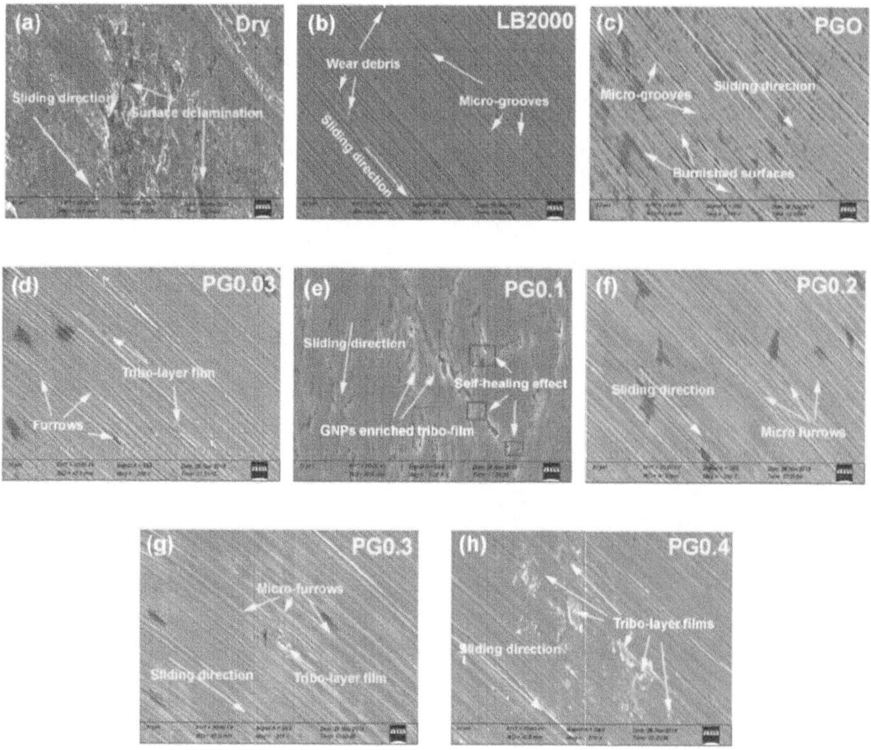

FIGURE 3.48 Surface morphological characterisation depicting various surface defects for different machining conditions. (Reproduced with permission from Ref. (Ibrahim et al., 2020). Copyright Elsevier, 2020.)

At 0.1 wt% of GnP in palm oil, there was a minimum COF of 0.0612 when compared with dry/flood/commercial + MQL mode of lubricant (Figure 3.48).

The unidirectional CFRP material used in this paper is vinyl resin carbon fibre composites, which are used as workpiece material with the application of CNT nanoparticles with palm oil base fluid as nanofluid (Gao et al., 2020). The results showed that the mean surface roughness Ra values in the feed and fibre direction while using nanofluid were reduced by 17.70% and 12.68% in comparison with dry processing. Similarly, Rz values in the feed and fibre direction while using nanofluid were reduced by 20.78% and 25.06% in comparison with dry processing, and for RMS values, the feed and fibre direction while using nanofluid was reduced by 25.40% and 11.43% in comparison with dry processing. The introduction of CNT nanoparticles into the spaces formed during the grinding of the grinder wheel, along with grain reduction from fibre fracture, contributed to a decrease in the surface roughness of the workpiece (Figures 3.49 and 3.50).

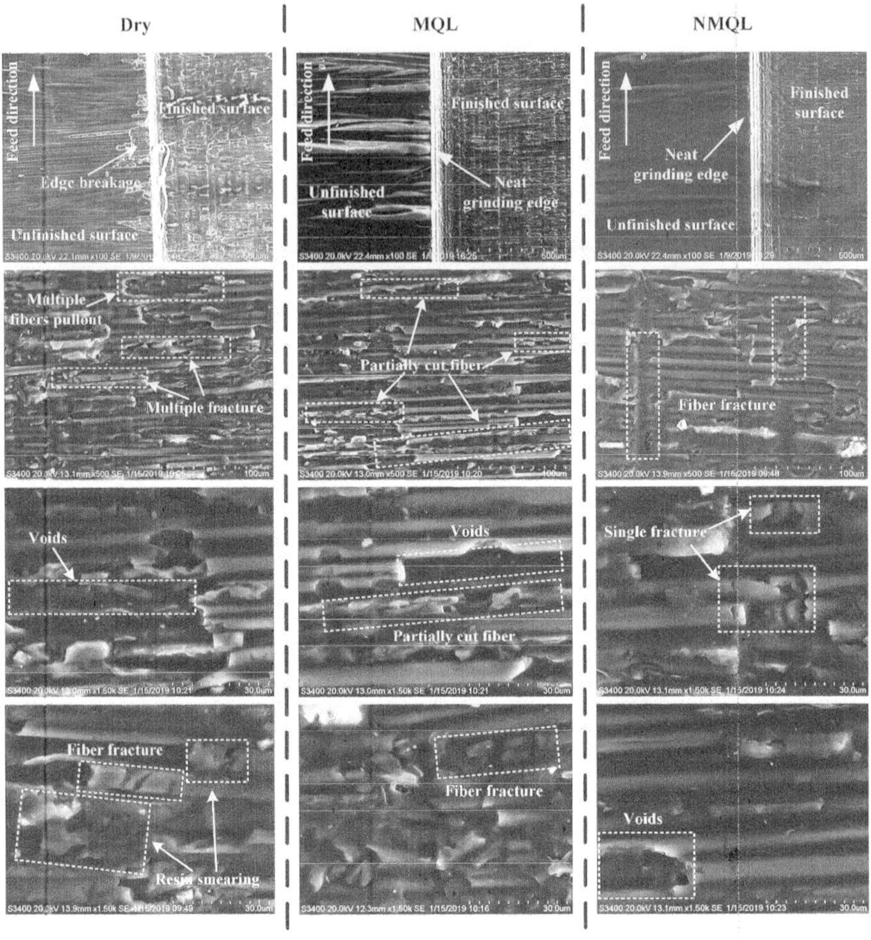

FIGURE 3.49 Surface morphological characterisation depicting different damage mechanisms for different machining conditions. (Reproduced with permission from Ref. (Gao et al., 2020). Copyright Elsevier, 2020.)

3.6 CONCLUSIONS

A comprehensive exploration of the utilisation of nanocutting fluids in various machining operations such as turning, milling, and drilling is presented, delving into detailed examinations of their impact on surface morphology, cutting forces, and tool wear. The discussion also briefly covers the influence of different types of nanofluids on machining processes. The following conclusions can be drawn from the aforementioned discourse:

- Nanocutting fluids demonstrate enhanced lubrication and cooling properties, significantly reducing tool wear and extending tool life during machining operations.

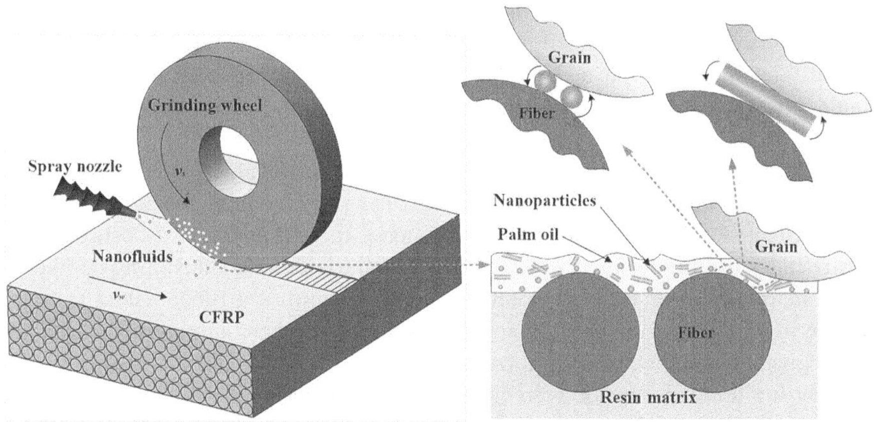

FIGURE 3.50 Illustration depicting the utilisation of nanofluid in the grinding process of a composite material, including a depiction of the underlying mechanism. (Reproduced with permission from Ref. (Gao et al., 2020). Copyright Elsevier, 2020.)

- The introduction of nanoparticles into cutting fluids alters their tribological behaviour, influencing friction, heat dissipation, and surface finish.
- Proper dispersion and stabilisation of nanoparticles within the cutting fluid are critical to achieving consistent and effective tribological performance.
- The choice of nanoparticle material and concentration is pivotal, affecting not only machining efficiency but also environmental and health considerations.
- Nanocutting fluids, particularly those with tailored nanoparticles, exhibit promising potential for sustainable and eco-friendly machining practices due to reduced resource consumption and waste generation.
- Tribological investigations showcase the intricate interplay between nanoparticle characteristics, cutting parameters, and tool–material interactions, shedding light on optimisation opportunities for specific machining applications.
- Surface characterisation techniques, such as AFM and SEM, provide insights into the alterations in tool surface morphology and chip formation mechanisms influenced by nanocutting fluids.
- The effectiveness of nanocutting fluids varies across various machining processes and materials, necessitating further research to establish guidelines for optimal usage in diverse industrial applications.
- Challenges persist regarding the stability and long-term performance of nanocutting fluids, emphasising the need for continued exploration and innovation in this field.
- Overall, nanocutting fluids present a paradigm shift in machining tribology, holding immense promise to revolutionise manufacturing processes by enhancing efficiency, precision, and sustainability.

The advancement of nanocutting fluids holds the potential to revolutionise traditional manufacturing processes. Their ability to enhance machining efficiency, precision, and sustainability presents a significant paradigm shift in industrial practices. These fluids offer multifaceted benefits, including extended tool life, improved surface finish, and reduced energy consumption, contributing to enhanced productivity and cost-efficiency in manufacturing. The evolution of nanocutting fluids aligns with the industry's pursuit of sustainable manufacturing. By reducing resource consumption, waste generation, and environmental impact, these fluids embody a shift toward eco-friendly machining practices. Their potential to optimise machining operations while maintaining or improving product quality signifies a fundamental transformation in manufacturing methodologies. Furthermore, the continuous research and development in nanotechnology and tribology continue to unveil new possibilities for nanocutting fluids. Innovations in nanoparticle engineering, fluid formulations, and cutting technologies aim to overcome current limitations and expand the applicability of these fluids across diverse machining processes and materials.

Moreover, the integration of smart technologies and predictive analytics into machining systems, coupled with advancements in nanocutting fluid technology, holds promise for highly efficient and adaptive manufacturing processes. Real-time monitoring of machining parameters, combined with the use of tailored nanocutting fluids, could revolutionise the precision, speed, and quality of manufactured components. In conclusion, the potential of nanocutting fluids to revolutionise manufacturing processes is evident. However, continued interdisciplinary research, technological advancements, and comprehensive understanding are required to harness their transformative capabilities fully and pave the way for the next era of machining and manufacturing.

REFERENCES

Gajrani, Kishor Kumar, P. S. Suvin, Satish Vasu Kailas, and Ravi Sankar Mamilla. 2019. "Thermal, Rheological, Wettability and Hard Machining Performance of MoS_2 and CaF_2 Based Minimum Quantity Hybrid Nano-Green Cutting Fluids." *Journal of Materials Processing Technology* 266 (October 2018): 125–39. https://doi.org/10.1016/j.jmatprotec.2018.10.036.

Gao, Teng, Changhe Li, Dongzhou Jia, Yanbin Zhang, Min Yang, Xiaoming Wang, Huajun Cao, Runze Li, Hafiz Muhammad Ali, and Xuefeng Xu. 2020. "Surface Morphology Assessment of CFRP Transverse Grinding Using CNT Nano-fluid Minimum Quantity Lubrication." *Journal of Cleaner Production* 277. https://doi.org/10.1016/j.jclepro.2020.123328.

Gaurav, Gaurav, Abhay Sharma, G. S. Dangayach, and M. L. Meena. 2020. "Assessment of Jojoba as a Pure and Nano-Fluid Base Oil in Minimum Quantity Lubrication (MQL) Hard-Turning of Ti–6Al–4V: A Step towards Sustainable Machining." *Journal of Cleaner Production* 272: 122553. https://doi.org/10.1016/j.jclepro.2020.122553.

Ibrahim, Ahmed Mohamed Mahmoud, Wei Li, Hang Xiao, Zhixiong Zeng, Yinghui Ren, and Mohammad S. Alsoufi. 2020. "Energy Conservation and Environmental Sustainability during Grinding Operation of Ti–6Al–4V Alloys via Eco-Friendly Oil/Graphene Nano Additive and Minimum Quantity Lubrication." *Tribology International* 150 (April): 106387. https://doi.org/10.1016/j.triboint.2020.106387.

Jia, Dongzhou, Changhe Li, Yanbin Zhang, Min Yang, Yaogang Wang, Shuming Guo, and Huajun Cao. 2017. "Specific Energy and Surface Roughness of Minimum Quantity Lubrication Grinding Ni-Based Alloy with Mixed Vegetable Oil-Based Nano-fluids." *Precision Engineering* 50: 248–62. https://doi.org/10.1016/j.precisioneng.2017.05.012.

Khan, Muzammil S., Mahendra Singh Sisodia, Shubham Gupta, M. Feroskhan, S. Kannan, and Karunamurthy Krishnasamy. 2019. "Measurement of Tribological Properties of Cu and Ag Blended Coconut Oil Nano-fluids for Metal Cutting." *Engineering Science and Technology, an International Journal* 22 (6): 1187–92. https://doi.org/10.1016/j.jestch.2019.04.005.

Kilincarslan, Enes, Sena Kabave Kilincarslan, and M. Huseyin Cetin. 2021. "Evaluation of the Clean Nano-Cutting Fluid by Considering the Tribological Performance and Cost Parameters." *Tribology International* 157 (February): 106916. https://doi.org/10.1016/j.triboint.2021.106916.

Kumar, Anil, Sudarsan Ghosh, and Sivanandam Aravindan. 2019. "Experimental Investigations on Surface Grinding of Silicon Nitride Subjected to Mono and Hybrid Nano-fluids." *Ceramics International* 45 (14): 17447–66. https://doi.org/10.1016/j.ceramint.2019.05.307.

Li, Benkai, Changhe Li, Yanbin Zhang, Yaogang Wang, Dongzhou Jia, Min Yang, Naiqing Zhang, Qidong Wu, Zhiguang Han, and Kai Sun. 2017. "Heat Transfer Performance of MQL Grinding with Different Nano-fluids for Ni-Based Alloys Using Vegetable Oil." *Journal of Cleaner Production* 154: 1–11. https://doi.org/10.1016/j.jclepro.2017.03.213.

Li, Ming, Tianbiao Yu, Rongchuang Zhang, Lin Yang, Zhelun Ma, Baichun Li, Xue Zhi Wang, Wanshan Wang, and Ji Zhao. 2020. "Experimental Evaluation of an Eco-Friendly Grinding Process Combining Minimum Quantity Lubrication and Graphene-Enhanced Plant-Oil-Based Cutting Fluid." *Journal of Cleaner Production* 244: 118747. https://doi.org/10.1016/j.jclepro.2019.118747.

Lv, Tao, Xuefeng Xu, Aibing Yu, and Xiaodong Hu. 2021. "Oil Mist Concentration and Machining Characteristics of SiO_2 Water-Based Nano-Lubricants in Electrostatic Minimum Quantity Lubrication-EMQL Milling." *Journal of Materials Processing Technology* 290 (October 2020): 116964. https://doi.org/10.1016/j.jmatprotec.2020.116964.

Musavi, S. H., B. Davoodi, and S. A. Niknam. 2019. "Effects of Reinforced Nanoparticles with Surfactant on Surface Quality and Chip Formation Morphology in MQL-Turning of Superalloys." *Journal of Manufacturing Processes* 40 (August 2018): 128–39. https://doi.org/10.1016/j.jmapro.2019.03.014.

Ni, Jing, Zhi Cui, Lihua He, Yongfeng Yang, Zhiqian Sang, and M. M. Rahman. 2021. "Reinforced Lubrication of Vegetable Oils with Nanoparticle Additives in Broaching." *Journal of Manufacturing Processes* 70 (September): 518–28. https://doi.org/10.1016/j.jmapro.2021.09.005.

Padmini, R., P. Vamsi Krishna, and G. Krishna Mohana Rao. 2016. "Effectiveness of Vegetable Oil Based Nano-fluids as Potential Cutting Fluids in Turning AISI 1040 Steel." *Tribology International* 94: 490–501. https://doi.org/10.1016/j.triboint.2015.10.006.

Pal, Amrit, Sukhpal Singh Chatha, and Hazoor Singh Sidhu. 2020. "Experimental Investigation on the Performance of MQL Drilling of AISI 321 Stainless Steel Using Nano-Graphene Enhanced Vegetable-Oil-Based Cutting Fluid." *Tribology International* 151 (March): 106508. https://doi.org/10.1016/j.triboint.2020.106508.

Qu, Shuoshuo, Yadong Gong, Yuying Yang, Wenwen Wang, Chunyou Liang, and Bing Han. 2020. "An Investigation of Carbon Nano-fluid Minimum Quantity Lubrication for Grinding Unidirectional Carbon Fibre-Reinforced Ceramic Matrix Composites." *Journal of Cleaner Production* 249: 119353. https://doi.org/10.1016/j.jclepro.2019.119353.

Rapeti, Padmini, Vamsi Krishna Pasam, Krishna Mohana Rao Gurram, and Rukmini Srikant Revuru. 2016. "Performance Evaluation of Vegetable Oil Based Nano Cutting Fluids in Machining Using Grey Relational Analysis-A Step towards Sustainable Manufacturing." *Journal of Cleaner Production* 172: 2862–75. https://doi.org/10.1016/j.jclepro.2017.11.127.

Rosnan, Rosmahidayu, Muhamad Nasir Murad, Azwan Iskandar Azmi, and Islam Shyha. 2019. "Effects of Minimal Quantity Lubricants Reinforced with Nanoparticles on the Performance of Carbide Drills for Drilling Nickel-Titanium Alloys." *Tribology International* 136 (February): 58–66. https://doi.org/10.1016/j.triboint.2019.03.029.

Salimi-Yasar, Hamid, Saeed Zeinali Heris, Mehdi Shanbedi, Ahmad Amiri, and Ali Kameli. 2017. "Experimental Investigation of Thermal Properties of Cutting Fluid Using Soluble Oil-Based TiO_2 Nano-fluid." *Powder Technology* 310: 213–20. https://doi.org/10.1016/j.powtec.2016.12.078.

Sharma, Anuj Kumar, Rabesh Kumar Singh, Amit Rai Dixit, and Arun Kumar Tiwari. 2017. "Novel Uses of Alumina-MoS_2 Hybrid Nanoparticle Enriched Cutting Fluid in Hard Turning of AISI 304 Steel." *Journal of Manufacturing Processes* 30: 467–82. https://doi.org/10.1016/j.jmapro.2017.10.016.

Sinha, Manoj Kumar, Rajeshkumar Madarkar, Sudarsan Ghosh, and P. Venkateswara Rao. 2017. "Application of Eco-Friendly Nano-fluids during Grinding of Inconel 718 through Small Quantity Lubrication." *Journal of Cleaner Production* 141: 1359–75. https://doi.org/10.1016/j.jclepro.2016.09.212.

Şirin, Emine, Turgay Kıvak, and Çağrı Vakkas Yıldırım. 2021. "Effects of Mono/Hybrid Nano-fluid Strategies and Surfactants on Machining Performance in the Drilling of Hastelloy X." *Tribology International* 157 (November 2020). https://doi.org/10.1016/j.triboint.2021.106894.

Şirin, Şenol, Murat Sarıkaya, Çağrı Vakkas Yıldırım, and Turgay Kıvak. 2021. "Machinability Performance of Nickel Alloy X-750 with SiAlON Ceramic Cutting Tool under Dry, MQL and HBN Mixed Nanofluid-MQL." *Tribology International* 153 (September 2020). https://doi.org/10.1016/j.triboint.2020.106673.

Şirin, Şenol, and Turgay Kıvak. 2019. "Performances of Different Eco-Friendly Nano-fluid Lubricants in the Milling of Inconel X-750 Superalloy." *Tribology International* 137 (April): 180–92. https://doi.org/10.1016/j.triboint.2019.04.042.

Yi, Shuang, Guangxian Li, Songlin Ding, and John Mo. 2017. "Performance and Mechanisms of Graphene Oxide Suspended Cutting Fluid in the Drilling of Titanium Alloy Ti-6Al-4V." *Journal of Manufacturing Processes* 29: 182–93. https://doi.org/10.1016/j.jmapro.2017.07.027.

Zhang, Xianpeng, Changhe Li, Yanbin Zhang, Yaogang Wang, Benkai Li, Min Yang, Shuming Guo, Guotao Liu, and Naiqing Zhang. 2017. "Lubricating Property of MQL Grinding of Al2O3/SiC Mixed Nano-fluid with Different Particle Sizes and Microtopography Analysis by Cross-Correlation." *Precision Engineering* 47: 532–45. https://doi.org/10.1016/j.precisioneng.2016.09.016.

4 Tribology of Metal Additively Manufactured Materials

Jitender Kumar Chaurasia, Vijayavarman Chakkravarthy, Jibin T. Philip, Nida Naveed, and Jinoop Arackal Narayanan

4.1 INTRODUCTION

Additive manufacturing (AM) is a method of adding materials to make components from digital data, in a layer-by-layer way, which is unlike subtractive manufacturing and formative manufacturing practices (International Organization for Standardization/ASTM International, 2017). AM introduced by Charles Chuck Hull in the year 1986 paved the way to solve most of the above issues associated with conventional manufacturing processes. AM was initially used for fabricating models and prototypes for various industrial and academic applications. However, with the advent of technology, AM progressed to become a "true manufacturing platform that develops the form, fit and function of a component" (Gisario et al., 2019). AM is one of the disruptive technologies, which is being used to build highly complex structures by exploiting its design freedom at reduced time and cost. AM is also attractive due to its ability to build highly customized components, paving the way for the fabrication of "mass-customized" products (Mahamood and Akinlabi, 2016). AM reduces material wastage (Ford and Despeisse, 2016), lead time (Kumar and Krishnadas Nair, 2017), and product development cycle (Frutos, 2015) and also provides manufacturing flexibility. It is seen that several industries are attracted to the use of AM, which is mainly due to the various freedoms it offers. The freedoms offered by AM are as follows:

a. **Shape design freedom:** AM provides the opportunity for "feature-based design and manufacturing" (Paul et al., 2013) and allows the fabrication of complex geometries with undercuts, lattice structures, honeycomb structures, etc. This aids in the fabrication of components with reduced weight and material usage. Further, fabrication of the entire component is possible as a single unit without joining separate parts.
b. **Material design freedom:** AM provides the opportunity to add multiple materials in a single component, mainly due to selective material fusion/deposition principle. AM also permits the fabrication of components with

DOI: 10.1201/9781003359333-4

multi-materials in one layer or different materials at different layers. This gives the freedom to join metallurgically incompatible materials by bringing a buffer material compatible with both materials in between them.

c. **Logistics freedom:** AM aids in reducing the gap between the manufacturer and the consumer with the availability of cyber technologies. It enables the design, manufacturing, and property evaluation of the components at distant and remote locations.
d. **Post-processing freedom:** Post-processing is the final stage in AM used for achieving the required properties (Bahnini et al., 2018). During AM, the building strategy can be selected according to the post-processing requirements.

In the engineering sector, metal AM (MAM) changed AM from a prototyping process to a method for manufacturing near-net-shaped end-use components. MAM is an advanced version of AM that transformed AM into industrial manufacturing for building high-performance functional end-use components. Commercially, high-energy source-based MAM processes are most commonly used to build functional components. The various energy sources used in MAM are laser beam, electron beam, and arc (Debroy et al., 2018). The types of feedstocks used in MAM are wire and powder. Another way to classify the process is the methodology used for material feeding. It can be of two types: powder bed-based (pre-placed bed) or dynamic feeding-based. Based on this, MAM can be classified primarily as Directed energy deposition (DED) and powder bed fusion (PBF). DED and PBF use high-power energy sources for melting the raw materials and depositing them as per the required geometry layerwise. DED uses a dynamic material feeding approach, while PBF uses a bed of powder as raw material. DED uses a laser beam, an electron beam, or an arc as the energy source, while PBF uses a laser beam or an electron beam to melt the raw material (DebRoy et al., 2018). Figure 4.1 presents the schematic of the DED and PBF techniques. Table 4.1 presents a comparison between PBF and DED processes.

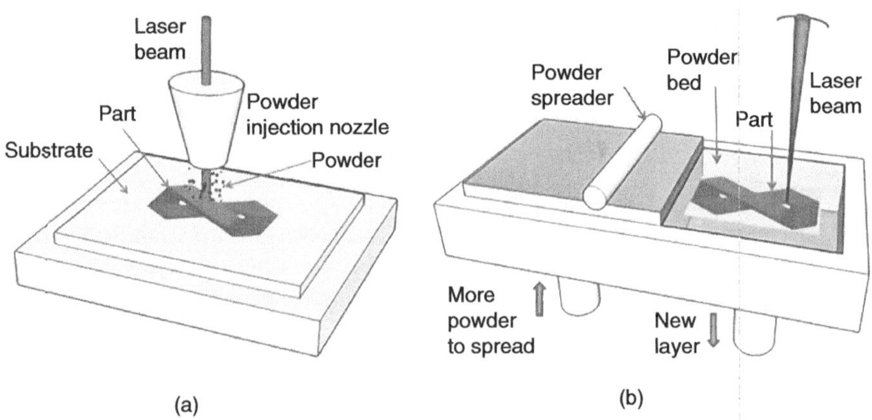

FIGURE 4.1 Schematic of MAM processes. (a) DED and (b) PBF (DebRoy et al., 2018). (Reproduced with permission from DebRoy et al., 2018). Copyright Elsevier, 2018.)

TABLE 4.1
Comparison between PBF and DED Process

Feature/Technique	Powder Bed Fusion	Directed Energy Deposition
Parameter	Power, scan speed, hatch spacing, layer thickness	Power, scan speed, material feed rate, hatch spacing
Typical power requirement (W)	100's of W	≥1 kW
Beam spot size	Micron scale	mm scale
Cooling rate	10^5–10^7 K/s	10^3–10^4 K/s
Main features	Precision, complex shaped components	High build rate, near-net-shaped manufacturing, material design freedom

It is required to understand the structure–property relation of MAM-built components in as-built and post-processed conditions to qualify the material for various applications. The tribological properties of MAM components are significant for applications that require excellent surface properties. "Tribology" is a Greek word that stands for *tribos*, meaning rubbing. It is the science that deals with surfaces in relative motion. Understanding engineering surfaces is a very important aspect of solving scientific issues related to friction, contact mechanics, and component design. Therefore, different aspects of surfaces are studied under tribology, and this includes friction, wear, and lubrication. In the last decade, MAM processes have come up as an alternate route for near-net or net shape part manufacturing techniques. Indeed, tribological performance is one of the concerned prerequisites of MAM components. Most of the MAM components that are designed for tribological applications will undergo repetitive or reciprocating contact load in the presence or absence of a harsh environment. As an effect, the component undergoes wear through the most frequent wear mechanisms, which include adhesion, delamination, abrasion, and corrosion (Bahshwan et al., 2020). Thus, considering the importance of analysing the tribological behaviour of MAM components, the present chapter explains the tribological behaviour of MAM parts in detail with a focus on the process parameters, metallurgy, process conditions, and post-processing.

4.2 EFFECT OF PROCESS PARAMETERS ON TRIBOLOGY BEHAVIOUR

Due to the unique way of part fabrication, i.e., layer-by-layer manufacturing, the quality of the part fabricated highly depends on the process parameters being used during the process. Laser power, scan speed, feed rate, and hatch spacing are some of the important process parameters that affect the performance of parts. There are several studies available where researchers have explored the influence of process parameters on metallurgical and mechanical characteristics. However, the one major aspect that has been least studied is the tribological behaviour of MAM parts. This section presents a critical review of the existing work focused on the tribology behaviour of MAM parts.

Engineering materials such as stainless steel (SS) 316, Inconel (IN) 625, IN718, and Ti-6Al-4V are the most widely used materials in the MAM process. Because of the application of these materials in critical and harsh places, surface integrity is an equally important aspect of the strength of the given material. In the MAM process, one of the biggest challenges is to have a good surface finish. MAM processes, due to the unique philosophy of layer-by-layer manufacturing, often result in a poor surface finish. One way to improve surface finish is to select an optimized process parameter. However, this can alter the grain structure and mechanical characteristics of the component being fabricated. Porosity in the additively manufactured parts, which is affected by process parameters, is another challenge that can significantly alter the tribology behaviour of the materials. To this end, Sun et al. did an investigation on SLMed SS316L where the effect of scan speed on the tribological behaviour was evaluated (Sun et al., 2014). The dry sliding test results of the SLMed 316L show a decrease in the wear resistance by 6–17 times compared to the wrought SS316L. A reduction in the volume percentage of porosity was found suitable for reducing the wear rate (Li et al., 2018). In addition to this, the hardness of the additively manufactured part is found to have a significant impact on the wear behaviour (Gu et al., 2012). As reported by Gu et al., an increase in hardness due to the increase in laser power does not significantly affect both coefficient of friction (CoF) and wear rate. Figure 4.2 presents the variation in hardness and CoF with process parameters. The SLM parts fabricated at a lower laser power were found to give similar results of CoF and wear rate as the samples fabricated using high laser power. The porosities observed due to the lack of fusion were found to improve the wear resistance. During the wear study, the debris was to get filled in the pores. At the same time, due to the high contact pressure between the relative surfaces, the debris was found to get welded to the base material, leading to an increase in the wear resistance (Gu et al., 2012). The layer-by-layer manufacturing also leads to anisotropy, especially along the build-up direction. Studies claim an insignificant impact of the build orientation on both COF and wear rate. However, at higher loads, a marginal impact of wear anisotropy on wear rate can be observed.

Repair is one sector where AM-related techniques are found useful. Sandhu et al. attempted to repair IN625 using cladding by shielded metal arc welding (Sandhu and Shahi, 2016). The clads were deposited at low (80 A) and high currents (105 A). The findings suggest that low welding current gives improved wear endurance as compared to the parts tested at a high welding current. From the EDS analysis, it was found that a rise in the welding current leads to a reduction in Cr and Ni content in the clad areas, which results in an improvement in the wear rate as compared to the samples where a low welding current was used. Comparing the AM-processed IN625 with conventional wrought IN625, it was found that AM-processed IN625 shows a better resistance towards wear (Gao and Zhou, 2018). AM processed IN625, due to the non-isothermal solidification, is known to have hard and brittle precipitates of Mo and Nb. The presence of these precipitate phase(s) increases the hardness of the material, leading to improved wear resistance. The presence of tribo-layers (a mixture of Fe_2O_3, Fe_3O_4, and Cr_2O_3) in the as-deposited sample is found to show a significant part in increasing wear resistance (Gao and Zhou, 2018).

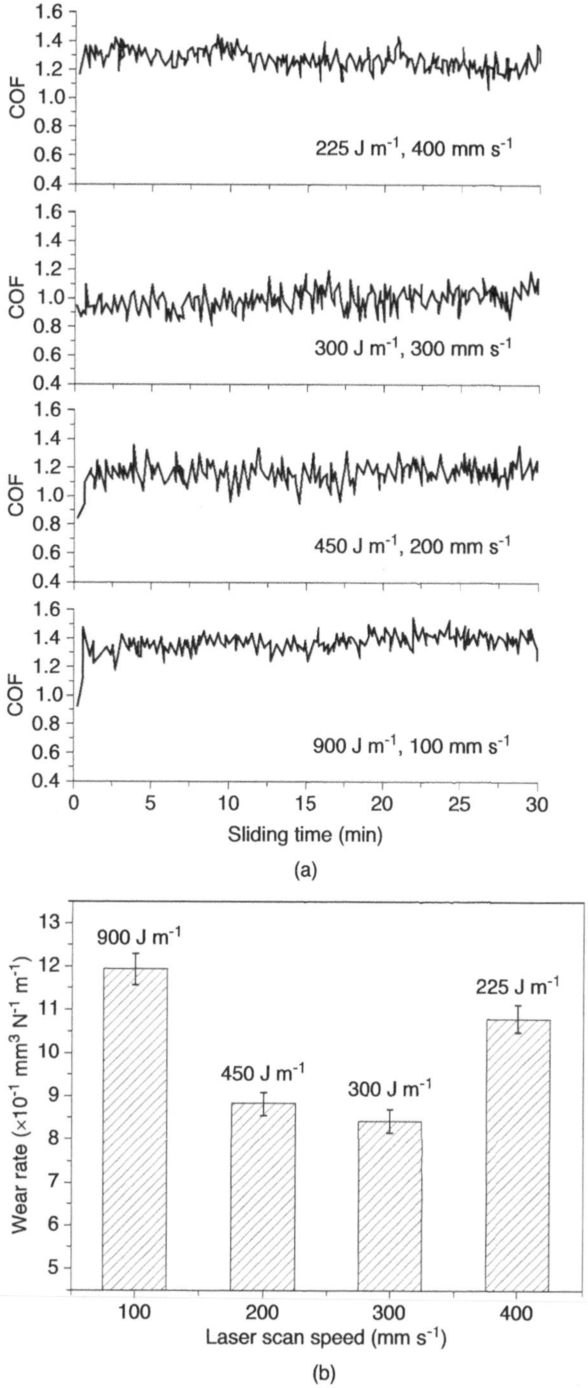

FIGURE 4.2 Effect of process parameters and build direction on CoF and wear rate (Gu et al., 2012). (Reproduced with permission from Gu et al., 2012). Copyright Elsevier, 2012.)

Build orientation in MAM processes is another important parameter that needs to be decided during the process planning stage. Build orientation is found to have a substantial influence on the microstructure and mechanical properties. A sample built in the horizontal direction can have a 1.5 times higher Young's modulus than the sample built in the vertical direction (Yadroitsev et al., 2009). Yan et al. used the SLM technique to deposit samples in 0°, 45°, and 90° orientations (Yan et al., 2021). The tribology results indicate that the COF and wear rate were lower in the 0° than in the other two orientations. As mentioned earlier, build direction can alter the resulting microstructure. The microstructure in the 0° build orientation was found to have fine columnar dendrites and equiaxed grain dispersed on the surface. The sample fabricated at 45° orientation is found to show the least wear resistance and the highest COF values, as it contains the coarsest microstructure among all cases. The 0° oriented sample shows several grooves and furrows indicating a clear abrasive wear mechanism (refer to Figure 4.3). However, on the other side, the sample with a 45° build orientation shows a wider wear track with deep grooves, indicating a delamination deformation mechanism on the worn surface. The Samples fabricated at 90° orientation show small cracks and creator, which usually occur due to the ploughing effect. In conclusion, it can be said that the abrasive and micro-ploughing wear methods are the dominating wear mechanisms in the SLM-fabricated IN625. Also, the effect of build orientation has a substantial impact on the surface wear performance, and therefore it must be a consideration during process planning.

Manufacturing of nanocomposite material using the MAM process is an area of interest for researchers to enhance the surface characteristics of a particular base part. In this, the reinforcement is first mixed with the metal matrix in a given

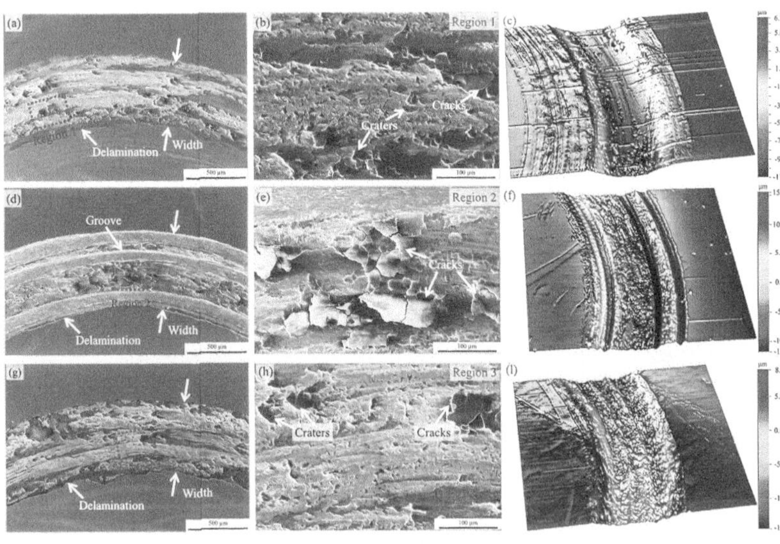

FIGURE 4.3 Wear morphology of (a, d, and g) sample fabricated at 0°, 45°, and 90°, respectively, (b, e, h) magnified view of the respective regions, and (c, f, and i) three-dimensional surface topography of the worn surfaces of sample built at 0°, 45°, and 90°, respectively (Yan et al., 2021). (Reproduced with permission from (Yan et al., 2021). Copyright Elsevier, 2021.)

proportion using the planetary micro mill. The mixed powder is then fed into the machine setup as feedstock material. Materials, such as Ti, despite having good mechanical strength, show a limited wear resistance towards applications involving abrasive and erosion phenomena (Gu and Zhang, 2013). To this end, Gu et al. have attempted to understand the influence of linear laser energy density (LLED) on the part density and wear behaviour of Ti reinforced with TiC. LLED is a parameter applied to define the total laser energy being delivered to the powder layer for melting. Mathematically, LLED is defined as Equation 4.1.

$$\text{Linear laser energy density} = \frac{P}{v} \left({}^{\text{J}}\!/_{\text{m}} \right) \tag{4.1}$$

where P is the laser power (W) and v is the scan speed (m/s).

The SLMed TiC/Ti structures showed the highest density of 98% at an LLED (1200 J/m) with a homogenous mixture of TiC/Ti. However, at lower LLED, the part density was found close to 90% (300 J/m). The significant reduction in the part density can be attributed to the balling effect, which usually arises at lower laser energy. Also, the part density was found to play a significant role in controlling the wear rate and CoF. The high porosity observed at low LLED leads to an increase in both wear rate and CoF. However, as the LLED increases, a significant fall in the wear rate and CoF can be observed. Interestingly, with the further increase in the laser energy input, both wear rate and CoF were found to increase, which can be ascribed to grain coarsening. The use of high LLEL increases the melt pool temperature; as a result, the thermal gradient within the melt pool reduces, which in turn lowers the cooling rate during solidification, thereby resulting in coarse grain size (Chaurasia et al., 2021). The coarsening of nanostructured TiC leads to an easy split of these particles during the dry sliding test. This leads to fracturing of the surface, offering reduced wear resistance. However, a comparison of the improvement in the wear resistance of SLMed TiC/Ti over SLMed Ti is missing in the study.

4.3 EFFECT OF MICROSTRUCTURE AND GRAIN ORIENTATION ON TRIBOLOGICAL PERFORMANCE

The tribological performance of polycrystalline metallic materials in terms of friction and wear is significantly influenced by grain orientation and microstructures of the initial and the sub-surfaces (Lakshmanan et al., 2021). In the context of tribological performance, it is challenging to single out the role of microstructure, grain size, and grain orientation (Lakshmanan et al., 2020). Many contributing variables such as contact load, contact geometry, contact environment, temperature, sliding direction, sliding speed, grain dislocation density, and hardness ratio of component and counterpart also need to be accounted (Chakkravarthy et al., 2022). In this section, the importance of microstructure, grain size, and grain orientation in determining the wear resistance of MAM components is discussed.

The final microstructure and grain orientation are influenced by both the temperature gradient and the cooling rate. It has been well accepted that the direction of the greatest heat flow in MAM components is typically directed from the top surface towards the bottom surface (Chakkravarthy and Jerome, 2020a). Furthermore, solidification happens parallel to the direction of the greatest temperature gradient. In a typical MAM process, the epitaxial and competitive grain development processes are the two basic solidification mechanisms. For instance, epitaxial grain development in a MAM component can come from the substrate or the partially remelted grains of earlier sublayers. Competitive grain growth occurs by penetrating different layer thicknesses and developing coarser columnar grains in a certain crystallographic direction. However, the non-equilibrium solidification processes of the MAM process can lead to the evolution of complicated microstructures across various zones. Similar to welding, typical MAM microstructures combine cellular, columnar, dendritic, and equiaxed morphologies that are dictated by the local temperature gradient (G) and grain growth velocity (R) (Kumar et al., 2022). It is worth mentioning that each microstructure has unique wear resistance capabilities. However, the microstructure and grain size are determined by the cooling speeds. Amongst different MAM technologies, the equipment that uses powder feedstock materials can provide extremely high cooling rates of between 10^5 and 10^6 K/s (Idury et al., 2021). Therefore, the majority of PBF components have finer grain sizes as a result of faster cooling. Finer grains provide greater yield strength and microhardness following the classical Hall–Petch empirical relationship. An increase in microhardness increases the material's resistance to wear following the well-known Archard's phenomenon. Unlike powder-based MAM, wire-based MAM techniques, such as wire arc additive manufacturing (WAAM) processes, experience exceptionally slow cooling rates, which leads to the evolution of coarser columnar grains that require post-processing to be used in applications requiring wear resistance. Yang et al. (2019) conducted a detailed wear study on laser PBF-built Stainless Steel 316L and noted the microstructural-dependent scratching resistance. For comparison, columnar and cellular microstructures were considered for wear study as shown in Figure 4.4. Due to the change in the local temperature gradient, a noticeable variation in the microstructure of the top side and front side was noticed. They concluded that for the same scratching load, columnar grains offer superior wear resistance compared to cellular grains. It was reasoned out that the microstructure experiences tangential force and the normal force during sliding. Particularly, the cells in the cellular structures have a smaller cross-section compared to that of columnar structures. Therefore, due to the smaller cross-section, the cellular structure quickly attains the maximum internal stresses during sliding. As an effect, the cellular grains quickly exceed the ultimate shear strength and undergo massive plastic deformation compared to that of columnar grains. However, while sliding, the direction of scratching should also be considered to ascertain the dislocation glide.

To provide in-depth insights, Bahshwan et al. (2020) performed a detailed study evaluating the tribological performance of laser PBF-built 316L stainless steel. They pointed out the significance of crystallographic orientation and texture in determining wear resistance. The most prevalent textures in face centered cubic (FCC) materials are Brass, Copper, Cube, Goss, and S. The initial cube texture exhibits the least plastic anisotropy among these textures, while the component with the predominant initial Goss texture is more likely to show high plastic anisotropy. It is well accepted that in an FCC system, components with <111> have roughly 2.4-fold times greater

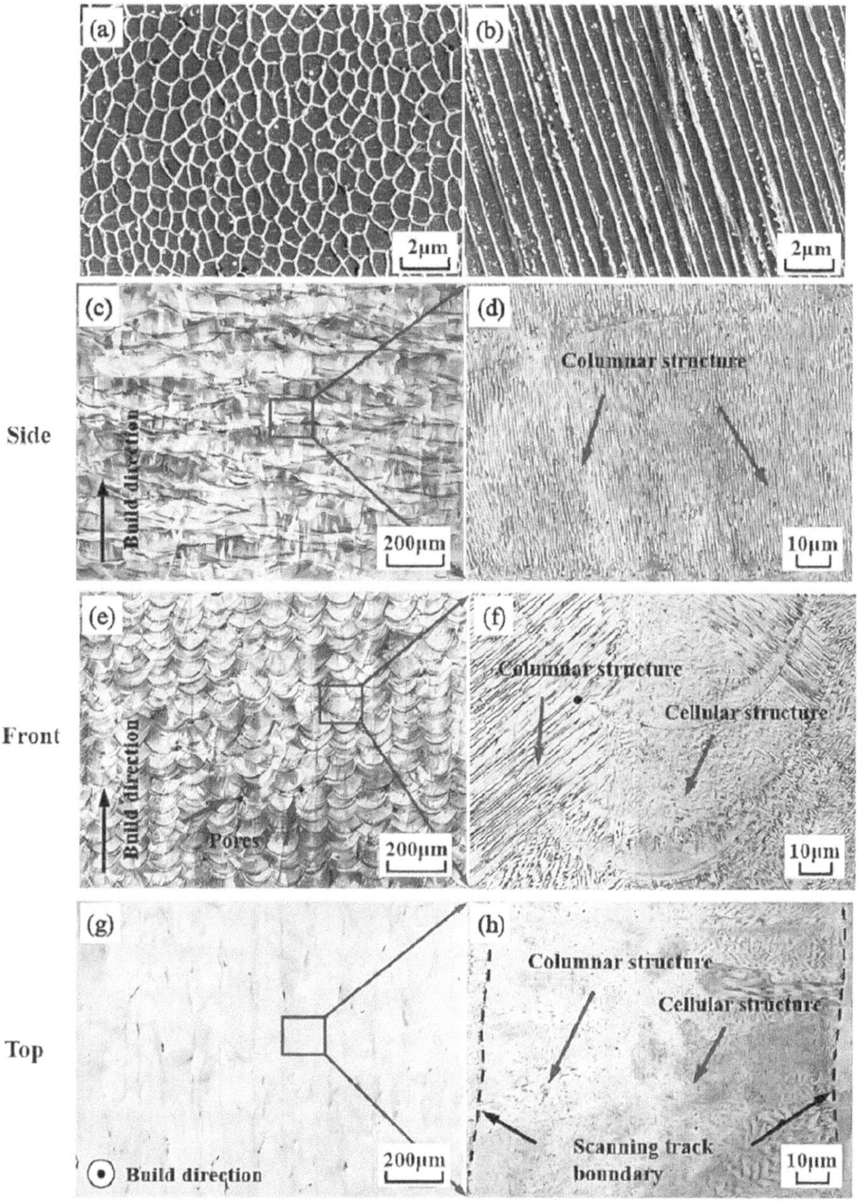

FIGURE 4.4 Microstructure of SLM-fabricated SS316L showing (a) cellular and (b) columnar structures and different structures of the (c and d) side view, (e and f) front view, and (g and h) top side; (d), (f), and (h) are the close magnifications of (c), (e), and (g), respectively (Yang et al., 2019). (Reproduced with permission from (Yang et al., 2019). Copyright Elsevier, 2019.)

Young's moduli than those of <001> texture. Consequently, to have crucial control over the amount of elastoplastic anisotropy, tailoring texture is inevitable. Using WAAM, Chakkravarthy and Jerome (2020a,2020b) created preferentially textured Al4043 components and investigated texture-dependent wear anisotropy. By adjusting the welding torch's orientation from (10° to 15°), a component with a preferred texture was developed. The component fabricated using a 15° welding torch angle showed a preferential texture of <111> and the component fabricated using a 10° torch angle was noted with <101> preferentially textured grains. Both the preferentially textured components were subjected to wear analysis as depicted in Figure 4.5.

Based on friction coefficient, surface roughness, and wear morphology, it was concluded that <111> textured grains exhibit superior wear resistance than <101> textured grains as shown in Figure 4.6.

During sliding, the contact shear stress is hindered by the grain boundaries and results in increased stress concentration at the sliding interface. In this context, during sliding, the <111> textured grains attenuate the contact shear stress at the sliding interface by undergoing faster work hardening as compared to <110> texture grains. It could be reasoned out that in the FCC system, the main deformation mechanisms are lattice rotation and dislocation slipping that occur more prominently in the {111} <110> slip system and less prominently in <111> and <001> textures (Kumar et al., 2022). The wear resistance of the sample is generally ascertained by its ability to restrain slipping during scratching. During scratching, the dislocation nucleation is activated along the slip plane to attenuate the local shear stress between the counterpart and the intender. During the initial phase of dislocation nucleation, the slip planes are active, and the grains undergo initial lattice rotation followed by which dislocation slipping would take place to reduce the local shear stress. However, continuous sliding of the intender increases the maximum

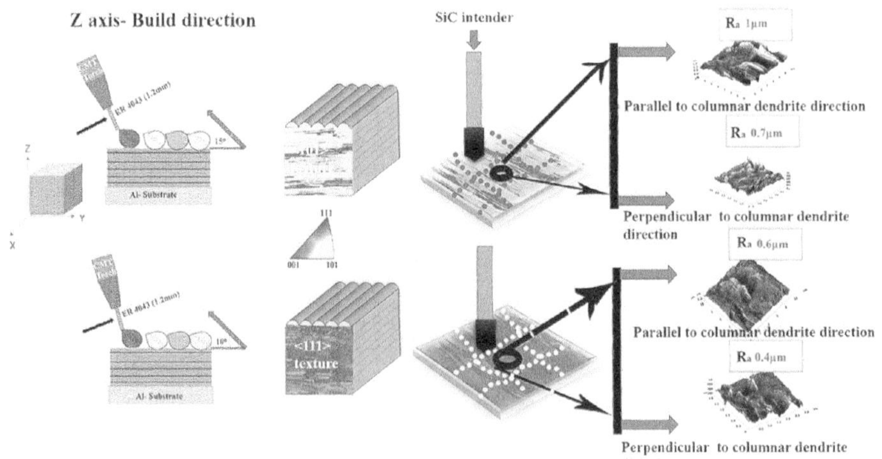

FIGURE 4.5 Texture-dependent wear resistance showing poor wear resistance in <101> textured grains and superior wear resistance in <111> textured components (Chakkravarthy and Jerome, 2020b). (Reproduced with permission from Chakkravarthy and Jerome, 2020b). Copyright Elsevier, 2020.)

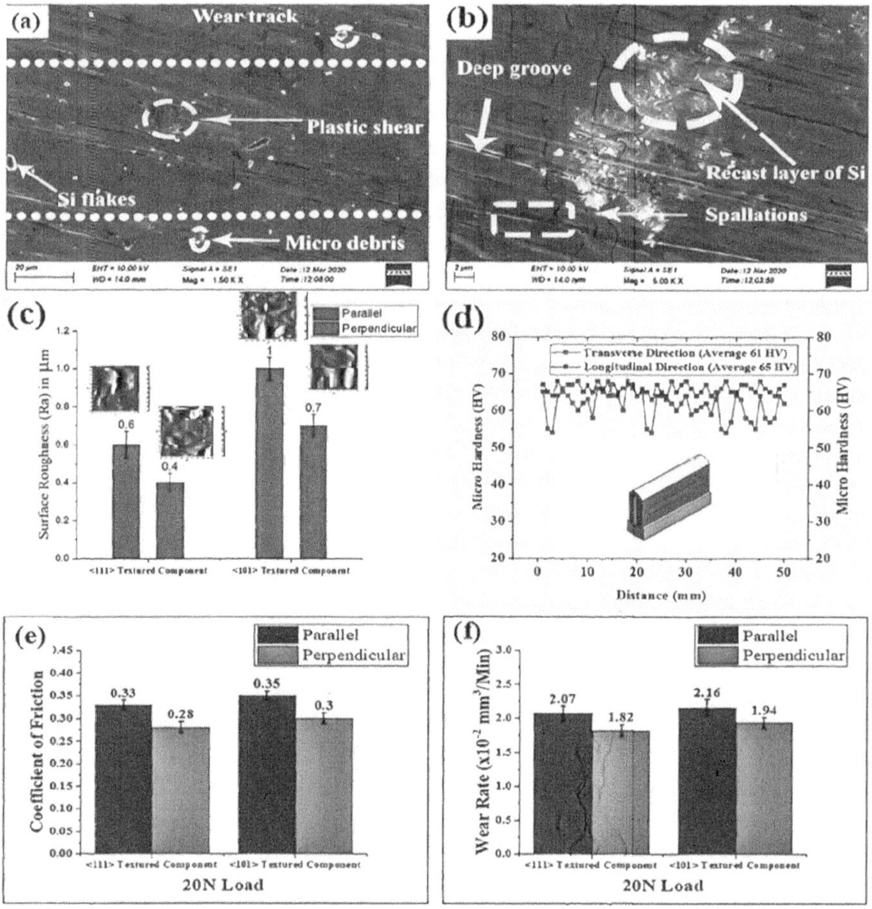

FIGURE 4.6 (a) SEM worn morphology of <111> textured component, (b) magnified view of worn surface, (c) friction coefficient graph of <111> and <101> textured components at a constant load of 20N, (d) wear rate histogram of <111> and <101> textured components (Chakkravarthy and Jerome, 2020b). (Reproduced with permission from Chakkravarthy and Jerome, 2020b). Copyright Elsevier, 2020.)

shear stress. Once after maximum lattice rotation and critical value of dislocation, the slipping becomes limited, under such circumstances, the grains undergo work hardening. But the rate of work hardening depends upon the number of active slip systems. The higher the active slip system, the greater the plastic deformation and the lesser the wear resistance. It is because during scratching the maximum forces act along the slip direction. In comparison with <110> textured grains, the <111> textured grains have a limited slip system. Therefore, during sliding, the sample containing <111> textured grains undergoes limited lattice rotation. Once after reaching the maximum lattice rotation, there will be a limited slip system to activate further plastic deformation. Under such circumstances, the <111> textured grains utilize the sliding load and quickly undergo dynamic strain hardening. As an effect, the contact shear stress across the grain boundaries was attenuated and adhesive wear of <111> textured grain was suppressed. This could be characterized by shallow pits and grooves. While <101> textured grains have a more active slip system due to which the plastic deformation is more prominent and the work hardening rate is much slower than <111> textured grains (Lakshmanan et al., 2021). As an effect, the <110> textured sample shows deep intense grooves with scale, spallation, and recast lumped debris, indicating intense adhesive wear. Further, the nature of slipping could be best explained through the critically resolved shear stress (CRSS) phenomenon. The lower the CRSS, the greater will be the metal deformation ability and the lower will be the wear resistance. In a hexagonally closely-packed (HCP) system, the three most notable orientations are basal {0001}, pyramidal {10–11}, and prismatic {10–10}. Interestingly, the wear response of each orientation can be very different. For instance, in AZ91 (Mg —9 wt% Al—1 wt% Zn) Mg alloy, the basal orientation was found to have a higher CRSS compared to that of pyramidal and prismatic (Kumar et al., 2021). Therefore, during scratching, the pyramidal and prismatic-oriented grains could easily undergo slipping. During scratching, the higher CRSS of {0001} basal-oriented grains undergoes limited slipping and easy activation of extension twinning as indicated in Figure 4.7.

The extension twinning arrests the dislocation motion of the intender and protects the chipping and pile-up of grain boundaries. Another striking example that falls under the HCP system is the titanium alloys, which are known for a high strength-to-weight ratio, biocompatibility, and superior corrosion resistance (Ye et al., 2022). One of the most widely used Ti alloys is the Ti-6Al-4V alloy, which contains a majority of α-phase (HCP) and a remanent β-phase (BCC) (Chakkravarthy et al., 2023). During plastic deformation, slip systems get activated based on the CRSS. The CRSS could be best explained through the Schmidt factor phenomenon. The greater the Schmidt factor, the greater will be the ability to undergo plastic deformation. Generally, Ti alloys undergo complex deformation modes. It has four main slip systems occurring in three glide planes: (1) <a> Basal <11–20> {0001}, (2) <a> Prismatic <11–20> {10-10}, (3) <a> Pyramidal <11–20> {10–11}, and (4) <c+a> on the pyramidal <11–23> {10–11} (Gong and Wilkinson, 2009). Among these slip systems, the most favourable slip system of Ti alloys is <a> on the prismatic plane. A recent study by

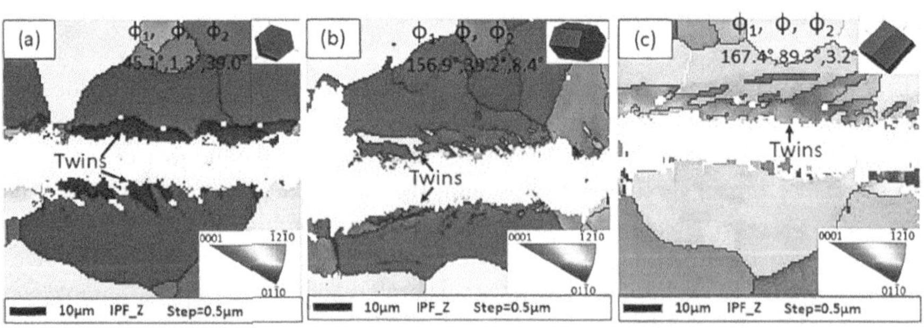

FIGURE 4.7 EBSD IPF maps of AZ91 Mg alloy (sample normal) to the scratched region for (a) basal, (b) pyramidal, and (c) prismatic orientations at room temperature (Kumar et al., 2021). (Reproduced with permission from Kumar et al., 2021) Copyright Elsevier, 2021.)

Liu (2023) performed a detailed indentation study and revealed superior hardness and elastic moduli while indenting perpendicular to a basal plane (i.e., indentation load parallel to the c-axis), while low hardness and elastic moduli were noted in the prismatic plane (i.e., indentation load perpendicular to the c-axis) (Hua et al., 2021). It is worth noting that at room temperature the basal slip system has ~23% greater CRSS than the prismatic slip system. Therefore, higher CRSS of the basal slip system arrest the dislocation against indentation and thereby offer superior hardness. However, at elevated temperature, the basal and prismatic have almost equal CRSS values. While the highest CRSS in Ti-6Al-4V alloy is noted for twinning, which is higher than basal and prismatic, but less than that as first order ⟨c+a⟩ slip. During scratching conditions, the β-Ti undergoes slipping in {110} <1–11> and while α-phase undergoes slipping in {0001} <11–20> parallel to the (sliding direction) (Hua et al., 2022). It is known that the deformation abilities vary following the phase composition where the α phase is hard to deform compared with the β phase. Greater dispersed α-precipitates are more favourable for accommodating the dislocations of β matrix. Therefore, the greater the β/α interface, the greater would be the resistance against the dislocation movement. Su et al. (2022) conducted a tribological analysis on an LPBF-built Ti-6Al-4V bulk sample. They observed microstructure-dependent wear resistance. They pointed out that the finer the width of α-lathe, the greater will be the wear resistance. Also, they noted that the lamellar β/α interface showed superior wear resistance to the bimodal α + α' microstructure. They reasoned out that the brittle nature of the α'-martensitic phase undergoes micro-cracks and flake-like debris during repetitive scratching.

4.4 EFFECT OF POST-PROCESSING ON TRIBOLOGICAL BEHAVIOUR OF MAM COMPONENTS

AM techniques offer complete design freedom because of the principle of the layer-by-layer manufacturing process. There is no doubt that these technologies are going to be industrial game changers when it comes to the fabrication of intricate/complex shape parts and functionally graded materials. However, AM processes do suffer from some drawbacks such as high porosity (leads to poor mechanical properties), tensile residual stresses (leads to unexpected failure or loss of dimensional accuracy of the part upon external loading), anisotropic mechanical properties (leads to variability in the parts' mechanical properties), and poor surface finish (results in poor fatigue and tribological behaviour). These drawbacks result in the partial failure well before the intended service life, as mentioned. Also, the two major sectors that are more likely to adopt MAM techniques are space and defence. These sectors usually have stringent part qualification criteria. To widen the MAM processes in these sectors, it is therefore required to look for possible solutions. The first solution to avoid these drawbacks is to use the optimal process parameters during the fabrication stage. However, it is sometimes difficult to find one optimal process parameter value, which can fix all drawbacks. For example, during fabrication using the laser powder bed fusion process, an increase in scan speed leads to fine microstructure, improving the mechanical properties, but at the same time, it can also lead to partial melting of powder particles on the top surface, leading to high surface finish and improper fusion. Similarly, to achieve a good surface finish in the as-built part, a lower slicing thickness is preferred. However, a lower slicing thickness can significantly improve the build time and cost of the part. In recent times, the post-processing of AM parts has gained significant interest in the research community. Post-processing of MAM parts is indeed a thrust area of research, which widely looks into ways to improve the surface finish, introduce compressive stresses for improving fatigue life, and hinder the micro-cracks for high mechanical properties. Processes such as heat treatment, laser peening, abrasive finishing, drag finishing, machining, and polishing are some of the post-processing techniques being used for MAM-fabricated parts. In this section, a review of the effect of post-processing techniques on tribological behaviour is discussed in detail.

Turning is the most simplistic and easiest way of post-processing, which can improve the surface finish. The turning operations carried out on SS316 cylindrical samples built using a selective laser melting process showed a fall in surface roughness from a Ra value of 7 to 1 μm (Kaynak and Kitay, 2018). Turning is a plastic deformation process that is found to change the microstructure of the outer region up to a depth of 10–20 μm. The microstructure of the post-processed sample shows a much smaller grain size than that of as-fabricated samples, which has resulted in up to a 23% increase in the surface hardness after post-processing. The significant increase in the hardness is expected to improve the tribology behaviour of SLMed parts. However, a clear physical understanding of the change in wear behaviour of SLMed parts followed by turning operations is still required.

Laser shock peening (LSP) is one of the most widely used post-processing techniques. In this process, LSP carried on the IN718 cylindrical sample fabricated using the direct metal laser sintering technique, showing a 1.7 times improvement in the specific wear rate over the as-build condition (Jinoop et al., 2019). AM parts are known to have tensile residual stress on the surface, as a result of AM parts showing poor surface characteristics. Post-processing of AM-built parts using LSP can be advantageous as this technique introduces compressive residual stresses on the surface. These compressive stresses do not allow material to plough or spall from the surface during the wear test, as a result of which the overall increase in the wear resistance can be observed. In addition, laser power (P) and the number of shots (n) are found to play an important role in improving the wear properties. Studies show that a rise in both P and n can improve wear resistance (Jinoop et al., 2019). However, an increase in the number of shots is found to be more significant than increasing laser power when it comes to improving specific wear rates.

Hot isostatic pressing (HIP) is a thermo-mechanical processing technique. In this process, the built parts are compressed under high temperatures. Depending on the material, the operating temperature during HIP is likely to range from 100°C to 2000°C and the compression load can vary from a few tens to 200 MPa. The HIP process is very similar to the hot-pressing technique. However, the major difference lies in the type of loading being used and therefore the reader should not be confused. In hot pressing, the applied load is uniaxial whereas in HIP the load is applied in isostatic (equal from all sides). In most cases, a pressured argon gas is used to create the required isostatic pressure. Ti-6Al-4V samples fabricated using the Electron Beam

Melting (EBM) technique show a marginal improvement in the fatigue limit (from 180 to 200 MPa) for 5×10^6 cycles after HIP. It is also found that the HIP process has the least effect on both surface finish and fatigue life.

Laser remelting is another post-processing technique that is used to tailor the surface properties of MAM parts. Laser remelting uses a laser to melt the top surface of the parts at a high scanning speed to obtain higher cooling rates. This results in a finer grain structure, higher surface hardness, and improved surface quality. Researchers have investigated the effect of laser remelting on tribological behaviour significantly. One of the recent studies by Karimi et al. investigated the effect of laser remelting on the surface properties of PBF-built Ti-6Al-4V. It was noted that the surface quality of the parts increased significantly with an increase in the number of remelting cycles. In addition, the surface hardness and wear resistance increased with an increase in the number of remelting cycles. Figure 4.8 presents the variation in wear behaviour at different process conditions, where SV, DV, TV, and HV indicate as-built SLM sample, double remelting, triple remelting, and heat-treated single melting, respectively. The improvement in wear resistance is seen with an increase in the number of melting steps from SV to TV for both the transverse cross-section and longitudinal cross-section surfaces. Table 4.2 presents the variation in mechanical properties at different conditions (Karimi et al., 2022).

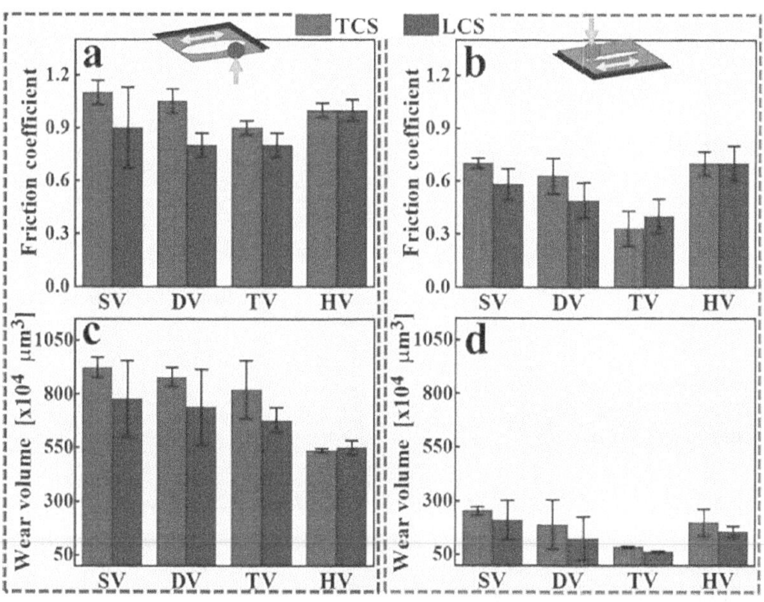

FIGURE 4.8 Wear behaviour at different process conditions for PBF-built Ti-6Al-4V (Karimi et al., 2022). (Reproduced with permission from Karimi et al., 2022) Copyright Elsevier, 2022.)

TABLE 4.2
Variation in Mechanical Properties at Different Conditions (Karimi et al., 2022). (Reproduced with Permission from (Karimi et al., 2022). Copyright Elsevier, 2022.)

	Vertically Oriented Specimens			Horizontally Oriented Specimens			
	SV	MV	HV	SH	DH	TH	HH
Ultimate Strength (MPa)	1208	1198	1043	1155	1133	1063	1047
Yield Strength (MPa)	1103	1108	1012	1051	1143	997	1008
Area Reduction (%)	3.3	2.81	11.3	2.7	1.28	0.8	4.9
Modulus (GPa)	123	145	147	128	127	136	147

4.5 TRIBOLOGY OF MAM PARTS IN ORTHOPAEDIC APPLICATIONS

One of the core areas of applications where metal AM techniques are coming up in a good way is the fabrication of orthopaedic implants. Several materials are used in orthopaedic applications as shown in Figure 4.9. The advantage of AM to fabricate parts directly from CAD modelling is best suitable for patient-specific implants. These implants when fitted in a patient are subject to different kinds of loads. As the patient improves and starts day-to-day activity, these implants will have a certain relative motion with other body tissues and fluid. As a result of relative motion, the implants undergo wear and tear, and therefore it becomes critical to understand the tribo-corrosive behaviour of these implants (Malik et al., 2022). As mentioned in

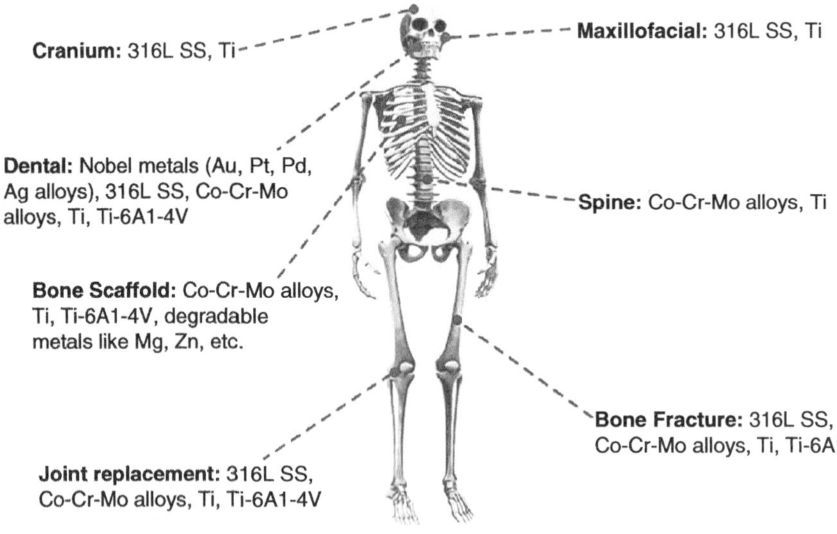

FIGURE 4.9 Materials being used as body implants (Bai et al., 2019). (Reproduced with permission from Bai et al., 2019) Copyright Elsevier, 2019.)

the last section, friction and wear lead to material disintegration in terms of debris. This debris is not a major concern if the system is open, such as mechanical parts. However, when it comes to body implants, this debris tends to accumulate near and around, reducing the intended life span of implants. The accumulation of this debris may also add to the annoying sound when a person walks (Malik et al., 2022).

Under the hood of available AM-tailored alloys, Ti-6Al-4V and SS316L are among the widely used material systems for body implants. Both SS316L and Ti-6Al-4V materials are known to have good mechanical properties and bio-compatibility. However, the wear performance of both of these alloys is very poor. As a result, researchers used some post-processing techniques such as coatings and surface treatments (Caminero et al., 2021; Gorsse et al., 2017; Saravanan et al., 2018).

There are studies discussing the various tribological aspects of additively manufactured Ti-6Al-4V (Roudnicka et al., 2020). It was found that the AM processing route does not have a significant effect on the wear rate. However, the refined microstructure and the induced stresses during the AM process have resulted in a slight reduction in specific wear. In the metal AM technique, especially LPBF and LDED, due to high cooling rates during the solidification, the sample is found to show a hard martensite phase. These samples show a comparatively lower value of specific wear (Roudnicka et al., 2020). Ti-6Al-4V is always a material of choice for manufacturing hip implants; however, the poor wear resistance and stiffness mismatch with other body parts are some of the known key issues limiting its widespread use (Fousova et al., 2017). The use of specific coatings and surface post-processing is a possible route to counter these issues. To this end, researchers have used PVD coatings of titanium nitride, titanium carbonitride, amorphous carbon, and silver containing amorphous carbon on Ti6Al7Nb alloy for biomedical use. These coatings have been found to show a significant improvement in the tribological performance against Al_2O_3. Other than the tribological performance, it is equally important to look into the adhesion strength of these coatings. The findings show TiN to have the highest adhesion strength among the different coatings (Tillmann et al., 2022).

4.6 EFFECT OF LUBRICATION ON MAM COMPONENTS

It is a well-established fact that improvement in hardness leads to improved wear resistance. However, the high hardness of the material can lead to severe wear loss in the counter body. This situation is again infeasible to be used anywhere, as there will be continuous demand to change the counterpart. These kinds of situations can be resolved by either using an appropriate lubrication medium (Viat et al., 2015) or by forming a self-lubrication film.

The deployment of solid lubricants is a technique to improve the tribological properties of metallic alloys. For example, the common solid lubricants that are used to improve tribological characteristics include graphite, MoS_2, Mo- oxide, Mo- fluoride, and soft metals. In one of the studies by Qin et al., the tribological behaviour of TC4 alloy was studied with and without bionic coating. TC4-SnAgCu bionic coating was prepared using the laser PBF technique with different concentrations of solid lubricants such as TC4 with SnAgCu concentration varying from 5 to 15 wt%. It was seen that the addition of bionic coating reduced the friction in a higher load range and improved the material's hardness and wear resistance. It was also noted that the lubricant moved to the worn surface and a lamellar lubrication film was generated. This film aids in reducing surface damage and constraining the development of vibration and noise due to friction. The composition of the lubricant also plays a significant role in controlling the lubricating film. In the above case, 10% of the solid lubricant was more favourable for forming the lubricating film on the worn surface, with the best friction and noise drop at 20 N load (Qin et al., 2021).

A novel strategy was developed by Wang et al. to advance the wear behaviour of TC4 alloy by merging surface texture and solid lubricant. The morphology and composition of the worn surface are presented in Figure 4.10. It was performed by preparing a coating with a honeycomb structure using PBF with a specified thickness, and polytetrafluoroethylene (PTFE) was fused as a solid lubricant to develop a composite layer using a traditional hot-pressing technique. The hybrid coating developed using the above technique displays steady low COF and negligible wear properties under several operating conditions. The coating layer has a uniform, chemically stable, and constantly replenishable lubricant film under frictional heat and frictional force due to the elevated thermal expansivity of PTFE and its continuous extrusion due to trapped abrasive debris. The surface of the coupling ball creates an exclusive transfer film owing to the adsorption and partial defluorination reaction of PTFE. This distinct friction interface is accountable for the composite coating's excellent wear resistance (Wang et al., 2022).

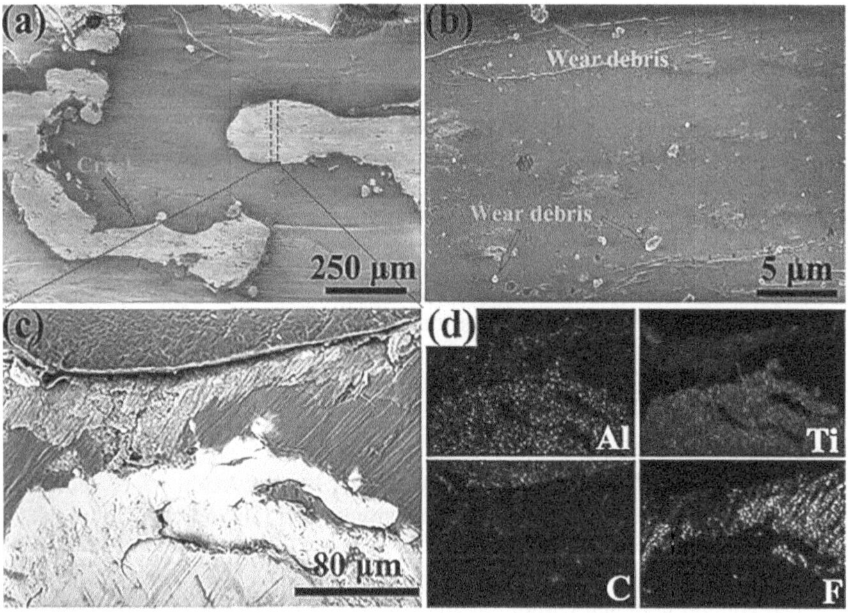

FIGURE 4.10 (a) and (b) Hybrid coating wear surface. (c) Cross-section of worn surface and (d) elemental mapping of the cross-section (Wang et al., 2022). (Reproduced with permission from Wang et al., 2022). Copyright Elsevier, 2022.)

4.7 CONCLUSIONS

MAM transformed AM from a process for prototyping to a process for fabricating near-net-shaped end-use components. It is necessary to understand the microstructure and mechanical properties of MAM-built components in as-built and post-processed conditions to qualify the material for various applications. The tribological properties of MAM components are significant for applications that require excellent surface properties.

Laser power, scan speed, feed rate, and hatch spacing are some of the important MAM process parameters that affect the performance of parts. In the MAM process, one of the biggest challenges is to have a good surface finish, which affects tribological behaviour. Studies also show that the hardness of the MAM part is found to have a significant impact on the wear behaviour. Build direction can alter the resulting microstructure, and the microstructure in the 0° build orientation was found to have fine columnar dendrites and equiaxed grain dispersed on the surface. The sample fabricated at 45° orientation is found to show the least wear resistance and the highest COF values, as it contains the coarsest microstructure among all cases. Manufacturing of nanocomposite material using the MAM process is an area of interest for researchers to improve the surface characteristics of a given base material.

Processes such as heat treatment, laser peening, abrasive finishing, drag finishing, machining, and polishing are some of the post-processing techniques being used for MAM-fabricated parts. LSP is one of the advanced post-processing techniques that is used to improve the fatigue behaviour of LDED-built components. The wear behaviour can be tweaked by varying the shock peening process parameters.

In addition, MAM has shown great progress in biomedical engineering. The poor wear resistance and stiffness mismatch with other body parts limit the usage of the commonly used Ti-6Al-4V material. Researchers have used PVD coatings of titanium nitride, titanium carbonitride, amorphous carbon, and silver containing amorphous carbon on Ti6Al7Nb alloy for biomedical use.

MAM is a widely expanding field and new research is being developed across the globe daily. New materials, new processes, and new post-processing treatments are being developed continuously to improve the properties of MAM components. The tribological behaviour of MAM-built components also depends on the microstructure and texture. The tribological behaviour of MAM parts can be improved and qualified for applications using various processing and post-processing techniques.

REFERENCES

Bahnini, I., Rivette, M., Rechia, A., Siadat, A., and Elmesbahi, A. (2018). "Additive manufacturing technology: the status, applications, and prospects." *Int. J. Adv. Manuf. Technol.*, 97(1), 147–161.

Bahshwan, M., Myant, C. W., Reddyhoff, T., and Pham, M. S. (2020). "The role of microstructure on wear mechanisms and anisotropy of additively manufactured 316L stainless steel in dry sliding." *Mater. Des.*, 196, 109076.

Bai, L., Gong, C., Chen, X., Sun, Y., Zhang, J., Cai, L., Zhu, S., and Xie, S. Q. (2019). "Additive manufacturing of customized metallic orthopedic implants: Materials, structures, and surface modifications." *Metals (Basel).*, 9(9), 1004.

Caminero, M. Á., Romero, A., Chacón, J. M., Núñez, P. J., García-Plaza, E., and Rodríguez, G. P. (2021). "Additive manufacturing of 316L stainless-steel structures using fused filament fabrication technology: mechanical and geometric properties." *Rapid Prototyp. J.*, 27(3), 583–591.

Chakkravarthy, V., and Jerome, S. (2020a). "Printability of multiwalled SS 316L by wire arc additive manufacturing route with tunable texture." *Mater. Lett.*, 260, 126981.

Chakkravarthy, V., and Jerome, S. (2020b). "Fabrication of preferentially oriented Al4043 alloy and its wear anisotropy." *Mater. Lett.*, 280, 128578.

Chakkravarthy, V., Lakshmanan, M., Manojkumar, P., and Prabhakaran, R. (2022). "Crystallographic orientation and wear characteristics of TiN, SiC, Nb embedded Al7075 composite." *Mater. Lett.*, 306, 130936.

Chakkravarthy, V., Oliveira, J. P., Mahomed, A., Yu, N., Manojkumar, P., Lakshmanan, M., Zhang, L., Jerome, S., Prabhu, T. R., and Narayan, R. L. (2023). "Effect of abrasive water jet peening on NaCl-induced hot corrosion behavior of Ti–6Al–4V." *Vacuum*, 111872

Chaurasia, J. K., Jinoop, A. N., Parthasarathy, P., Paul, C. P., Bindra, S., and Bontha, S. (2021). "Study of melt pool geometry and solidification microstructure during laser surface melting of Inconel 625 alloy." *Optik (Stuttg).*, 246(July), 167766.

DebRoy, T., Wei, H. L., Zuback, J. S., Mukherjee, T., Elmer, J. W., Milewski, J. O., Beese, A. M., Wilson-Heid, A., De, A., and Zhang, W. (2018). "Additive manufacturing of metallic components – Process, structure and properties." *Prog. Mater. Sci.*, 92, 112–224.

Ford, S., and Despeisse, M. (2016). "Additive manufacturing and sustainability: an exploratory study of the advantages and challenges." *J. Clean. Prod.*, 137, 1573–1587.

Fousova, M., Vojtech, D., Kubasek, J., Jablonska, E., and Fojt, J. (2017). "Promising characteristics of gradient porosity Ti-6Al-4V alloy prepared by SLM process." *J. Mech. Behav. Biomed. Mater.*, 69, 368–376.

Frutos, G. M. ınez de. (2015). "Product development process for additive manufacturing Considerations on the design of final products for 3D printing." Thesis, Instituto Superior Técnico Lisboa.

Gao, Y., and Zhou, M. (2018). "Superior mechanical behavior and fretting wear resistance of 3D-printed Inconel 625 superalloy." *Appl. Sci.*, 8(12), 2439.

Gisario, A., Kazarian, M., Martina, F., and Mehrpouya, M. (2019). "Metal additive manufacturing in the commercial aviation industry: A review." *J. Manuf. Syst.*, 53, 124–149.

Gong, J., and Wilkinson, A. J. (2009). "Anisotropy in the plastic flow properties of single-crystal α titanium determined from micro-cantilever beams." *Acta Mater.*, 57(19), 5693–5705.

Gorsse, S., Hutchinson, C., Gouné, M., and Banerjee, R. (2017). "Additive manufacturing of metals: a brief review of the characteristic microstructures and properties of steels, Ti-6Al-4V and high-entropy alloys." *Sci. Technol. Adv. Mater.*, 18(1), 584–610.

Gu, D., Hagedorn, Y. C., Meiners, W., Meng, G., Batista, R. J. S., Wissenbach, K., and Poprawe, R. (2012). "Densification behavior, microstructure evolution, and wear performance of selective laser melting processed commercially pure titanium." *Acta Mater.*, 60(9), 3849–3860.

Gu, D., and Zhang, G. (2013). "Selective laser melting of novel nanocomposites parts with enhanced tribological performance: Nanocrystalline TiC/Ti nanocomposites parts were built via SLM technology and the densification, microstructures, microhardness and tribological Performance were." *Virtual Phys. Prototyp.*, 8(1), 11–18.

Hua, K., Tong, Y., Zhang, F., Wang, C., Kou, H., Wu, H., and Wang, H. (2022). "Dependence of fretting wear on the microstructure characteristics and impact on the subsurface stability of a metastable β titanium alloy." *Tribol. Int.*, 165, 107351.

Hua, K., Zhang, Y., Zhang, F., Kou, H., Li, X., Wu, H., and Wang, H. (2021). "Microstructure refinement and enhanced wear-resistance modulated by stress relaxation processing in a metastable β titanium alloy." *Mater. Charact.*, 181, 111505.

Idury, K. S. N. S., Chakkravarthy, V., and Narayan, R. L. (2021). "Mechanical behavior of laser powder bed fusion processed inconel 625 alloy." *Trans. Indian Natl. Acad. Eng.*, 6(4), 975–990.

International Organization for Standardization/ASTM International. (2017). *ISO/ASTM 52900: 2017-06 Additive Manufacturing—General Principles—Terminology*. The International Organization for Standardization and ASTM International.

Jinoop, A. N., Subbu, S. K., Paul, C. P., and Palani, I. A. (2019). "Post-processing of laser additive manufactured inconel 718 using laser shock peening." *Int. J. Precis. Eng. Manuf.*, 20(9), 1621–1628.

Karimi, J., Antonov, M., Kollo, L., and Prashanth, K. G. (2022). "Role of laser remelting and heat treatment in mechanical and tribological properties of selective laser melted Ti6Al4V alloy." *J. Alloys Compd.*, 897, 163207.

Kaynak, Y., and Kitay, O. (2018). "Porosity, surface quality, microhardness and microstructure of selective laser melted 316l stainless steel resulting from finish machining." *J. Manuf. Mater. Process.*, 2(2), 36.

Kumar, D., Gosvami, N. N., and Jain, J. (2021). "Influence of temperature on crystallographic orientation induced anisotropy of microscopic wear in an AZ91 Mg alloy." *Tribol. Int.*, 163, 107159.

Kumar, L. J., and Krishnadas Nair, C. G. (2017). "Current trends of additive manufacturing in the aerospace industry BT". In *Advances in 3D Printing & Additive Manufacturing Technologies*. D. I. Wimpenny, P. M. Pandey, and L. J. Kumar, eds., Singapore: Springer Singapore, 39–54.

Kumar, S. P., Chakkravarthy, V., Mahalingam, A., Rajeshshyam, R., Sriraman, N., Marimuthu, P., Narayan, R. L., and Babu, P. D. (2022). "Investigation of crystallographic orientation and mechanical behaviour in laser-welded stainless steel 316L additive components." *Trans. Indian Inst. Met.* 76(2), 527–535.

Lakshmanan, M., SelwinRajadurai, J., Chakkravarthy, V., and Rajakarunakaran, S. (2020). "Wear and EBSD studies on (SiC/NiTi) reinforced Al7075 composite." *Mater. Lett.*, 272, 127879.

Lakshmanan, M., Selwin Rajadurai, J., Chakkravarthy, V., and Rajakarunakaran, S. (2021). "Tribological investigations on h-BN/NiTi inoculated Al7075 composite developed via ultrasonic aided squeeze casting." *Mater. Lett.*, 285, 129113.

Li, H., Ramezani, M., Li, M., Ma, C., and Wang, J. (2018). "Effect of process parameters on tribological performance of 316L stainless steel parts fabricated by selective laser melting." *Manuf. Lett.*, 16, 36–39.

Liu, Z. (2023). *Multiscale Mechanical Behaviours of a Near-Alpha Titanium Alloy Made by Additive Manufacturing* (Doctoral dissertation, University of Toronto (Canada)).

Mahamood, R. M., and Akinlabi, E. T. (2016). "Achieving mass customization through additive manufacturing BT." In *Advances in Ergonomics of Manufacturing: Managing the Enterprise of the Future*. C. Schlick and S. Trzcieliński, eds., Cham: Springer International Publishing, 385–390.

Malik, A., Rouf, S., Ul Haq, M. I., Raina, A., Valerga Puerta, A. P., Sagbas, B., and Ruggiero, A. (2022). "Tribo-corrosive behavior of additive manufactured parts for orthopaedic applications." *J. Orthop.*, 34, 49–60.

Paul, C. P., Bhargava, P., Kumar, A., Pathak, A. K., and Kukreja, L. M. (2013). "Laser rapid manufacturing: Technology, applications, modeling and future prospects." In *Lasers Manufacturing*. P. Davim, ed., Hoboken, NJ: Wiley Online Books, 1–67.

Qin, S., Shi, X., Xue, Y., Huang, Q., and Zhang, K. (2021). "Tribological behaviors and friction-induced vibration and noise performance of TC4 with bionic coating prepared by laser additive manufacturing." *J. Mater. Eng. Perform.*, 30(12), 9455–9469.

Roudnicka, M., Bayer, F., Michalcova, A., Kubasek, J., Hamed Alzubi, E. G., and Vojtech, D. (2020). "Biomedical titanium alloy prepared by additive manufacturing: Effect of processing on tribology." *Manuf. Technol.*, 20(6), 809–816.

Sandhu, S. S., and Shahi, A. S. (2016). "Metallurgical, wear and fatigue performance of Inconel 625 weld claddings." *J. Mater. Process. Technol.*, 233, 1–8.

Saravanan, M., Devaraju, A., Venkateshwaran, N., Krishnakumari, A., and Saarvesh, J. (2018). "A review on recent progress in coatings on AISI austenitic stainless steel." *Mater. Today Proc.*, 5(6), 14392–14396.

Su, J., Jiang, F., Li, J., Tan, C., Xu, Z., Xie, H., Liu, J., Tang, J., Fu, D., Zhang, H., and Teng, J. (2022). "Phase transformation mechanisms, microstructural characteristics and mechanical performances of an additively manufactured Ti-6Al-4V alloy under dual-stage heat treatment." *Mater. Des.*, 223, 111240.

Sun, Y., Moroz, A., and Alrbaey, K. (2014). "Sliding wear characteristics and corrosion behaviour of selective laser melted 316L stainless steel." *J. Mater. Eng. Perform.*, 23(2), 518–526.

Tillmann, W., Lopes Dias, N. F., Kokalj, D., Stangier, D., Hein, M., Hoyer, K. P., Schaper, M., Gödecke, D., Oltmanns, H., and Meißner, J. (2022). "Tribo-functional PVD thin films deposited onto additively manufactured Ti6Al7Nb for biomedical applications." *Mater. Lett.*, 321, 132384.

Viat, A., Fouvry, S., Barros Bouchet, M. I. De, and Pin, L. (2015). "Influence of carbon-based solid lubricant on fretting wear response for alumina-based ceramics versus cobalt superalloy contact." *Surf. Coat. Technol.*, 284, 327–333.

Wang, B., Lai, W., Li, S., Huang, S., Zhao, X., You, D., Tong, X., Li, W., and Wang, X. (2022). "Self-lubricating coating design strategy for titanium alloy by additive manufacturing." *Appl. Surf. Sci.*, 602, 154333.

Yadroitsev, I., Pavlov, M., Bertrand, P., and Smurov, I. (2009). "Mechanical properties of samples fabricated by selective laser melting." *14èmes Assises Eur. du Prototypage Fabr. Rapide*, 625(February 2016), 24–25.

Yan, X., Gao, S., Chang, C., Huang, J., Khanlari, K., Dong, D., Ma, W., Fenineche, N., Liao, H., and Liu, M. (2021). "Effect of building directions on the surface roughness, microstructure, and tribological properties of selective laser melted Inconel 625." *J. Mater. Process. Technol.*, 288, 116878.

Yang, Y., Zhu, Y., Khonsari, M. M., and Yang, H. (2019). "Wear anisotropy of selective laser melted 316L stainless steel." *Wear*, 428–429, 376–386.

Ye, J., Chen, X., Luo, H., Zhao, J., Li, J., Tan, J., Yang, H., Feng, B., Zheng, K., and Pan, F. (2022). "Microstructure, mechanical properties and wear resistance of Ti particles reinforced AZ31 magnesium matrix composites." *J. Magnes. Alloy.*, 10(8), 2266–2279.

5 Tribo-Behavioural Enhancement through Surface Texturing

Krishnkant Sahu, Adesh Kumar Tomar, and Sandeep Kumar Dwivedi

5.1 INTRODUCTION

As the demand for materials continues to surge across diverse functions and applications, there has been a notable upswing in the adoption of surface texturing techniques. This surge is particularly evident in the realm of enhancing the tribological performance of materials and tribo pairs. Texturing surfaces has emerged as a crucial focus of investigation in the field of material science, showcasing its remarkable capability to fine-tune tribological properties tailored to specific applications. An improvement in tribological performance was found by different researchers when they closely and carefully examined the impact of texturing on materials in diverse tribological conditions. With the intention of improving tribological properties like friction and wear, much extensive research has focused on material texturing techniques. Surface texturing employs methods like forming, micro grooves, micro dimples, and micro channels to alter the topography. Different types of surface texturing influence shapes such as micro-pits, nano-dots, micro-pillars, micro-dimples, micro-grooves, and inverted pyramid shapes have been analysed and fabricated. Laser surface texturing (LST) is by far the most advanced technology that can even produce nano-level texture features over the surfaces (Etsion et al., 1999; Etsion, 2005). Besides the load capacity of smooth surface bearings, these micro-bearings contribute to the generation of an additional load-bearing ability by generating their own localized hydrodynamic action. Additionally, this micro-texture collects wear debris and serves as a lubricant reservoir for sliding contacts using either boundary or mixed lubrication. The application of textured surface bearings operating in mixed and hydrodynamic bearings significantly enhances the capacity to carry load and reduces the frictional coefficient. Research has been carried out by various researchers on textured surface shapes and geometries for various tribological functions and applications. Furthermore, the impact of texturing parameters like the area fractions, distance, shapes, spacing, geometries, dimensions, width, and tuning of microstructure depth has been thoroughly investigated. The most important reason for friction is the relative motion among the tribo pairs.

Enhancing the durability of mechanical components has been a primary focus, with a key emphasis on reducing friction. In 1967, the Organization for Economic

DOI: 10.1201/9781003359333-5

Cooperation and Development proposed a committee to figure out the root cause of the economic losses that occurred due to friction and wear of the mechanical components. The Committee's report revealed that the wear of mechanical components resulted in a substantial economic loss (3%–5% of GDP) due to improper tribological practices. Therefore, friction and wear can be controlled effectively by adequate lubrication between mechanical components.

Friction is the main parameter that comes under tribology. Even though there are a lot of ways to reduce friction, surface texturing is a possible and promising method. Several methods to fabricate surface texturing include atomic layer deposition, two-photon lithography, ball burnishing, micro-milling, LST, ultrasonic-assisted grinding, electrical discharge machining (EDM), and chemical etching. Particularly in bearing applications, the texture pattern protects the surfaces from dents (scratches) by imparting lubricant reservoirs, wear debris traps, and micro-hydrodynamic bearing. This dent contributes to sufficient lubrication among the interacting surfaces.

It is sure that certain textured surfaces may have a positive influence on tribological performance, but it is generally agreed that, based on specific necessities of applications, texture optimization should be performed. In the cutting-edge field of applying surface texturing to mechanical components like journal bearings, sealing, piston-ring and cylinder liners, and cutting tools, success hinges on meeting application requirements, conducting rigorous testing, and ensuring superior tribological performance. In many mechanical applications, the micron-scale partial texturing shapes improve performance compared to full texturing. In the experimental test, it has been observed that for the case of journal bearings, seals, cutting tools, and piston-ring & cylinder liners, friction reduces up to 18%, 65%, 34.5%, and 82%, respectively.

One of the main aims of tribologists is to design and develop energy-efficient journal-bearing/tribo pairs. To attain this objective, tribologist engineers across the globe are actively pursuing their research activities to analyse textured surfaces, which enhance the tribological performance. In recent decades, the use of textured surface bearings has emerged as a viable means to enhance tribological performance. The idea/concept of applying texture features actually came from the special surface textures (scales) observed on the skins of marine animals and plants (Bhushan, 2009; Dean and Bhushan, 2010). For instance, the textured scales and aerodynamic surface of fish help in reducing frictional drag. Also, the natural texture on the surface of the frogs and lizards helps them to walk/crawl over wet surfaces or walls. The use of dimples over the surface of a golf ball (Jearl, 1979) results in a significant reduction in the frictional drag acting during its flight. The dimpled surface golf ball was reported to be flying four times the distance covered by a smooth surface golf ball. A textured surface essentially consists of intentionally induced micro-features (dimples, grooves, cross-hatch, etc.). Such micro-features possess a deterministic character. These micro-features, in contrast to micro-roughness, are identical in dimensions to each other and systematically arranged over the bearing surfaces. Contrary to the above, micro-roughness over the bearing surface possesses a random character and occurs on a smaller scale as compared to the textured features. The use of textured surfaces has been reported to enhance the efficiency of tribo systems, by improving the capacity to carry loads and reducing the amount of friction between the sliding surfaces (Etsion et al., 1999; Etsion, 2005; Yadav and Sharma, 2014; Gropper et al., 2016; Yadav and Sharma, 2016). Over the past two decades, numerous researchers have conducted both numerical and experimental studies to explore the viability of textured surfaces in designing energy-efficient tribo-systems.

Firstly, Hamilton et al. (1966) carried out an experimental examination regarding the effect of surface micro-regularities on rotary seals. As a result of the hydrodynamic action, these micro-asperities on the seal surface generate fluid film pressure, which produces more load-carrying capacity. A comprehensive investigation and experimental analysis were undertaken to examine how micro-asperities impact the load capacity and oil leakage rate of seals/bearings (Anno et al., 1969). Mathematically, the micro-asperities were modelled as peaks and valleys over the bearing surface. The author reported that peaks and valleys generate almost a similar magnitude of load-supporting capacity. However, the valleys were reported to be offering load-supporting capacity with a negligible oil leakage rate. Some promising results regarding the use of engineered (textured) surfaces have been delivered (Hamilton et al., 1966). However, the lack of manufacturing machines during that time, to generate micro-scale characteristics on the bearing surfaces with enhanced dimensional precision, resulted in very little research in the area of textured surface bearing (Anno et al., 1969). Over the course of time, advancement in the computing machines and improvement in numerical technique, for example, finite-difference method (FDM) (Sharma and Pandey, 2008; Kango and Sharma, 2010; Fesanghary and Khonsari, 2011; Etsion, 2013; Woloszynski et al., 2013), finite volume method (FVM) (Dobrica et al., 2010; Xiong and Wang, 2012; Woloszynski et al., 2013), finite element method (FEM) (Hajjam and Bonneau, 2007; Giacopini et al., 2010; Woloszynski et al., 2013, 2015b), spectral element method (Woloszynski et al., 2013, 2015b), and finite cell method (Pei et al., 2011), provided a new impetus and life to studies related to the numerical simulation of textured surfaces. FEM (Woloszynski et al., 2013, 2015a), compared to other numerical techniques, is reported to provide a higher convergence speed. The FEM is also noticed to require a smaller number of nodes than other numerical techniques to obtain a desired level of accuracy in the numerical simulation of textured surface bearings. The FDM approach requires a greater number of nodes and computational time, while numerically simulating the textured surface bearing. Still, FDM is a preferred choice of most of the researchers, owing to its ease of implementation. FVM and FEM techniques (Woloszynski et al., 2013; Woloszynski et al., 2015a) are reported to have good capabilities for handling complex geometries, which include intricate shapes/discontinuities such as a dimple in the film thickness. In recent times, advanced manufacturing techniques (Etsion, 2005; Venkatesan and Stephens, 2005; Yang et al., 2009; Coblas et al., 2015; Costa and Hutchings, 2015; Ibatan et al., 2015; Gropper et al., 2016) have been developed, which have the ability to create micro- or even nano-level features on the bearing surface with a high degree of dimensional control, repeatability, and accuracy. As a result, many research efforts have been carried out by investigators across the world, which deal with numerical and experimental examination of textured surfaces.

5.2 SURFACE TEXTURING TECHNIQUES

Innovative fabrication techniques and advancements in additive manufacturing have made intricate texture designs increasingly achievable with high precision, efficiency, repeatability, and cost-effectiveness. Beyond traditional shapes like circular dimples and grooves, the realm of texture characteristics now embraces the creation of mixed or overlapped motifs with varying scales and geometries on the same functional surface. The selection of the texturing technique for particular applications must adhere to minimum requirements like productivity and safety of the substrate material, compatibility with the samples' dimensions, accuracy, compatibility with the essential grain size and shape, and cost-effectiveness.

Normally, the most common types of textured surface such as discrete and continuous textures are used according to their need and applications. Discrete textures feature shapes like circles, rectangles, triangles, spheres, and ellipses. Continuous textures encompass straight lines, sinusoidal patterns, positive full-wave patterns, arrays of curved lines, and crosshatch patterns. The design and direction of surface textures play a crucial role in altering the performance of tribo-contacts, affecting aspects such as pressure regeneration and load capacity. Here, finding the crucial (optimal) dimensional relationships between the many geometrical parameters of texture surfaces is a difficult task. These geometrical characteristics include pitch, array, and angle as well as the length, width, and depth of the texture surface. In addition, the density of texture surfaces plays a significant role in enhancing the bearing (tribo-contacts) performance (Brizmer et al., 2003; Tala-Ighil et al., 2007; Tala-Ighil et al., 2011; Sharma and Yadav, 2014). Extensive research has been dedicated to exploring surface texturing. The objective is to ascertain the fundamental impact of surface texturing on tribological performance, employing various techniques for creating textured surfaces. The salient features of the above-mentioned texturing techniques are explored in the following subsections (Figure 5.1).

5.2.1 Laser Surface Texturing Technique

The Laser Surface Texturing (LST) technique is probably the most advanced and convenient method (Figure 5.2). This is due to the laser texturing method, which allows for short processing times, is exceptionally fast, offers good shape and size management, and is environmentally friendly.

The intensity and pulsation of the laser radiation may be controlled very accurately to create spherical or custom-shaped pores with selected dimensions and depth. This technique excels in producing a multitude of micro-dimples on tribological surfaces. In this technique, CO_2 and Nd:YAG laser systems are generally used for surface texturing (Etsion, 2004a,b; Sudeep et al., 2015). LST can create diverse textures by treating specific areas, with each feature produced separately. This involves either moving the sample using a translation stage or scanning the laser beam across the sample surface with a galvanometric scanner, targeting specific areas for precision. A shutter mechanism stops the laser beam between texture generation locations. Crucial laser parameters, like focusing system, pulse duration, fluence, and wavelength, significantly influence texture and morphology. Direct laser writing involves

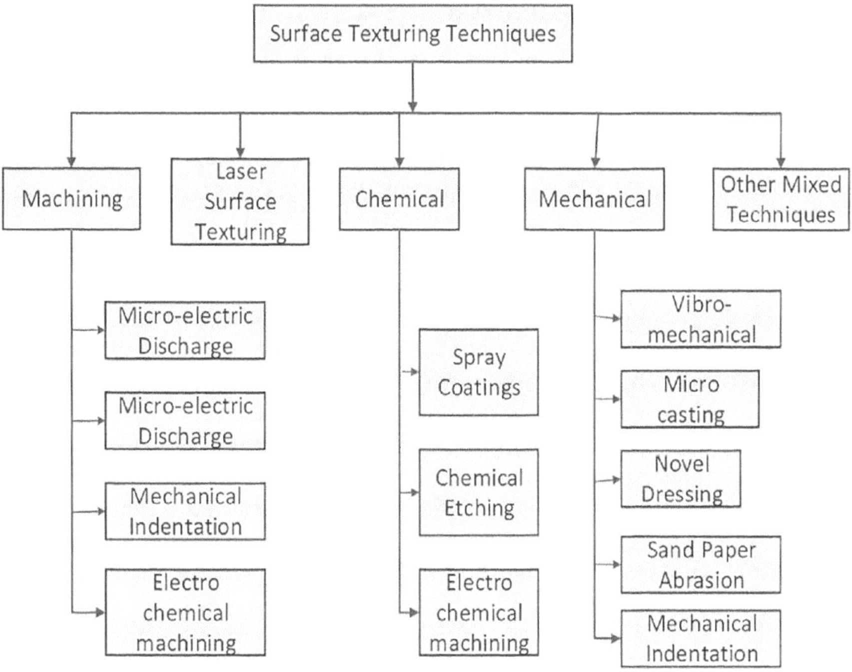

FIGURE 5.1 Various surface texturing techniques. (Venkatesan and Stephens, 2005; Coblas et al., 2015; Costa and Hutchings, 2015; Singh et al., 2021.)

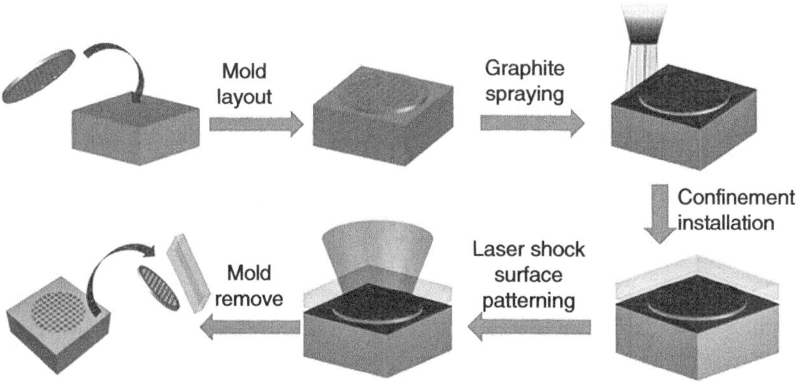

FIGURE 5.2 Laser surface texturing technique. (Reproduced with permission from Copyright *Journal of Manufacturing Processes*, Elsevier (Mao et al., 2018).)

focusing and scanning the laser directly over the sample. However, laser texturing has drawbacks, including heat-affected zones, burns, and high costs. Additional procedures, such as grinding, can be employed to eliminate dimensional flaws like periodic textures and can also be advantageous in managing hydrophobicity.

5.2.2 Vibro-Mechanical Texturing Technique

The Vibro-Mechanical Texturing (VMT) technique is an improved texturing technique, which generally relies on the standard turning procedure or operation. In a typical computer-numeric-controlled (CNC) lathe, this operation is carried out, which is governed by a tool positioning state piezoelectric actuator. In the VMT, an oscillated single-point cutting tool is used, and this tool engages with the surface of a rotating workpiece.

Due to this oscillation, the cutting tool locally removes surface material and forms dimples on workpiece. Dimples of different sizes can be produced by controlled vibratory motion because the dimple size is regulated on the basis of the tunable oscillation tool path. A continuous spiral groove is produced around the perimeter of the workpiece surface as a result of this technique. Surface texturing is made simple with VMT technology, which has a number of benefits over current methods, including high-precision texturing, cost-effective manufacturing, and little induced surface damage (Greco et al., 2009; Costa and Hutchings, 2015). VMT or vibro machining, an effective surface texturing technique, creates micro-texture patterns on components in tribological interfaces. Enhancing hydrodynamic motion, this method improves the tribological performance of lubricated contact systems.

The VMT approach is a standard turning process with a sophisticated tool positioning mechanism, enabling controlled tool vibration. VMT is a contemporary method for producing small-scale dimples or textures, rooted in the turning process. In this method, a single-point cutting tool is mounted on a CNC lathe machine and spins to engrave or texture the rotating workpiece surface. An FTS system or Fast Tool Servo mechanism is used to control the oscillatory motion of the tool, and dimples are produced by this oscillatory movement. It is a precise and very neat technique of producing dimples with the least surface variation and damage. This technique is generally employed for creating micro-textures on cylindrical surfaces. Tool motion has been controlled to cut grooves on the surface. Varieties of textures and dimples are created with the help of tunable oscillations.

5.2.3 Chemical Etching Texturing Technique

Chemical etching is an economical/inexpensive method that modifies the desired topography without changing the substrate's mechanical properties; however, the procedure consumes more time and involves material removal, which diminishes the overall specimen size. In the industry, the etching procedure is often referred to as chemical milling. In this procedure, acids or other chemicals are employed to dissolve undesirable materials like glass, metals, and semiconductor materials. This method has been used in a wide range of metals with metal removal depths up to 12 mm (0.5 inch). This method is frequently applied to the production of printed circuit boards and semiconductors, as well as in the aerospace sector to remove a thin film of material from huge aircraft parts, components, and missile skin panels. With this method, the substrate surface can be imparted micro- or nanoscale textures.

The surface of the substrate is typically washed with deionized water after being roughened with grating paper. The surface is treated with hydrochloric acid (HCl), sodium hydroxide (NaOH), and acetic acid (CH3COOH) to create micro- and nano-scale textures. Submersion in a silane solution reduces surface energy. However, limitations such as substrate size, texture accuracy, and prolonged etching time hinder its widespread use in industries (Singh et al., 2021).

5.2.4 Mechanical Indentation Technique

Mechanical indentation texturing technique is a simple and convenient method for the incorporation of micro-dimples on the tribo-contacts. This method involves pressing a hard indenter into the part's surface, which is responsible for the plastic deformation and creates the desired surface texture shape. In this technique, the mild steel rod is used in which steel balls are welded in a sequence called indenters. The rod with indenters is pressed with the help of a hydraulic device or screw jack arrangement; due to this action, micro-dimples are created on the surface (Coblas et al., 2015; Costa and Hutchings, 2015).

5.2.5 Photolithography Texturing Technique

In micro-fabrication, photolithography (also known as optical lithography) is a technique used to remove specific areas of a thin film or a substrate's bulk. It offers precise control over the size and shape of the texture and can accurately and simultaneously create patterns across the entire surface. In this method, a photomask is utilised to transfer a geometric pattern on the photoresist using ultraviolet light. This process completed in the several steps such as cleaning, followed by coating, exposure, and developing, and then etching and banking. The Novalac resin was applied in three steps using a spin coater, with each step operating at different speeds and durations, resulting in a layer thickness of 2.0–2.5 μm. To create a pattern on the surface, the photoresist is exposed to bright light, and a specific etchant is used for the etching process. In areas without photoresist, the etchant removes the top layer of the substrate. Furthermore, the banking process has been used to remove the photoresist from the substrate and form the required surface texture shape (Venkatesan and Stephens, 2005; Coblas et al., 2015).

5.2.6 Abrasive Jet Machining Technique

In the abrasive jet machining (AJM) technique, firstly, a mask is placed on the sample to be textured. Additionally, a jet nozzle directs the high-velocity air containing tiny abrasive particles. The primary benefit of this approach is that the area around the dimples still has a smooth surface, negating the need for deburring (Coblas et al., 2015; Costa and Hutchings, 2015). AJM is a method employed to create spherical shapes with dimples on a silicon nitride ceramic plate. The dimples have a diameter of 5 μm and a width ranging from 40 to 120 μm. The ceramic plate is paired with cylindrical steel pins (Wakuda et al., 2003).

5.2.7 LIGA Texturing Technique

The LIGA texturing technique can be used to manufacture either positive or negative asperities on the surface. In this process, a substrate is affixed with a polymethyl methacrylate (PMMA) layer at a designated asperity height. Following that, a mask is applied and subjected to X-ray radiation. On the substrate, microstructures are electroplated using the template. Once the PMMA has received the required quantity of radiation, it causes the molecular weight to decrease and the last molecules to disintegrate. The substrate is left with the microstructures that ultimately determine the shape and texture when the residual PMMA is removed. Contrary to the LST method, this method allows for the precise creation of vast textured fields. The resultant texture may exhibit a cross-section characterized by a standard shape, with dimensions depth ranging from 800 to 2 nm and a diameter ranging from 2 to 1000 nm (cylindrical, triangular, square, hexagonal, etc.) (Venkatesan and Stephens, 2005; Coblas et al., 2015; Costa and Hutchings, 2015). With the help of this technique, Stephens et al. (2004) created a hexagonal array of positive asperities on the thrust bearing surface.

5.2.8 Novel Dressing Technique

The novel dressing technique also sometimes called pattern grinding is an efficient method for generating regular surface texturing. A quick, efficient, and low-cost technique called "Grinding with the wheel shaped" enables the creation of a wide variety of regular patterns (Stępień, 2009; Costa and Hutchings, 2015). This method was first demonstrated by Stepien and Szafarczyk (1989) in 1989.

In this technique, the grinding wheel has a helical groove for the creation of the desired texture surface. Making use of textured tools is one solution to these technological issues and a means to improve the level of control over the grinding operation. Similar methods are under investigation for precision machining operations like turning, drilling, and milling, but their application in grinding remains unexplored. The primary challenge lies in identifying a reliable and cost-effective technology capable of imparting textures to the surface of a grinding wheel.

The utilization of textured cutting tools can enhance the performance of mechanical components by generating tailored surface patterns. The grooved surfaces creation for hydrodynamic bearings is one example. The goal of these designed patterns is to maximise load capacity and dynamic bearing stability. Surface grooving patterns created for axial bearings are shown in Figure 5.3. This section presents an innovative dressing method that simplifies the creation of pre-configurable texture patterns on a grinding wheel's surface. This approach is applicable for both the production of patterned parts and the improvement of conventional grinding processes. Any CNC cylindrical grinding machine can implement it at a moderate cost and with minimal changes to the normal or standard cycle times. The textured wheel can also be used to improve grinding efficiency or transmit geometric patterns to ground components to enhance their functional performance (Oliveira et al., 2010).

Upon comparing these methods, LST has garnered significant interest due to its excellent controllability, superior accuracy, and high flexibility compared to other

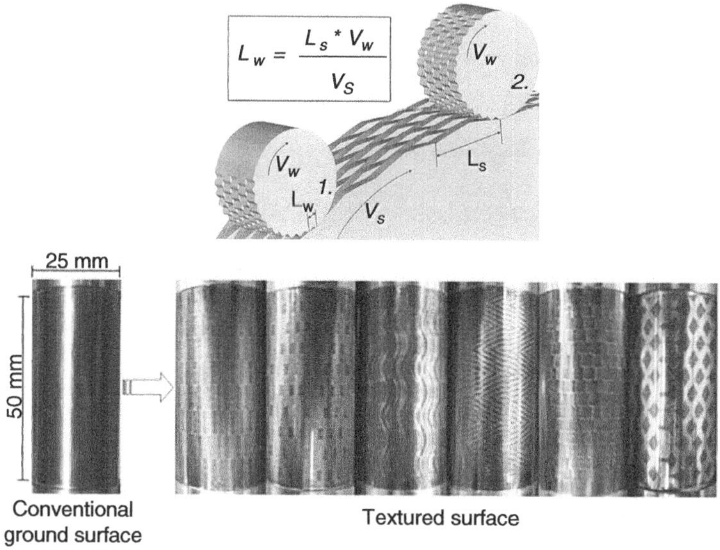

FIGURE 5.3 Surface texturing by a novel dressing technique. (Reproduced with permission from Copyright CIRP Annals, Elsevier, 2010, (Oliveira et al., 2010).)

methods. Parameters such as laser power intensity, pulse number, and type influence tribological performance, resulting in improved wear resistance and superior frictional coefficient through the use of LST.

5.3 INFLUENCE OF TEXTURE ATTRIBUTES ON TRIBOLOGICAL PERFORMANCE

Operating condition of the machine as well as the geometrical parameters of the surface texture influence the tribological performance of the tribo pairs. This section presents the main factors influencing the tribological behaviour.

In Figure 5.4, various geometries and shapes of a surface texture are presently studied by various researchers/investigators. Main geometries are cellular arrangement, texture depth, orientation, bottom shape profile, surface texture to direction of motion/sliding, and texture density. The optimization of surface texture is very important for obtaining an enhanced tribological performance. It is insisted that if the surface texture is not properly optimised, it may negatively impact a bearing's tribological performance, such as a decrease in the value of film thickness performance (Brizmer et al., 2003; Sharma and Yadav, 2014). Past researchers looked into how the surface texture's shape and orientation affected tribological performance. The spherical, circular, triangular, rectangular, and elliptical shapes of texture are the ones that are most frequently examined. The circular dimples' surface texture is the most popular among them since it is simple to manufacture using the available manufacturing techniques. In circular surface texturing, the minimum frictional torque is observed compared to other textured surfaces, whereas for the circular texturing the enhanced

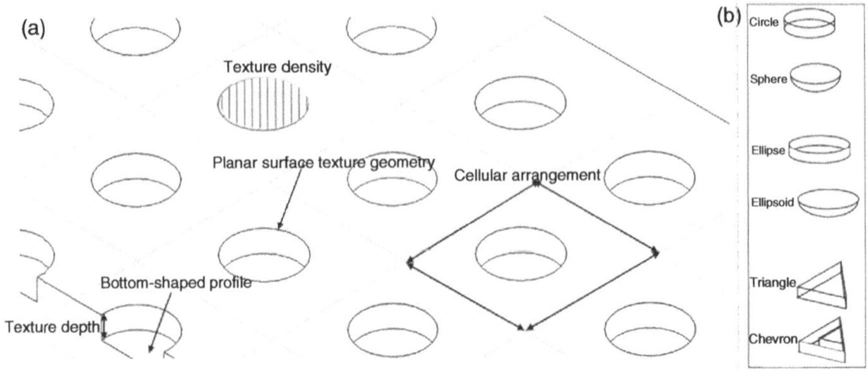

FIGURE 5.4 Various geometries of surface texture. (Reproduced with permission from Copyright Surface and Coatings Technology, Elsevier, Ibatan et al., 2015.)

value of stiffness coefficient and stability is observed as compared to the other textured surfaces. The influence of surface texture orientation with surface texture shapes like triangle, circle, and ellipse in hydrodynamic lubrication behaviour was mathematically studied (Yu et al., 2010). Texture orientation and geometric shape significantly impact the load-bearing capacity of tribological systems. Triangular textures, with their side and major axis perpendicular to the sliding direction, show a 26.3% greater load capacity in comparison with circular dimples. Detailed surface optimization was conducted by Vivek and Sharma (Kumar and Sharma, 2018). The micro-texture is provided in the form of a regular array of micro-dimples along radial and circumferential directions. These micro-dimples have spherical, conical, cylindrical, and rectangular shapes. Each of the shapes considered has an identical topographical area. The micro-dimples over the bearing surface have been characterized by dimple non-dimensional attributes, namely radius, depth, textured length (number of dimples), and area density of textured surface. Each dimple shape is further characterized into three more configurations, i.e., Conf I, Conf II, and Conf III, having the same number of micro-dimples, dimple radius, and dimple depth. Configuration I consists of full-section dimples, whereas Configuration II and Configuration III consist of half-section dimples oriented along the circumferential/rotational direction of the runner. Firstly, we conducted a parametric study to examine how the attributes and configuration of dimples affect the load-carrying capacity. These numerical simulations yield the optimum values of dimple attributes at which maximum load carrying can be obtained from the textured surface bearing, under given operating conditions. The optimum value of dimple depth and texture length, non-0.6 times film thickness and 6 mm, respectively, is observed.

The various shapes and orientations of the herringbone groove influence the performance. Of the several geometric forms of grooves taken into account in the literature (Sahu and Sharma, 2019), it was found that a rectangular groove provides a substantial improvement (–25.94%) in frictional torque. Incorporating a rectangular groove over the bearing surface provides more stiffness and stability threshold speed margin than parabolic and triangular grooves. Because of increases in the bearing's effective recess area, direct stiffness coefficient values have increased. Another important parameter while designing a surface texture is the orientation angle of the groove. Therefore, for achieving an optimal bearing performance, it was suggested that an appropriate selection of groove angles, groove shapes, and groove depth is essential.

The bottom shapes of textured surfaces significantly impact tribological performance. According to the literature, micro-step or micro-wedge bottom shapes generally result in higher film thickness compared to flat bottom surfaces. The study examined the performance of six texture shapes (chevron, circular, spherical, ellipsoidal, triangular, and elliptical) in slider bearings, considering optimized textured density and geometry (Moghbelli et al., 2014). The ellipsoidal shape achieved the maximum load capacity. Despite the better functioning of the ellipsoid, it was seen that spherical texture is used for real-world applications due to its ease of manufacturing process. Texture shapes featuring a curved bottom profile or round edge exhibit superior bearing stiffness and lower frictional coefficients compared to those with straight or flat bottoms. Among conventional texture shapes, ellipsoidal shapes demonstrate the best performance. Furthermore, texture density has a greater influence on load-carrying capacity and friction than the texture aspect ratio. Texture aspect ratio is the depth-to-diameter ratio of the texture feature, while texture density is the ratio of the texture-covered area to the entire unit cell area.

Previous research highlights the essential roles of geometry and texture shape in maximizing load capacity and reducing friction. Although most studies concentrated on regular geometric shapes, there is potential for a globally optimal texture shape that surpasses well-known shapes in improving tribological performance. Researchers have recently devised a novel mathematical optimization method called the Sequential Quadratic Programming-based model, to find the optimal texture shape for both unidirectional and bidirectional sliding. This approach starts with an arbitrary texture shape, comparing its performance to conventional shapes. Unlike existing literature, which lacks a standard approach, this method provides a systematic way to optimize surface texture, eliminating the need for trial and error. There is still a gap in the understanding of how surface texturing works, as evidenced by the numerous researchers who found various ideal geometrical properties in numerical experiments.

5.4 MATHEMATICAL EXPRESSION FOR SOME IMPORTANT TEXTURED SHAPES

Here are more detailed descriptions of the mathematical expressions for some important textured shapes.

5.4.1 Spherical-Textured Shape

The non-dimensional expression for the nominal fluid film thickness of a spherical-shaped textured surface, as depicted in Figure 5.5a, can be formulated as follows:

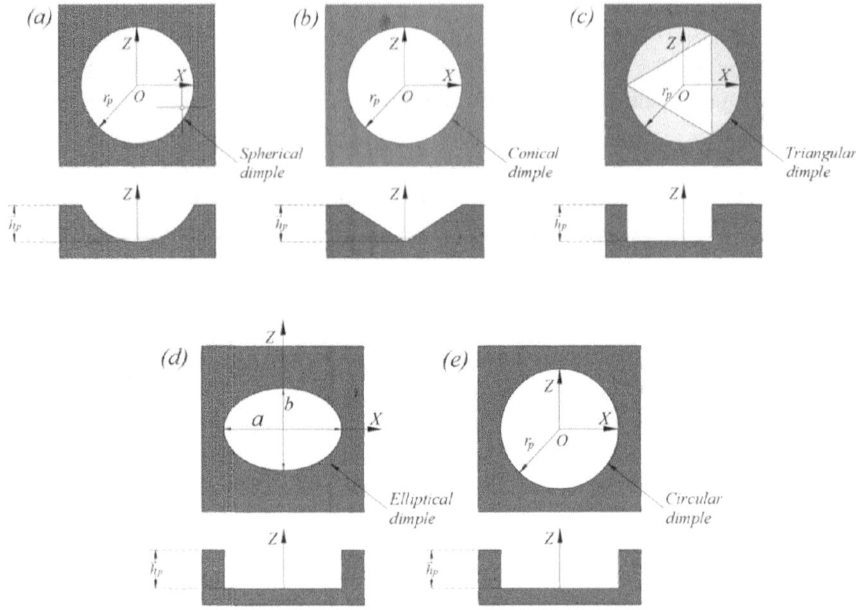

FIGURE 5.5 Geometry of (a) spherical dimple shape, (b) conical dimple shape, (c) elliptical dimple shape, (d) elliptical dimple shape, and (e) circular dimple shape.

$$\bar{h}_{\text{texture}} = \begin{cases} \sqrt{\left(\dfrac{\bar{h}_p}{2} + \dfrac{\bar{\delta}^2}{2\bar{h}_p}\right)^2 - \bar{\delta}^2\left(\bar{x}_l^2 + \bar{z}_l^2\right)} + \left(\dfrac{\bar{h}_p}{2} - \dfrac{\bar{\delta}^2}{2\bar{h}_p}\right) & \text{at } \sqrt{\bar{x}_l^2 + \bar{z}_l^2} < 1 \\ 0 & \text{at } \sqrt{\bar{x}_l^2 + \bar{z}_l^2} \geq 1 \end{cases} \tag{5.1}$$

where $\bar{\delta} = \frac{r_p}{c}$, $\bar{x} = \frac{x}{r_p}$, $\bar{z} = \frac{z}{r_p}$; $\bar{x}_l$ and $\bar{z}_l$ are the Local Cartesian coordinates of dimple; r_p is the base radius of dimple; $\bar{\delta}$ is the dimensionless dimple radius; $\bar{h}_p = \frac{h_p}{c}$ is the dimensionless depth of texture, h_p is the depth of dimple, and c is the radial clearance.

5.4.2 Conical Dimple Shape

The expression defining the nominal fluid film thickness for a conical-textured surface, as illustrated in Figure 5.5b, is given by the following equation:

$$\bar{h}_{\text{texture}} = \begin{cases} \left[1 - \sqrt{\left(\bar{x}_l^2 + \bar{z}_l^2\right)}\right]\bar{h}_p & \text{at } \sqrt{\bar{x}_l^2 + \bar{z}_l^2} < 1 \\ 0 & \text{at } \sqrt{\bar{x}_l^2 + \bar{z}_l^2} \geq 1 \end{cases} \tag{5.2}$$

5.4.3 Triangular Dimple Shape

The expression for the nominal fluid film thickness of a triangular-textured surface, as depicted in Figure 5.5c, is given by the following equation:

$$\bar{h}_{\text{texture}} = \begin{cases} \bar{h}_p \text{ at } \left(-\frac{3}{4} \leq \bar{x}_l \leq \frac{3}{4}\right) \text{and} \left(-\frac{1}{\sqrt{3}}\bar{x}_l - \frac{\sqrt{3}}{4}\right) \leq \bar{z}_l \leq \left(-\frac{1}{\sqrt{3}}\bar{x}_l - \frac{\sqrt{3}}{4}\right) \\ 0 \text{ (else)} \end{cases} \tag{5.3}$$

5.4.4 Elliptical Dimple Shape

The expression for the nominal fluid film thickness of an elliptical-textured surface, as depicted in Figure 5.5d, is given by the following equation:

$$\bar{h}_{\text{texture}} = \begin{cases} \bar{h}_p & \text{at } \left(\frac{a}{b}\bar{x}_l^2 + \frac{b}{a}\bar{z}_l^2\right) \leq \sqrt{\bar{x}_l^2 + \bar{z}_l^2} \\ 0 & \text{at } \left(\frac{a}{b}\bar{x}_l^2 + \frac{b}{a}\bar{z}_l^2\right) > \sqrt{\bar{x}_l^2 + \bar{z}_l^2} \end{cases} \tag{5.4}$$

The values of a and b represent the semi-major and semi-minor axes of the ellipse, respectively.

5.4.5 Circular Dimple Shape

The expression for the nominal fluid film thickness of a circular-textured surface, as depicted in Figure 5.5e, is given by the following equation:

$$\bar{h}_{\text{texture}} = \begin{cases} \bar{d}_t & \text{at } \sqrt{\bar{x}_l^2 + \bar{z}_l^2} < 1 \\ 0 & \text{at } \sqrt{\bar{x}_l^2 + \bar{z}_l^2} \geq 1 \end{cases} \tag{5.5}$$

5.5 CASE STUDIES ON TEXTURED SURFACES

Here, some case studies (Sahu and Sharma, 2019; Sharma and Sahu, 2021) have been used to demonstrate the influence of various texture shapes, texture attributes, and various texture orientations.

5.5.1 Influence of Orientation Angle of Texture on Maximum Fluid Film Pressure

Figure 5.6a–f illustrates the impact of herringbone groove (continuous texture) angle on pressure distribution for an external load, $\overline{W}_o = 1.5$. The figures demonstrate how textured surfaces influence fluid film pressure distribution, with a higher groove angle leading to an increased maximum fluid film pressure $\left(\overline{p}_{\max}\right)$. Notably, a specific groove angle (40°) yields the maximum $\overline{p}_{\max}$ after a decline occurs. Potentially due to the lubricant layers undergoing sudden expansion beyond this angle, resulting in a drop in fluid film pressure. Further, the figure presents the magnetic field's impact on groove angles in fluid film pressure generation under identical operating/geometric conditions. For instance, at a 40° groove angle, a 52.77% improvement in $\overline{p}_{\max}$ is observed compared to a 10° groove angle.

5.5.2 Influence of Texture Shapes on Discrete Textured Surfaces (Micro-Dimple Surfaces)

Figure 5.7 depicts the variation of frictional torque $\left(\overline{T}_f\right)$ with the applied external load for different surface textures, such as triangular, rectangular, spherical, and circular dimples. As the value of the applied load $\left(\overline{W}_o\right)$ increases, the magnitude of the frictional torque increases continuously. The incorporation of surface texturing in the journal-bearing system significantly reduces the frictional torque. The maximum frictional torque occurs in the absence of texture, while the circular surface texture yields the minimum frictional torque compared to other textured surfaces.

This could be attributed to the higher presence of lubricant in the contact area due to the textured surface profile. Bearing designers can leverage these trends for $\overline{T}_f$ based on the numerically computed results.

$$\left[\overline{T}_f\Big|_{\text{smooth}} > \overline{T}_f\Big|_{\text{triangular}} > \overline{T}_f\Big|_{\text{rectangular}} > \overline{T}_f\Big|_{\text{spherical}} > \overline{T}_f\Big|_{\text{circular}}\right] \tag{5.6}$$

Figures 5.8 and 5.9 show the graphs of stiffness coefficients ($\overline{S}_{11}$ and $\overline{S}_{22}$) in relation to the applied load $\left(\overline{W}_o\right)$ for various textured shapes. It has been noticed that using textured film enhances the stiffness capabilities of the bearing system. The circular shape textured journal-bearing system exhibits higher values of $\overline{S}_{11}$ and $\overline{S}_{22}$ compared to other shapes. In contrast, non-textured journal-bearing systems show minimum values of $\overline{S}_{11}$ and $\overline{S}_{22}$ compared to textured bearings. This observation is attributed to the sudden expansion experienced by lubricant layers when entering a textured surface, leading to an alteration in stiffness coefficients. The results suggest that for

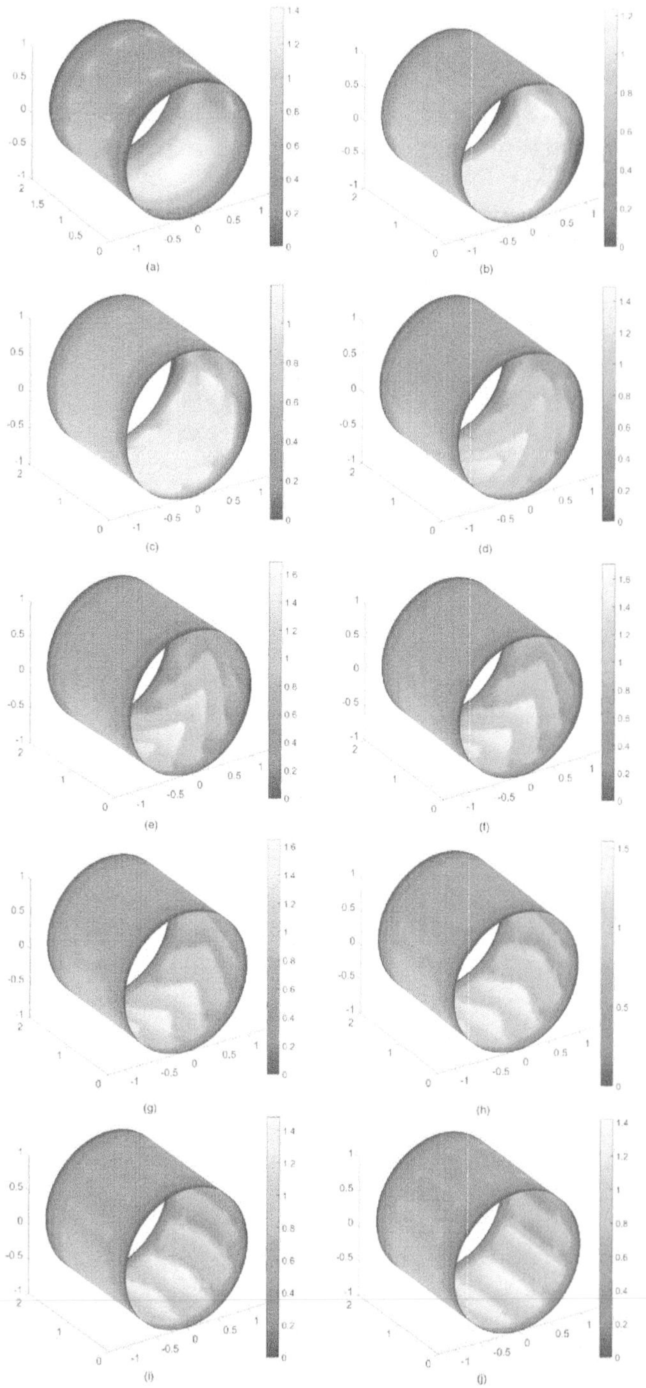

FIGURE 5.6 Influence of textured orientations on the surface of fluid film journal bearings (a) $\varphi=0°$, (b) $\varphi=10°$, (c) $\varphi=20°$, (d) $\varphi=30°$, (e) $\varphi=40°$, (f) $\varphi=50°$, (g) $\varphi=60°$, (h) $\varphi=70°$, (i) $\varphi=80°$, and (j) $\varphi=90°$. (Sharma and Sahu, 2021.)

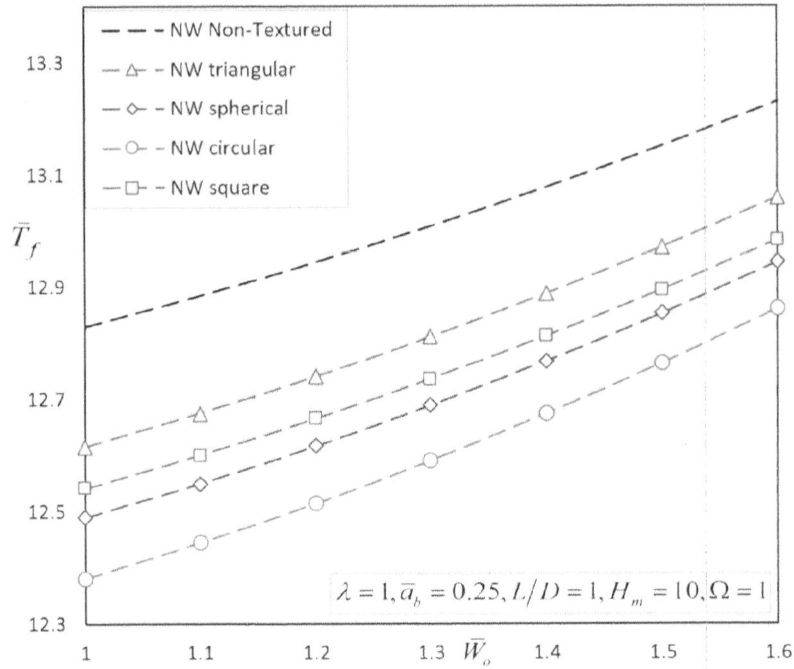

FIGURE 5.7 Influence of texture shapes on frictional torque $\left(\bar{T}_f\right)$.

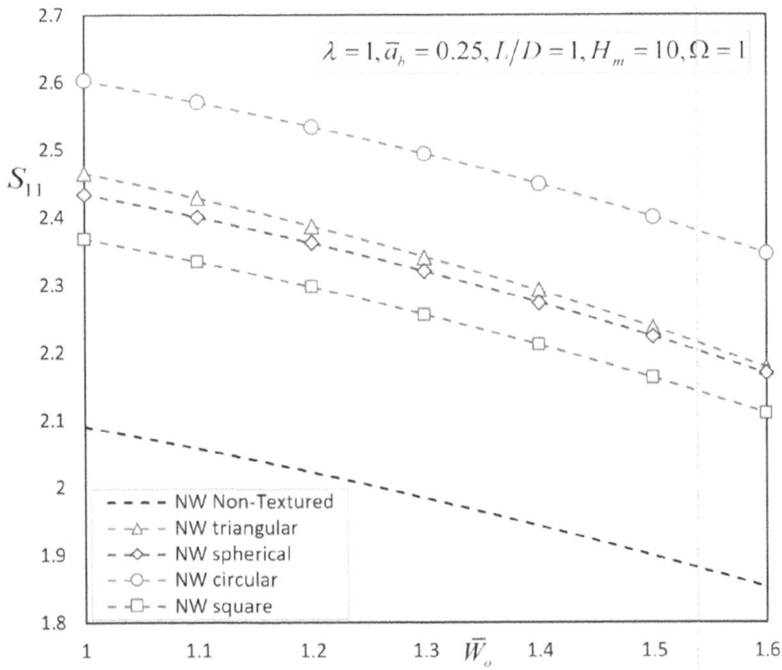

FIGURE 5.8 Influence of texture shapes on stiffness coefficient $\left(\bar{S}_{11}\right)$.

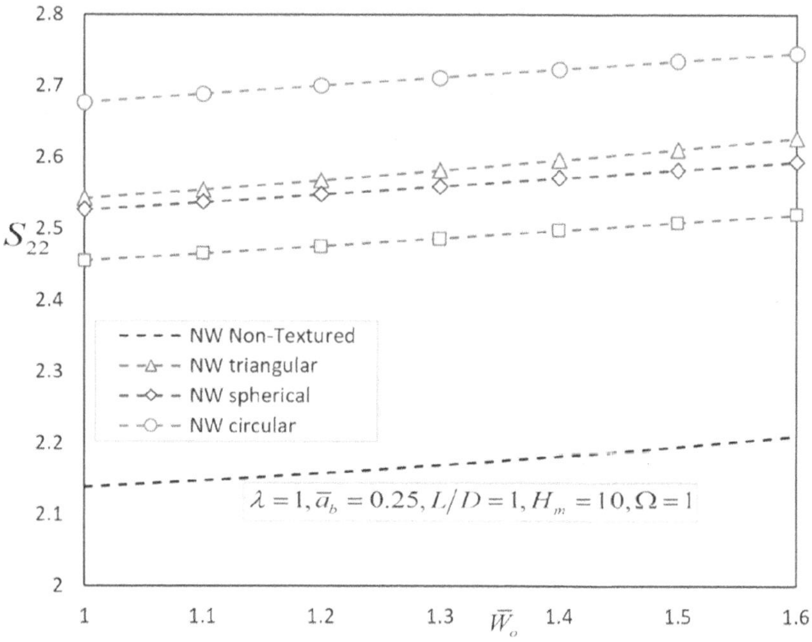

FIGURE 5.9 Influence of texture shapes on stiffness coefficient $\left(\bar{S}_{22}\right)$.

certain textured shapes, such as the circular one, lubricant films undergo a sudden expansion in grooves, resulting in a drop in fluid film pressure and consequently enhancing the stiffness capabilities. Bearing designers can leverage these trends to inform decisions regarding $\bar{S}_{11}$ and $\bar{S}_{22}$.

$$\left[\bar{S}_{11}\Big|_{\text{circular}} > \bar{S}_{11}\Big|_{\text{triangular}} > \bar{S}_{11}\Big|_{\text{spherical}} > \bar{S}_{11}\Big|_{\text{rectangular}} > \bar{S}_{11}\Big|_{\text{smooth}}\right] \tag{5.7}$$

$$\left[\bar{S}_{22}\Big|_{\text{circular}} > \bar{S}_{22}\Big|_{\text{triangular}} > \bar{S}_{22}\Big|_{\text{spherical}} > \bar{S}_{22}\Big|_{\text{rectangular}} > \bar{S}_{22}\Big|_{\text{smooth}}\right] \tag{5.8}$$

5.6 CONCLUDING REMARKS

The key textured geometries that influence the tribological properties are the depth of the textured surface, length, width, density of texture, orientation of surface texture with respect to the sliding direction, and cellular pattern of texture. Bottom profile/shape of surface texture plays an important role in the tribological performance, and studies indicate that bottom shapes with a micro-step or micro-wedge tend to produce greater film thickness in comparison to the flat bottom surface texture. Similarly, round edge texture shapes with curved bottom profiles have higher bearing stiffness and lower friction coefficient than flat or straight bottoms. From the viewpoint of textured shapes, the ellipsoidal shapes perform the best among all conventional texture shapes. From the viewpoint of full and partial texturing, partial surface texturing provides better performance as compared to full surface texturing. It is undisputable that certain surfaces with textured shapes may have a positive influence on tribological performance, but it is generally agreed that, based on specific necessities of applications, texture optimization should be performed. It is noted that if surface texturing is not optimised, it may have detrimental impacts on the tribological performance. Therefore, for achieving an optimum tribological performance, a proper selection of texture attributes is essential.

NOMENCLATURE

X, Y, Z:	Cartesian coordinates
C_{ij}:	Damping coefficients, Ns/mm $(i, j = 1,2)$
p, p_s, p_c:	Fluid film pressure, supply pressure, pressure at slots (N/mm^2)
h:	Fluid film thickness (nominal) (mm)
$\bar{T}_f$:	Friction torque (N mm)
X_J, Z_J:	Journal centre coordinates (mm)
R_J/R_b:	Journal/Bearing radius (mm)
L:	Length of the journal bearing (mm)
x_l, z_l:	Local coordinates of textures
$h_{\min}$:	Minimum fluid film thickness (mm)
c:	Radial clearance (mm)
r_p:	Radius of texture shape
S_{ij}:	Stiffness coefficient (N/mm) $(i, j = 1,2)$

DIMENSIONLESS PARAMETERS

$$\bar{C}_{ij} : C_{ij}\left(c^3/\mu R_J^4\right)$$

$$\bar{S}_{ij} : S_{ij}/p_s R_j^2$$

$$\bar{W}_o : W_o/p_s R_J^2$$

$$\bar{X}_J,\ \bar{Z}_J : (X_J, Z_J)/c$$

$$\bar{h}_{\min},\ \bar{h}, \bar{h}_0, \bar{h}_{\text{texture}} : (h_{\min},\ h, h_0, h_{\text{texture}})/c$$

$$\bar{p}, \bar{p}_{\max}, \bar{p}_c : (p, p_{\max}, p_c)/p_s$$

$$\bar{T}_f : T_f\left(1/p_s c R_J^2\right)$$

$$\alpha,\ \beta : (X, Y)/R_J$$

REFERENCES

Anno, J., Walowit, J. & Allen, C., 1969. Load support and leakage from microasperity-lubricated face seals. *Journal of Lubrication Technology,* 91 (4), 726–731.

Bhushan, B., 2009. *Biomimetics: Lessons from Nature–An Overview*. The Royal Society.

Brizmer, V., Kligerman, Y. & Etsion, I., 2003. A laser surface textured parallel thrust bearing. *Tribology Transactions,* 46 (3), 397–403.

Coblas, D.G., Fatu, A., Maoui, A. & Hajjam, M., 2015. Manufacturing textured surfaces: State of art and recent developments. *Proceedings of the Institution of Mechanical Engineers, Part J: Journal of Engineering Tribology,* 229 (1), 3–29.

Costa, H. & Hutchings, I.M., 2015. Some innovative surface texturing techniques for tribological purposes. *Proceedings of the Institution of Mechanical Engineers, Part J: Journal of Engineering Tribology,* 229 (4), 429–448.

Dean, B. & Bhushan, B., 2010. Shark-skin surfaces for fluid-drag reduction in turbulent flow: A review. *Philosophical Transactions of the Royal Society of London A: Mathematical, Physical and Engineering Sciences,* 368 (1929), 4775–4806.

Dobrica, M., Fillon, M., Pascovici, M. & Cicone, T., 2010. Optimizing surface texture for hydrodynamic lubricated contacts using a mass-conserving numerical approach. *Proceedings of the Institution of Mechanical Engineers, Part J: Journal of Engineering Tribology,* 224 (8), 737–750.

Etsion, I., 2004a. Improving tribological performance of mechanical components by laser surface texturing. *Tribology Letters,* 17 (4), 733–737.

Etsion, I., 2004b. State of the art in laser surface texturing. *ASME 7th Biennial Conference on Engineering Systems Design and Analysis* American Society of Mechanical Engineers, 585–593. Manchester, United Kingdom.

Etsion, I., 2005. State of the art in laser surface texturing. *Journal of tribology,* 127 (1), 248–253.

Etsion, I., 2013. Modeling of surface texturing in hydrodynamic lubrication. *Friction,* 1 (3), 195–209.

Etsion, I., Kligerman, Y. & Halperin, G., 1999. Analytical and experimental investigation of laser-textured mechanical seal faces. *Tribology Transactions,* 42 (3), 511–516.

Fesanghary, M. & Khonsari, M., 2011. A modification of the switch function in the Elrod cavitation algorithm. *Journal of Tribology,* 133 (2), 024501.

Giacopini, M., Fowell, M.T., Dini, D. & Strozzi, A., 2010. A mass-conserving complementarity formulation to study lubricant films in the presence of cavitation. *Journal of Tribology,* 132 (4), 041702.

Greco, A., Raphaelson, S., Ehmann, K., Wang, Q.J. & Lin, C., 2009. Surface texturing of tribological interfaces using the vibromechanical texturing method. *Journal of Manufacturing Science and Engineering,* 131 (6), 061005.

Gropper, D., Wang, L. & Harvey, T.J., 2016. Hydrodynamic lubrication of textured surfaces: A review of modeling techniques and key findings. *Tribology International,* 94, 509–529.

Hajjam, M. & Bonneau, D., 2007. A transient finite element cavitation algorithm with application to radial lip seals. *Tribology International,* 40 (8), 1258–1269.

Hamilton, D., Walowit, J. & Allen, C., 1966. A theory of lubrication by microirregularities. *Journal of Basic Engineering,* 88 (1), 177–185.

Ibatan, T., Uddin, M. & Chowdhury, M., 2015. Recent development on surface texturing in enhancing tribological performance of bearing sliders. *Surface and Coatings Technology,* 272, 102–120.

Jearl, W., 1979. More on boomerangs, including their connection with the dimpled golf ball. *Scientific American,* 180. 31.

Kango, S. & Sharma, R., 2010. Studies on the influence of positive roughness on the performance of journal bearing. *Journal of Tribology Surface Engineering,* 1, 243–62.

Kumar, V. & Sharma, S.C., 2018. Influence of dimple geometry and micro-roughness orientation on performance of textured hybrid thrust pad bearing. *Meccanica,* 53, 3579–3606.

Mao, B., Siddaiah, A., Menezes, P.L. & Liao, Y., 2018. Surface texturing by indirect laser shock surface patterning for manipulated friction coefficient. *Journal of Materials Processing Technology,* 257, 227–233.

Moghbelli, E., Banyay, R. & Sue, H.-J., 2014. Effect of moisture exposure on scratch resistance of PMMA. *Tribology International,* 69, 46–51.

Oliveira, J.F.G.D., Bottene, A. & Franca, T., 2010. A novel dressing technique for texturing of ground surfaces. *CIRP Annals,* 59 (1), 361–364.

Pei, S., Ma, S., Xu, H., Wang, F. & Zhang, Y., 2011. A multiscale method of modeling surface texture in hydrodynamic regime. *Tribology International,* 44 (12), 1810–1818.

Sahu, K. & Sharma, S.C., 2019. A simulation study on the behavior of magnetorheological fluid on herringbone-grooved hybrid slot-entry bearing. *Tribology Transactions,* 62 (6), 1099–1118.

Sharma, R. & Pandey, R., 2008. Thermohydrodynamic analysis of infinitely wide cycloidal profiled pad thrust bearing with rough surface and a comparison to plane profiled pad. *Lubrication Science,* 20 (3), 183–203.

Sharma, S.C. & Sahu, K., 2021. On the behavior of a ferrofluid-lubricated herringbone-grooved hybrid slot-entry bearing. *Proceedings of the Institution of Mechanical Engineers, Part J: Journal of Engineering Tribology,* 235 (11), 2295–2315.

Sharma, S.C. & Yadav, S.K., 2014. Performance analysis of a fully textured hybrid circular thrust pad bearing system operating with non-Newtonian lubricant. *Tribology International,* 77, 50–64.

Singh, S., Kango, S., Sharma, N. & Verma, R., 2021. Recent advances in the mechanical durability of superamphiphobic surfaces: A review. *Proceedings of the Institution of Mechanical Engineers, Part J: Journal of Engineering Tribology,* 235 (11), 2474–2499.

Stephens, L.S., Siripuram, R., Hayden, M. & McCartt, B. 2004. Deterministic micro asperities on bearings and seals using a modified LIGA process. *Journal of Engineering for Gas Turbines and Power,* 126 (1), 147–154.

Stępień, P., 2009. Regular surface texture generated by special grinding process. *Journal of Manufacturing Science and Engineering,* 131 (1), 011015.

Stepien, P. & Szafarczyk, M., 1989. Generation of regular patterns on ground surfaces. *CIRP Annals-Manufacturing Technology,* 38 (1), 561–566.

Sudeep, U., Tandon, N. & Pandey, R., 2015. Tribological studies of lubricated laser-textured point contacts in rolling/sliding reciprocating motion with investigations of wettability and nanohardness. *Tribology Transactions,* 58 (4), 625–634.

Tala-Ighil, N., Fillon, M. & Maspeyrot, P., 2011. Effect of textured area on the performances of a hydrodynamic journal bearing. *Tribology International,* 44 (3), 211–219.

Tala-Ighil, N., Maspeyrot, P., Fillon, M. & Bounif, A., 2007. Effects of surface texture on journal-bearing characteristics under steady-state operating conditions. *Proceedings of the Institution of Mechanical Engineers, Part J: Journal of Engineering Tribology,* 221 (6), 623–633.

Venkatesan, S. & Stephens, L.S., 2005. Surface textures for enhanced lubrication: Fabrication and characterization techniques. In *World Tribology Congress III.* American Society of Mechanical Engineers, 521–522.

Wakuda, M., Yamauchi, Y., Kanzaki, S. & Yasuda, Y., 2003. Effect of surface texturing on friction reduction between ceramic and steel materials under lubricated sliding contact. *Wear,* 254 (3), 356–363.

Woloszynski, T., Podsiadlo, P. & Stachowiak, G.W., 2013. Evaluation of discretisation and integration methods for the analysis of hydrodynamic bearings with and without surface texturing. *Tribology Letters,* 51 (1), 25–47.

Woloszynski, T., Podsiadlo, P. & Stachowiak, G.W., 2015a. Efficient solution to the cavitation problem in hydrodynamic lubrication. *Tribology Letters,* 58 (1), 18.

Woloszynski, T., Podsiadlo, P. & Stachowiak, G.W., 2015b. Evaluation of discretization and integration methods for the analysis of finite hydrodynamic bearings with surface texturing. *Proceedings of the Institution of Mechanical Engineers, Part J: Journal of Engineering Tribology,* 229 (4), 465–477.

Xiong, S. & Wang, Q.J., 2012. Steady-state hydrodynamic lubrication modeled with the Payvar-Salant mass conservation model. *Journal of Tribology,* 134 (3), 031703.

Yadav, S.K. & Sharma, S.C., 2014. Performance of hydrostatic tilted thrust pad bearings of various recess shapes operating with non-Newtonian lubricant. *Finite Elements in Analysis and Design,* 87, 43–55.

Yadav, S.K. & Sharma, S.C., 2016. Performance of hydrostatic textured thrust bearing with supply holes operating with non-Newtonian lubricant. *Tribology Transactions,* 59 (3), 408–420.

Yang, H., Ratchev, S., Turitto, M. & Segal, J., 2009. Rapid manufacturing of non-assembly complex micro-devices by microstereolithography. *Tsinghua Science & Technology,* 14, 164–167.

Yu, H., Wang, X. & Zhou, F., 2010. Geometric shape effects of surface texture on the generation of hydrodynamic pressure between conformal contacting surfaces. *Tribology Letters,* 37, 123–130.

6 Measurement and Control of Wettability and Surface Energy of Materials

Jibin T. Philip, Alphene G. Alex, and M. Gururaj Acharya

6.1 INTRODUCTION

The importance of understanding wettability on solid surfaces is necessary for fields like engineering, science, and technology, which are part of day-to-day life (Della Volpe et al., 2004; Marmur and Valal, 2010). The study of wetting spans over two centuries, originating with Thomas Young and Pierre Simon de Laplace's insights in 1805. They introduced the concept of interfacial energy, directly related to the number of molecules at an interface and thus the surface area (Paty, 1993, 1749–1827; Young, 1805; Rowlinson and Widom, 2013). Consequently, in air, drops assume a spherical shape, and this form persists when a liquid/vapor interface encounters a solid surface, provided the drops are sufficiently small to ignore gravitational flattening effects.

Young extended this concept to nondeformable surfaces like solids, enabling the calculation of the solid/liquid contact size or the contact angle (CA) between the liquid and the solid (Young, 1805). This understanding offers comprehensive descriptions of systems and facilitates control over drop behavior on substrates by manipulating surface tensions. Wettability, or the capacity of a droplet to adhere to a rigid surface, is measured by the equilibrium CA. There are four states of wettability commonly found on the surface of a material: hydrophilic, hydrophobic, superhydrophilic, and superhydrophobic. These states are used to quantify a solid's interaction with liquids, independently of the specific liquid used. The CA method is utilized to determine the wettability state, where a drop of liquid is placed on the solid surface in the presence of gas (Figure 6.1).

The CA measurement is widely used to understand the surface properties in many areas ranging from ink-jet to spray cooling technologies (Spori et al., 2008; Mishchenko et al., 2010). The understanding of the surface chemistry, morphology, and forces acting on the treated surfaces can be provided by CA ($\theta°$) (Zhao and Jiang, 2018). The parameter employed to evaluate the wettability of a solid surface is referred to as the CA. Hydrophilic solid surfaces have a CA, $\theta < 90°$, whereas

DOI: 10.1201/9781003359333-6

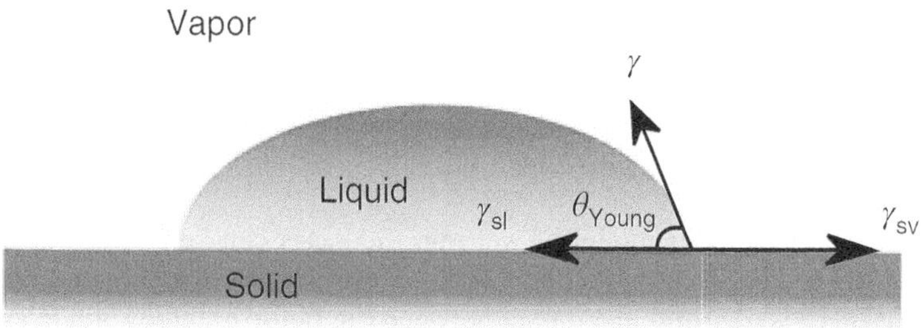

FIGURE 6.1 Contact angle analysis. (Reproduced with permission from Ref. (Huhtamäki et al., 2018). Copyright Springer Nature, 2018.)

hydrophobic solid surfaces have a CA, $\theta > 90°$. Young's equation can be used to explain the connection between the CA and interface stresses based on Equation 6.1 (Adamson and Gast, 1967).

$$\gamma_{sg} - \gamma_{sl} = \gamma_{lg}\cos\theta \tag{6.1}$$

where θ denotes the CA and γ_{sg}, γ_{sl}, γ_{lg} reflect the interface tensions of solid/gas, solid/liquid, and liquid/gas, respectively. For a smooth solid surface study, the shift between hydrophilicity and hydrophobicity was studied by new research on the structural states of water droplets that had taken place at CA = 65° (Guo et al., 2008; Berg et al., 1994; Vogler, 1998; Yoon et al., 1997; Patel et al., 2010).

Young's contact angle (YCA) is applicable only for ideal conditions, such as perfectly smooth and chemically homogeneous solid surfaces. On real solids, the angle may vary largely. This is for two reasons: (1) the CA may be equal to the YCA in rough surfaces when the surface tension is negligible and (2) due to contact angle hysteresis (CAH), a wide range of metastable CAs exists. Consequently, the stable CA and the resulting YCA estimate (determined by the Wenzel and Cassie equation) are not suitable.

The CA and the surface tension of liquids (which is a key player because of its influence of physical properties on the wettability of a solid surface) can be used to determine the surface free energy (SFE) of the material, which helps determine the hydrophilicity of the solid (Lu et al., 2022). Hydrophilicity and hydrophobicity are important properties of any solid surface. This article will discuss different theories and methods for the determination, evaluation, and manipulation of SFE, wettability, and interfacial surface tensions of solid surfaces by different CA-based methods.

6.2 THEORETICAL BACKGROUND

The two major theoretical models explaining the influence of surface roughness on wettability are discussed, i.e., the Wenzel and Cassie–Baxter models.

In the Wenzel model of wetting, for a homogenous but rough surface, the equation is as follows:

$$\cos\theta_W = \frac{r(\gamma_{sg} - \gamma_{sl})}{\gamma_{lg}} = r\cos\theta_Y \tag{6.2}$$

where r is the overall ratio of the true solid surface area to the nominal area. In this approach, the surface having inherent wettability may be enhanced by roughness. That is, a hydrophilic or hydrophobic surface becomes superhydrophilic or hydrophobic by increasing its surface roughness. The Wenzel model cannot be used for the research if the surface is highly rough or porous, and if the roughness factor value "r" is more than 1.

In the Cassie–Baxter model of wetting, the gases can be trapped in between the grooves or channels, which can act as a beanbag for the droplets and make sure that there is a minimum contact between the liquid and the solid (Cassie and Baxter, 1944). By including the inputs of the fractional area of the wet surface and the air pockets of this model, the CA (θ_{CB}) can be calculated using a variation of Wenzel's equation.

$$\cos\theta_{CB} = -1 + f_{sl}(1 + \cos\theta) \tag{6.3}$$

where f_{sl} is the fractional area of the solid–liquid interfaces under the droplet. This particular model calculates the increase in the hydrophobicity of the surface from the initial CA. Due to the absence of a gaseous trap and complete penetration of the liquid on the solid in the Wenzel state, the adhesion of droplets greatly differs compared to the Cassie–Baxter state. Figure 6.2 shows the graphical image of the Wenzel state and the Cassie–Baxter state. Thereafter, the obtained superhydrophobic surface must find a correlation between Wenzel and Cassie–Baxter methods.

Lafuma and Quere proposed a new method in which they introduced critical CA (θ_c), where the theory shows the formation of air trapped at the liquid–gas interface.

$$\cos\,\theta_c = \frac{f_{sl} - 1}{r - f_{sl}} \tag{6.4}$$

If $\theta > \theta_c$, grooves consist of entrapped air, and the liquid drop stays upon it. If the liquid adsorbs on the grooves due to capillary action, then it gives a good superhydrophilic state (Majhy et al., 2020). The Wenzel–Cassie state is a transition shift between the Wenzel and Cassie–Baxter states, where liquid partly fills the pores and air partially fills the surface structures. The transition can be completely changed to the Wenzel state via droplet press or vibrations on the surfaces (Sun et al., 2021).

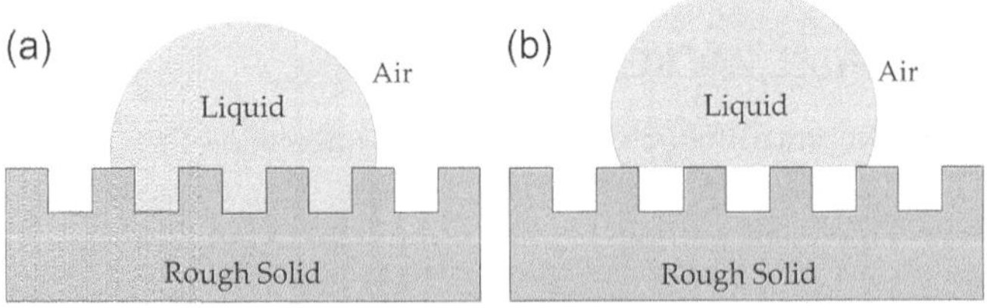

FIGURE 6.2 (a) A drop of liquid assumed in the Wenzel state and (b) a drop of liquid assumed in the Cassie–Baxter state. (Reproduced with permission from Ref. (Ramachandran and Nosonovsky, 2016). Copyright Springer Nature, 2020.)

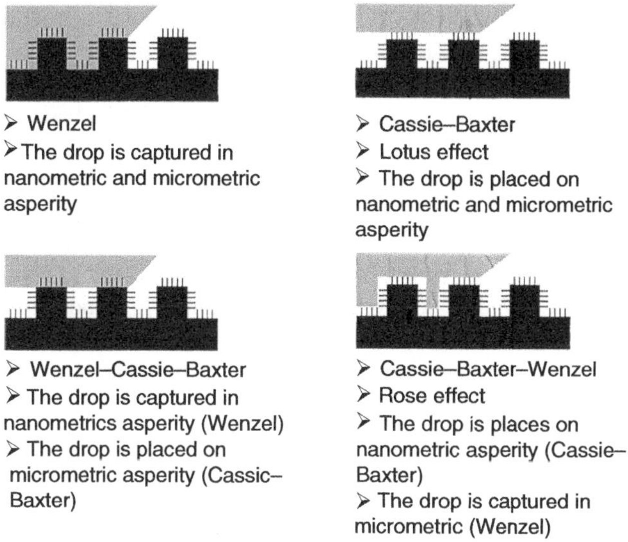

FIGURE 6.3 Schematics of the different wetting scenarios for the roughness surface models and the multiscale roughness surfaces. (Reproduced with permission from Ref. (Nosonovsky and Bhushan, 2012). Copyright Springer Nature, 2012.)

Figure 6.3 illustrates various scenarios observed on real heterogeneous and complex rough surfaces. The behavior of drops on such surfaces can be categorized into distinct models. In the first scenario, drops conform to either the Wenzel or Cassie–Baxter models when encountering roughness. Alternatively, drops exhibit a combination of both models: following the Wenzel model for nanometric scale roughness and the Cassie–Baxter model for micrometric scale roughness. In the second scenario, drops adhere to the Wenzel and Cassie–Baxter models for micrometric and nanometric scales, respectively. These behaviors are evident in natural superhydrophobic surfaces, such as lotus or rose leaves (Nosonovsky and Bhushan, 2012).

6.3 WETTING

The investigation of wetting phenomena stands at the intersection of chemistry, physics, and engineering disciplines. Surface chemistry emerges as a pivotal determinant of wetting behaviors, prompting extensive exploration into altering the surface chemistry of various solids to achieve desired wetting characteristics (Durian and Franck, 1987). Techniques such as plasma treatment (Wu, 1982) and silanization (Brzoska et al., 1992) are commonly employed to modify surface chemical properties, thereby influencing the interaction energy between the surface and liquids, vapors, or other solids. These chemical interactions, which operate at molecular scales, are often referred to as short-ranged interactions. Alongside surface chemistry, surface forces such as van der Waals or electrostatic forces play critical roles in determining fluid–surface interactions. Physicists and physical chemists extensively investigate these forces, recognizing the enduring significance of van der Waals forces over considerable molecular distances, thus termed "long-ranged." These forces dictate the equilibrium thickness of wetting films, impacting fluid spread over solid surfaces and the stability of soap films (Vrij, 1976).

Surface and interfacial interactions, typically occurring at nanometer scales for van der Waals or electrostatic interactions, are explored using advanced experimental techniques like atomic force microscopy and theoretical tools such as molecular dynamics. These methodologies provide new insights into longstanding challenges related to surface forces. Additionally, innovative concepts like patterned surfaces, surfactants, and non-Newtonian flow emerge to address technological challenges. Wetting phenomena, prevalent in both natural and technological contexts, involve the widespread coverage of solid substrates by fluid material. Understanding the surface forces governing wetting and the equilibrium surface coverage of substrates interfacing with liquid drops is crucial. Various wetting phase transition scenarios emerge based on the nature of surface forces, with recent progress enabling a direct correlation between critical exponents and underlying surface forces.

Thermal fluctuations, particularly notable in geometrically or chemically structured substrates and colloidal suspensions, alter adsorption singularities and influence wetting behavior relevant to microfluidics and nanofluidics applications. Wetting dynamics encompass the spreading of drops on wettable substrates and dewetting upon alterations in system parameters. Recent advancements illuminate moving contact line dynamics, intertwined with wetting thermodynamics. Moreover, research investigates wetting characteristics on rough surfaces, examining contact line anchoring and CAH resulting from surface irregularities. Innovations such as patterned surfaces, surfactants, and complex fluids provide avenues to customize wetting properties according to technological requirements. Two distinct types of continuous or critical wetting transitions have been identified, characterized by discontinuities in higher derivatives of SFE. Long-range critical wetting, driven by van der Waals forces, precedes a first-order wetting transition, resulting in a new intermediate wetting state termed frustrated complete wetting. Short-range critical wetting involves the continuous divergence of layer thickness from microscopic to macroscopically thick films.

6.3.1 Wetting without Roughness

6.3.1.1 Ideal Wetting

Surface engineering is a pivotal aspect of material science, particularly concerning the control of wettability in solid materials. The objective often revolves around achieving two extreme limits. The first limit entails complete wetting, where a liquid in contact with a solid effortlessly forms a film. In practical scenarios, such as a car windshield, this film sustains the glass's transparency and has the ability to flow under the influence of gravity when the car is stationary or due to air friction when in motion, effectively carrying away dust particles. Conversely, the second limit involves complete drying, wherein liquid drops maintain a spherical shape without establishing any contact with the substrate. This spherical shape facilitates easy evacuation, preventing liquid contamination of the solid surface. Determining the parameters conducive to both these situations is of considerable interest. The foundational principles were initially established for ideal solids characterized by flatness and chemical homogeneity.

As elucidated by Young and Laplace, surfaces possess a specific energy known as surface tension, denoted as γ_{IJ} for an interface between phases I and J (with S, L, and A representing solid, liquid, and air, respectively). This energy, expressed as force per unit length, acts along the IJ surface interface to minimize the corresponding positive surface energy. The liquid/air surface energy is denoted simply as γ_L. This framework involves the spreading of a film from a liquid reservoir onto a solid. The condition for this spreading is contingent on the solid/air surface tension (γ_{SA}), which propels the film, being greater than the sum of the solid/liquid (γ_{SL}) and liquid/air (γ) surface tensions.

If the difference between the solid–vapor (γ_{SA}) and solid–liquid (γ_{SL}) surface tensions, denoted as ($\gamma_{SA}-\gamma_{SL}$), exceeds the total surface tension (γ), the droplet tends to completely spread on the solid surface, resulting in a CA of null value ($\theta=0$). Marangoni initially identified this condition, expressed as $\gamma_{SA}>\gamma_{SL}+\gamma$, indicating that the solid reduces its surface energy by becoming wetted (Marangoni, 1871). Complete wetting occurs with solids possessing high surface energy, such as glass or noble metals. However, these solids often get contaminated by aerosols, leading to a decrease in their actual surface energy over time and, consequently, losing their ideality. An example is glass, which is initially wetted by water but tends to show only partial wetting in most cases as it gets polluted. Efficient cleaning of the glass with a strong acid allows it to transiently regain complete hydrophilicity.

Complete wetting is also observed with liquids having low surface tension ($\gamma<<\gamma_{SA}$), such as light alkanes or silicone oils. These liquids exhibit complete wetting on glass, steel, and most plastics. Surfactants, which reduce the liquid/vapor surface tension, are commonly added to paints to enhance their wettability and stabilize the paint film. When a droplet is placed on a solid surface that it completely wets, it spontaneously spreads. The final contact size (l) that the droplet attains on its substrate has been extensively discussed over the past two decades. The spreading parameter plays a pivotal role in determining the behavior of a drop on a solid surface. A positive S leads to drop spreading, while a negative S results in the formation of a small lens as denoted in Equation 6.5.

$$S=\gamma_{SA}-\gamma_{SL}-\gamma_L \tag{6.5}$$

When the difference between surface tensions ($\gamma_{SL}-\gamma_{SA}$) exceeds γ, the drop assumes a purely nonwetting state ($\theta=180°$). To illustrate, if we invert the fluids in a fully wetting scenario (θ in Equation 6.1 transforms into $\pi-\theta$), silicone oil, which typically wets most surfaces, will cause an air bubble introduced into a container filled with silicone oil to adhere to the container's surface at a CA of 180°. However, achieving the scenario where a solid remains completely unwetted by water, with air present, is challenging. Even on the most hydrophobic solids known, such as waxes or fluorinated materials like Teflon, water droplets exhibit CAs around 120° or 130°, significantly below the maximum of 180° (Johnson, 1964). In this review, one of our objectives is to elucidate various methods that enable us to approach this maximum limit and to assess the anticipated decrease in drop adhesion under such conditions. The drying parameter D can be defined as

$$D=\gamma_{SL}-\gamma_{SA}-\gamma_L \tag{6.6}$$

If γ_{SL} surpasses $\gamma_{SA}+\gamma_L$, the contact line experiences withdrawal due to surface forces until a film of air intervenes between the solid and the liquid, indicating $D>0$ and serving as the criterion for complete drying. This criterion aligns with $\cos\theta<-1$ in Equation 6.1. A notable scenario where this criterion is met is in systems achieving complete wetting, such as water on freshly cleaned glass. Inverting the positions of the liquid and the air results in immediate $D>0$, leading to the complete dewetting of glass when a water-filled glass forms a bubble of air at its bottom.

In specific circumstances, exemplified by the Leidenfrost effect, D is compelled to be zero. This occurs when water (or any volatile oil) is deposited on a solid with a temperature significantly higher than the liquid's boiling point, leading to the formation of a vapor film between the solid and the liquid. This vapor film rests on its own vapor, and considering the solid role played by the vapor in Equation 6.6, $D=0$ is established. However, this situation differs under standard temperatures, where complete drying of water (or any oil) on a flat solid is never observed, even though complete wetting can be achieved with most light oils on most solids. On the most hydrophobic solids (fluorinated materials), the CA never exceeds approximately 120°, corresponding to a negative parameter D of approximately $-\gamma/2$. Such situations, characterized by obtuse angles, are termed hydrophobic.

6.3.1.2 Ideal Wicking

Wicking is the spontaneous transport of a liquid through a porous medium as a result of capillary suction taking place at liquid–gas interfaces at the surface or within the porous medium. Porous materials, although typically not completely solid, exhibit characteristics that can be explored by examining cylindrical pores with a constant diameter.

If the surface energy of the solid is lower when wet than when dry, a liquid will infiltrate the tube. To quantify this, we can introduce a wicking parameter, denoted as W, which is calculated as the difference between the solid–air surface energy (γ_{SA}) and the solid–liquid surface energy (γ_{SL}).

$$W=\gamma_{SA}-\gamma_{SL} \tag{6.7}$$

The sign of W serves as an indicator for liquid penetration within the tube. Wicking is indicated when $W>0$, signifying a CA smaller than 90°. In this scenario, the meniscus formed by the liquid inside the tube exhibits curvature, resulting in a negative Laplace pressure beneath the surface. This depression facilitates the suction of the liquid into the tube. Notably, CAs tend to be acute, leading to the absorption of liquids by most sponges.

In cases where the system satisfies $S>0$, the condition $W>0$ is met, indicating a CA smaller than 90°. This situation warrants a thorough examination of the invasion mechanism, as depicted in Figure 6.4a. Here, a molecular-thick liquid film progresses along the tube walls, and the meniscus behind the film advances on a prewet tube. As this meniscus moves, it suppresses the liquid/air interface, favoring penetration. The wicking parameter, in this case, can be expressed as $W=\gamma$, representing the limit as the CA approaches zero.

Conversely, when $W<0$, wicking is not favourable, and the liquid is expelled from the pore. This concept supports the idea that a solid adorned with hydrophobic cavities can retain air, even when exposed to a liquid, approaching the Leidenfrost limit. Thus, surface roughness not only influences the wettability of a solid but, importantly, can impart novel properties, such as water repellence. Initially, we highlight that the inherent roughness of most solids can lead to contact line pinning and variability in the CA. Subsequently, we explore how specific types of roughness, particularly well-designed microstructures, can be implemented to regulate wettability, pinning, and, more broadly, hydrodynamic properties like slip.

6.3.1.3 Laplace Law

The movement of a meniscus within a pore can also be interpreted as a consequence of pressure, as initially proposed by Laplace (Paty, 1993, 1749–1827). Laplace demonstrated that pressure undergoes alteration when traversing a curved boundary, where the alteration is denoted by $P = \gamma C$, with C representing the total curvature of the boundary. In the case of a spherical boundary (with a radius denoted by R), the Laplace pressure simplifies to $P = 2\gamma/R$. In the scenario of a tube (Figure 6.4), the radius of curvature of the meniscus is determined as $r/\cos\theta$ (where r signifies the inner radius of the tube), resulting in a depression at the tube's forefront of $2\gamma\cos\theta/r$. This pressure exerts its influence over a surface area of πr^2, giving rise to a propelling force $F = 2\pi r\gamma\cos\theta$. This formula allows the reinstatement of the invasion criterion: liquid advancement occurs when $F > 0$, indicating $\theta < 90°$. In instances of complete wetting ($\theta = 0$), this force achieves its utmost magnitude, expressed as $F = 2\pi r\gamma$.

6.3.2 Wetting in Rough Solids

6.3.2.1 Contact Angle Hysteresis

Solids often exhibit inherent roughness, typically at a micrometric scale. Various fabrication processes, like lamination, can introduce striations or microgrooves. The compaction of grains results in roughness corresponding to the grain scale. Additionally, coating processes may induce roughness, especially when the coating film dewets, leading to micro drops on the surface. Conversely, molecularly flat solids are rare, and they usually form when a liquid film, either free or suspended on

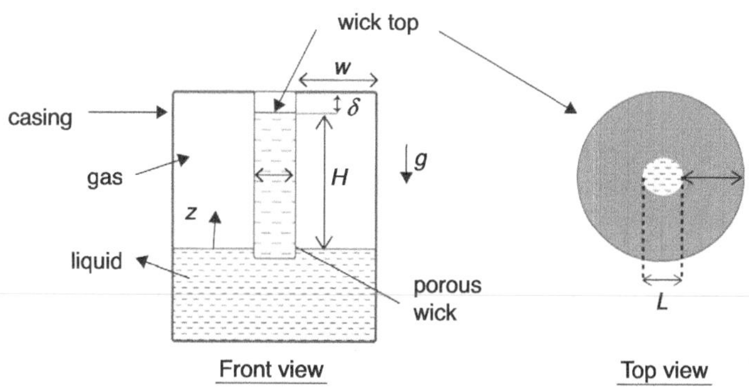

FIGURE 6.4 Schematics of the system that is to be studied. (Reproduced with permission from Ref. (Beyhaghi et al., 2014). Copyright John Wiley and Sons, 2014.)

another liquid, solidifies. In such cases, the roughness aligns with the thermal roughness of a liquid interface, typically around a few angstroms.

The presence of imperfections in a surface has significant implications for wettability, leading to two primary consequences. Firstly, these imperfections impact the measurement of the CA. When a drop, typically at the millimetric scale, comes into contact with numerous defects on the surface, it results in a shift in the CA. This phenomenon warrants discussion regarding the alteration of CAs due to such surface heterogeneities. Secondly, imperfections on the surface allow the contact line of the drop to adhere to these defects, resulting in the generation of multiple observed CAs [Adamson and Gast, 1967; Mishchenko et al., 2010; Zhao and Jiang, 2018]. This phenomenon is known as CAH.

CAH can be quantified through experimental procedures. A drop is deposited onto a given substrate and slowly fed using a syringe. In an ideal scenario where the angle is uniform, the length of the contact line would gradually increase as the drop is filled. However, in the presence of imperfections, the contact line remains pinned, causing the observed angle to increase inconsistently. Even upon ceasing to fill the drop, the angle remains fixed at its last value, reflecting one of the many possible observable angles.

Gibbs highlighted that defects on a solid surface can pin a contact line. Consequently, droplets on an incline remain stationary, with nonwetting and wetting defects at the front and rear contacts, respectively (Furmidge, 1962). This asymmetry in CAs creates a Laplace pressure difference between the front (high curvature) and the rear (smaller curvature), generating a force capable of opposing gravity if the droplet size is sufficiently small (Chow, 1983). Both chemical heterogeneities and roughness can serve as pinning sites. Visualizing a single defect even on a chemically homogeneous surface, the edge of the defect (characteristic angle ϕ) introduces flexibility in the CA. A Young angle θ is measured before the edge, and Young's angle $\pi-\phi+\theta$ is measured after the edge, taking the horizontal as the reference (ignoring the defect's existence with the naked eye). This implies that any angle between θ and $\pi-\phi+\theta$ is possible at the edge (Oliver et al., 1977). Consequently, a groove can arrest the front of a liquid drop, mimicking nonwetting behavior, while a tip acts in the opposite manner, resulting in a solid adorned with both defect types exhibiting both small and large apparent angles.

The determination of CAs is intricately linked to the procedural history of liquid deposition. When a drop is delicately placed, its spreading halts upon encountering predominantly nonwetting defects, restricting further exploration of the solid surface. Subsequently, as the drop undergoes evaporation, its configuration aligns with that of a drop affixed to wetting defects. To quantitatively assess the CAH, a systematic approach involves incrementally augmenting the drop's volume. Initially, the contact line remains adhered before abruptly surpassing a critical volume, causing the line to deepen and shift toward the subsequent set of pinning defects. The maximum angle observed during this process is denoted as the advancing contact angle (ACA) (θ_a). Conversely, extracting liquid from the drop results in its flattening until depinning occurs, retracting to the next wetting series, where the line halts and pins. The corresponding minimum angle, as shown in Figure 6.5, is termed the receding contact angle (RCA) (θ_r).

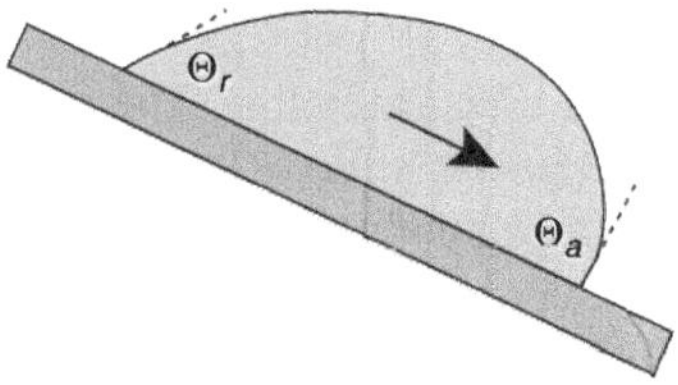

FIGURE 6.5 Hysteresis contact angle distribution. (Reproduced with permission from Ref. (Butt et al., 2022). Copyright Elsevier, 2022.)

The implications of CAH can be either advantageous or disadvantageous. Exploiting it, for instance, guides flow along a line of defects, following a predetermined path. On the contrary, when water drops adhere to window panes, they distort transparency and contribute to the degradation of the glass. Therefore, comprehending CAH is crucial. However, there remains an ongoing debate regarding the laws that connect the microscopic phenomena, such as pinning on a single defect, to macroscopic observations like the measurement of hysteresis, which averages over multiple defects. An illustrative calculation further elucidates this point. Broadly, the CAH $\Delta\theta = \theta_a - \theta_r$ exhibits a significant variation on a rough solid, ranging from nearly zero to a substantial value, approximately on the order of θ_a itself (Johnson Jr and Dettre, 1964).

6.3.3 The Wenzel Model

The influence of surface roughness extends beyond its impact on CAH, also influencing the typical or apparent angle, which often deviates significantly from the anticipated value defined by Equation 6.1. Wenzel (1936) initially recognized this phenomenon by introducing a geometrical argument hinging on the roughness factor, denoted as "*r*." This factor represents the ratio between the actual surface area and the apparent surface area of a rough surface. Upon placing a droplet on a rough surface (depicted in Figure 6.6), it tends to spread until reaching its equilibrium state, characterized by a CA denoted as θ^*, which may differ from Young's angle (θ). The underlying assumption of this model, as illustrated in Figure 6.6, postulates that as the contact line advances on the dry solid, it faithfully follows all the topological variations of the material. Consequently, each segment of the liquid/air interface undergoes replacement by a solid/liquid interface with an equivalent surface area.

The energy variation (dE) arising from an apparent displacement (dx) of the line can be expressed per unit length of the contact line as

$$dE = r(\gamma_{SL} - \gamma_{SA})dx + \gamma dx\cos\theta^* \tag{6.8}$$

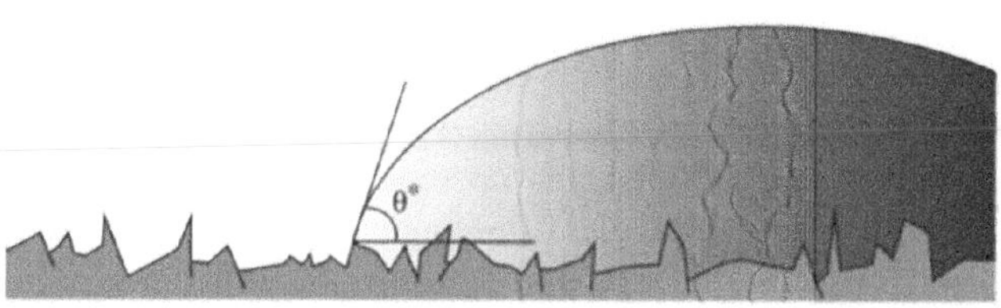

FIGURE 6.6 Infinitesimal spreading of a liquid wedge on a rough surface. (Reproduced with permission from Ref. (Bush et al., 2007). Copyright Elsevier, 2007.)

Here, the second term on the right side corresponds to the alteration of the liquid/vapor surface area during the droplet spreading. The roughness amplifies both solid energies, geometrically enhanced by a factor of "r." The minimum of E ($dE = 0$) yields Equation 6.1 under the assumption of a flat solid ($r = 1$). However, if the solid is not flat, the expression becomes (Wenzel, 1936)

$$\cos\theta^* = r\cos\theta \tag{6.9}$$

This formulation highlights the intricate interplay between surface roughness, CAs, and the geometric features of the solid substrate.

The Wenzel relation, represented by Equation 6.2, anticipates that surface roughness influences wettability. If the roughness factor, denoted as r, surpasses 1, a hydrophilic solid ($\theta < 90°$) tends to become even more hydrophilic when exhibiting roughness ($\theta^* < \theta$). Conversely, a hydrophobic solid ($\theta > 90°$) exhibits increased hydrophobicity with roughness ($\theta^* > \theta$). Although these trends are generally observable, adherence to Equation 6.3 is not consistently quantitative (refer to the following section for details). It is noteworthy that the Wenzel relation suggests peculiar aspects, such as an absence of limitations on the effect. The roughness factor can be arbitrarily large, seemingly implying the potential induction of complete wetting ($\cos\theta^* > 1$) or complete drying ($\cos\theta^* < -1$) with significant roughness ($r >> 1$). However, it is demonstrated that such behavior is not commonly observed due to unmet Wenzel assumptions.

Even when the Wenzel relation is expected to be valid, direct verification of its adherence is challenging. In this state, where the liquid conforms to the roughness, the pinning of the contact line becomes notably strong, both at the edges and along the defects. Furthermore, pushing a Wenzel drop results in the formation of cavities filled with liquid, contributing to the pinning of the drop by the liquid itself. Consequently, a Wenzel state is characterized by exceptionally low receding angles and substantial hysteresis ($\Delta\theta \sim \theta_a$). Under such conditions, extracting the singular angle θ^* or confirming Equation 6.3 becomes highly challenging. In-depth discussions regarding the validity of the Wenzel model can be found in modern references (Wolansky and Marmur, 1999; Wolansky and Marmur, 1998; Swain and Lipowsky, 1998), emphasizing the necessity for drops to be significantly larger than the defects when employing such an averaged model. In the opposite scenario, where the drop size is comparable to the defects, the liquid rearranges, leading to a dependence of the CA on the drop size (Wolansky and Marmur, 1998; Swain and Lipowsky, 1998; Checco et al., 2003).

6.3.4 Surface Texturing

In the preceding section, it was demonstrated that surface roughness not only alters the ideal characteristics of Young's equation, rendering the angle nonunique, but also impacts the observed apparent angle. Consequently, the tunability of wettability is achievable through the manipulation of roughness. The utilization of rough surfaces presents an opportunity to modulate the surface properties of solids. Moreover, it enables the induction of properties that would otherwise remain unattainable. This thematic exploration has garnered substantial attention over the past decade.

To delve into the catalysts of this burgeoning domain, three pivotal factors deserve mention. Firstly, toward the conclusion of the 1990s, researchers at the Kao Corporation in Japan unveiled that the application of fluorinated rough (fractal) surfaces resulted in exceedingly large CAs (Onda et al., 1996). While this revelation was not entirely groundbreaking, with similar outcomes documented in the 1940s (L. Gao and McCarthy, 2006), it had somewhat faded from collective memory. We will discuss this experiment in detail in Section 6.8.1. Secondly, during the same period, Neinhuis and Barthlott in Germany conducted a systematic analysis of structures on hydrophobic plant surfaces. Their investigation revealed the remarkable diversity of surface designs, implying that nature had optimized these patterns (Neinhuis and Barthlott, 1997). This line of inquiry extended to animals, uncovering new and captivating designs that were either rediscovered or discussed (Wagner et al., 1996). Subsequent efforts were directed at mimicking these natural patterns to comprehend their efficiency. Thirdly, the contemporary advancements in microfabrication techniques have empowered the construction of well-defined micro or nanostructures. This has spurred researchers to envision novel designs and optimize existing ones. The ensuing discussion encapsulates diverse findings related to these three pivotal factors.

6.4 SURFACE FREE ENERGY AND SURFACE TENSION

There is always confusion in understanding the terminology between surface energy and surface tension. A surface is the area of liquid and gas phases, and the interface is the area between them. These cannot be called different phases or separate phases since they are thin and change their properties from the bulk phase. All the neighboring molecules in the same phase draw the molecules to the surface. This results in the SFE creating an asymmetric force field. Surface tension (γ) is a phenomenon at the liquid–gas interface and is determined assuming that the molecules at the interface are less tightly bound to the surface of the liquid than in the substrate phase.

To attain the surface energy of the given solid surface, the best method is to use CA measurement. Young's equation provides for the theory of the CA of any liquid on a solid, which is discussed in the upcoming section.

6.4.1 Zisman Approach

W. A. Zisman worked on several liquids on different low-energy solid surfaces to calculate CAs (Zisman, 1963). The surface tension γ_l of the liquids is provided in the equation below, and the cosine of the CA for various liquids is determined (Figure 6.7).

$$\cos\theta = 1 - b(\gamma_l - \gamma_s) \tag{6.10}$$

where b is the slope of the regression line.

The equation was proposed by Zisman as the critical surface energy (γ_c), which is the surface tension of a reference liquid that wets the surface ($\gamma_l = \gamma_c$). The reference liquid values when lies in a straight line can be extrapolated to provide $\cos\theta = 1$ tending to the full wettability of the surface.

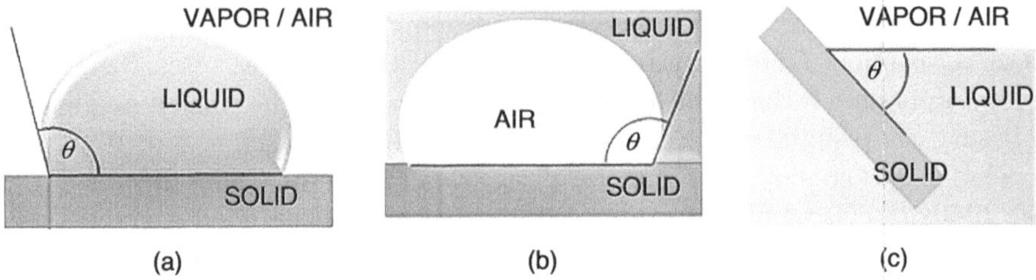

FIGURE 6.7 Methods for CA measurement: (a) sessile drop technique, (b) captive bubble method, and (c) tilting plate method of CA measurement. "θ" corresponds to the CA. (Reproduced with permission from Ref. (Liber-Kneć and Łagan, 2021). Copyright MDPI, 2021.)

The findings from Zisman's work lead to two types of calculations that are as follows (Gindl et al., 2001; Mittal, 2003):

- The two-liquid CA method, such as the geometric/harmonic mean equation, and the three-liquid CA method, such as the acid–base strategy, can be used to calculate the SFE values.
- One liquid CA method is a synonym for the equation of state methods.

6.4.2 Wu Approach

S. Wu (1971) proposed intermolecular energy between two materials composed of a dispersion component and a polar component. The SFE (or surface tension) is equivalent to the intermolecular energy. The SFE (γ) comprises a dispersion component (γ^d) and a polar component (γ^p). The harmonic mean equation is given as follows:

$$\gamma_l(1+\cos\theta_{\text{flat}}) = 4\left(\frac{\gamma_l^d\gamma_s^d}{\gamma_l^d+\gamma_s^d}+\frac{\gamma_l^p\gamma_s^p}{\gamma_l^p+\gamma_s^p}\right) \tag{6.11}$$

$$\gamma_{sl} = \gamma_{sv}+\gamma_{lv}-4\left(\frac{\gamma_l^d\gamma_s^d}{\gamma_l^d+\gamma_s^d}+\frac{\gamma_l^p\gamma_s^p}{\gamma_l^p+\gamma_s^p}\right) \tag{6.12}$$

or by geometric mean (S. Wu, 1971):

$$\gamma_{sl} = \gamma_s+\gamma_l-2\left(\sqrt{\gamma_l^d\cdot\gamma_s^d}+\sqrt{\gamma_l^p\cdot\gamma_s^p}\right) \tag{6.13}$$

The liquids used for testing affect the surface energy values on various substrates in the above equation. For the geometric and harmonic mean formulae to yield accurate surface-free energy values, both polar and nonpolar substances must be used (Owens, 1970).

6.4.3 Acid–Base Theory or Three-Liquid Method

This method is the most complex and was developed by Fowkes (1987), Van Oss et al. (W. Wu et al., 1995), and Good and Chaudhury (Oss et al., 1988). The equation for the interface tension of solid/liquid is given by

$$\gamma_{sl} = \gamma_s + \gamma_l - 2\left[\sqrt{\gamma_s^{lw}\gamma_l^{lw} + \gamma_s^{+}\gamma_l^{-} + \gamma_l^{+}}\right] \tag{6.14}$$

It is necessary to determine the CA of at least three liquids, including both polar and nonpolar liquids, in order to calculate the SFE components (γ_s^{lw}) and factors (γ_s^{+} and γ_s^{-}) of solids.

6.4.4 Neumann's Method

The method suggests that a relationship between $\gamma_{s,}\, \gamma_{l,}$ and γ_{sl} is possible and must exist and is called the equation of state.

$$\gamma_{sl} = f(\gamma_s, \gamma_l) \tag{6.15}$$

Further formulation of the equation of state gave a new equation as shown below:

$$\cos\theta_{\text{flat}} = -1 + 2\sqrt{\frac{\gamma_l}{\gamma_s}}.e^{-\beta(\gamma_l - \gamma_s)^2} \tag{6.16}$$

where θ_{flat} is the average CA of the liquid with a flat solid surface which is about to be measured to obtain SFE. γ_l is liquid's surface tension, γ_s represents SFE, and β is a constant of value 0.0001247 $(m^2/mJ^1)^2$. Finally, through this equation, a single observation of the CA of a liquid with a known surface tension (γ_l), the equation of state provides the SFE of a solid (γ_s). The polar component or the spread, however, cannot be measured.

6.4.5 Owens–Wendt–Rabel–Kaelble (OWRK) Method

In the method of OWRK, the assumption is that the surface free energies are obtained by the dispersive and polar liquid's SFE components. The OWRK model is also known as the expanded Fowkes model based on the Young–Dupree equation (Owens and Wendt, 1969).

By Young's equation, as shown in Equation 6.1:

$$\gamma_l \cdot \cos\theta = \gamma_s - \gamma_{sl} \tag{6.17}$$

where γ_s, γ_l, and γ_{sl} are the surface free energies of solid, liquid, and solid–liquid (mJ/m^2) and θ is the CA (°) between solid and liquid, respectively.

γ_l is a known value. θ is the CA (°). To calculate γ_s and γ_{sl}, we need the work of adhesion equation (Yao et al., 2016).

$$W_A = \gamma_s + \gamma_l - \gamma_{sl} \tag{6.18}$$

Combined with Young's equation, the following equation is obtained:

$$W_A = \gamma_l.(1+\cos\theta) \tag{6.19}$$

The OWRK model also termed the extended Fowke's model based on the Young–Dupre equation leads

$$\gamma_l(1+\cos\theta_{\text{flat}}) = 2\left(\sqrt{\gamma_l^d.\gamma_s^d} + \sqrt{\gamma_l^P\gamma_s^P}\right) \tag{6.20}$$

where γ_l^d and γ_l^p depict the liquid's dispersive and polar components, respectively, while γ_s^d and γ_s^p stand for the solid's relative contributions.

Combined with Young's equation, this can be rewritten as given below:

$$\sqrt{\gamma_s^d\gamma_l^d} - 2\sqrt{\gamma_s^p\gamma_l^p} = \frac{1}{2}\gamma_l(1+\cos\theta) \tag{6.21}$$

Because there are two unknowns, γ_s^d and γ_s^p two liquids with the known dispersive and polar components are needed to solve it.

Substituting this term for γ_{sl} in the following equation:

$$\gamma_{sl} = \gamma_s + \gamma_l - 2\left(\sqrt{\gamma_s^d\gamma_l^d} + \sqrt{\gamma_s^p\gamma_l^p}\right) \tag{6.22}$$

By plotting the above equation, the slope and the intercept from the graph will provide the surface energy of the solid. The value is calculated using the sum of the two SFE components.

We can also determine solid surface energy using Fowkes's theory, i.e., using the dispersive component.

$$\gamma_s^d = \left(\frac{\gamma_l}{4}\right)(\cos\theta + 1)^2 \tag{6.23}$$

γ_s^d can be calculated directly from CA data for the probe liquid on the solid.

According to the results of CA measurements, the OWRK method can be perfectly used for calculating SFE, the dispersive and polar components.

6.5 METHODS FOR CONTACT ANGLE MEASUREMENT

6.5.1 Sessile-Drop Method

Sessile-drop goniometry is involved by determining a contact angle (CA) of a water drop on a solid surface through the images by fitting a tangent procedure. Following this protocol while measuring CA is called the sessile-drop goniometry method. The straightforward approach is deceptive because most impact angle readings in the literature come from "static" or "placed" drop CAs (Figure 6.7). This method presumes that the drop put on the solid surface will have minimum energy and therefore a stable state which is called YCA. But usually, the drop obtained will be metastable as it could have any minimum energy within the hysteresis range as shown in Figure 6.8. Therefore, these "static or placed" CAs cannot be reproducible, and it fails to provide the necessary information about the interaction of the liquid/solid by CAH. The most used, reproducible, and measurable CAs are ACA and RCA. According to rising and declining drop volumes, they are the greatest and lowest angles in the CAH range, as shown in Figure 6.9.

6.5.2 Captive Bubble Method

The captive bubble method is another way to measure the CA, which is by introducing an air bubble in liquid where the flat solid substrate is submerged in liquid (Figure 6.7). The method is mainly known to be the "captive bubble method," which was developed by Taggart et al. (1929). Herein, a minute amount of air is exposed to the testing liquid for an air bubble to be submerged. The needle must be handled carefully to prevent the air bubble from floating away over the solid surface and disrupting the advancing angle of the balance. This method also ensures that the solid sample is at equilibrium with the atmospheric interfacial tension and with the liquid phase during the process. Due to the absence of any other airborne absorption, the substance is only minimally contaminated. Additionally, the temperature dependence of CA studies is possible only through this method (Wark and Cook, 1934). Figure 6.10 displays a graphic depiction of the captive bubble method.

6.5.3 Pendant Drop Tensiometry

The pendant drop is the relationship between the Laplace pressure at an interface with the interfacing circle and interfacial tension (γ). It follows the Young–Laplace equation at equilibrium.

$$\gamma\left(\frac{1}{R_1}+\frac{1}{R_2}\right)=\Delta P=\Delta P_0-\Delta\rho g z \tag{6.24}$$

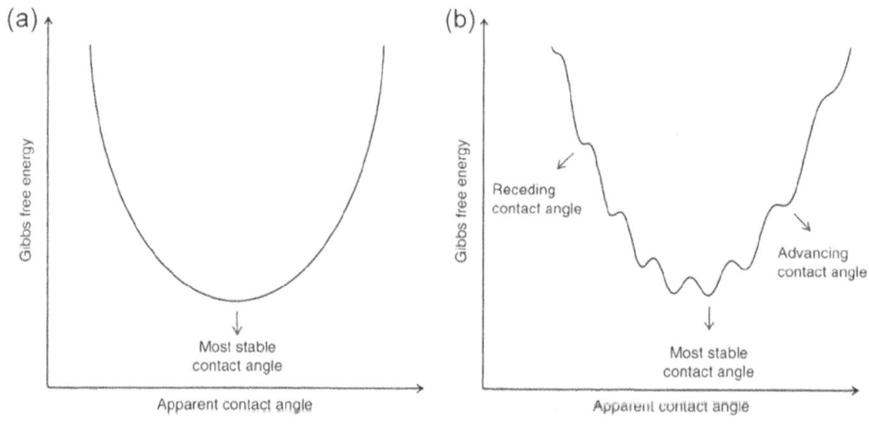

FIGURE 6.8 (a) The wetting system having a single free-energy minimum and is stable YCA and (b) the wetting system of a solid surface at minimum free energy and metastable CA. (Reproduced with permission from Ref. (Huhtamäki et al., 2018). Copyright Springer Nature, 2018.)

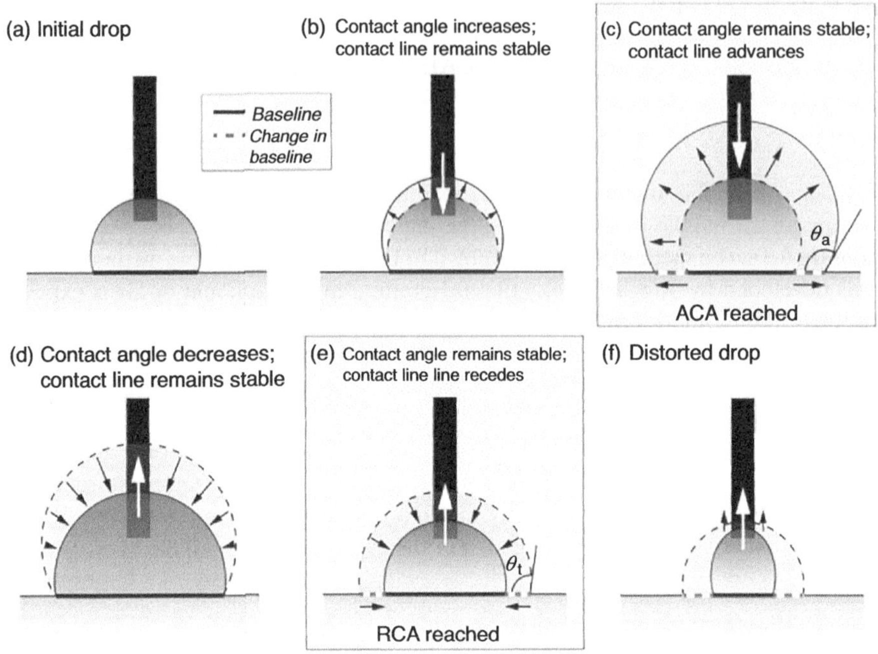

FIGURE 6.9 Different types of measurable CAs, namely, ACA and RCA, respectively. (Reproduced with permission from Ref. (Huhtamäki et al., 2018). Copyright Springer Nature, 2018.)

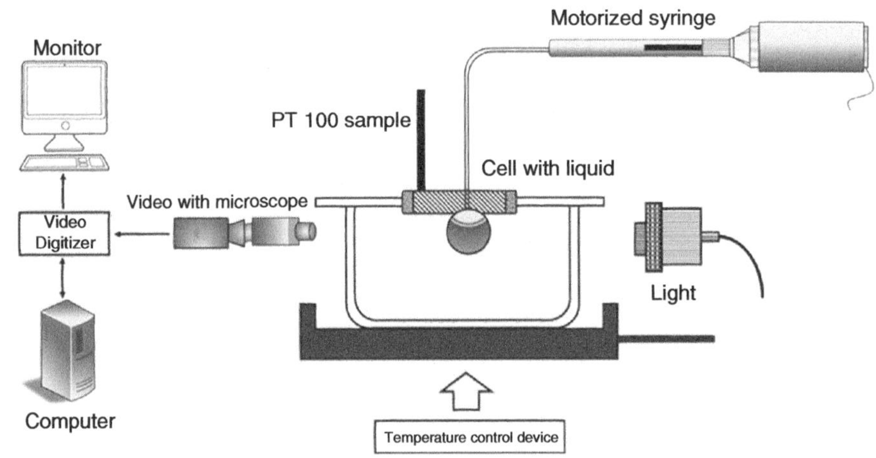

FIGURE 6.10 The captive bubble method's visual depiction. (Reproduced with permission from Ref. (Hebbar et al., 2017). Copyright Elsevier Books, 2017.)

where radii of the curvature are R_1 and R_2; the Laplace pressure across the contact is $\Delta P = P_{in} - P_{out}$; the density differential is defined as $\Delta\rho = \rho d - \rho$, where ρd and ρ are the drop density and the continuous phase density, respectively.

$$B_o = \frac{\Delta\rho g R_0^2}{\gamma} \tag{6.25}$$

The associated boundary conditions are as follows:

$$\bar{r} = 0, \bar{z} = 0, \phi = 0 \text{ at } \bar{s} = 0 \tag{6.26}$$

The dimensionless Bond number (B_o) is influenced by the pendant drop shape. The Bond number can be used to calculate the drop radius (R_0) at the tip and the interfacial tension (γ) is then easily derived from the above equation, which is the theoretical significance of the pendant drop techniques.

The Young–Laplace solution, when applied, can give us data like drop volume (V_d) and drop surface area (A_d), which is defined as

$$V_d = \Pi \int \bar{r}^2 \sin\phi d\bar{s}, \; A_d = 2\Pi \int \bar{r} d\bar{s} \tag{6.27}$$

This will determine the interfacial data. Also, the drop evaporation rate can be determined by tracking the drop volume over the time of experiments.

6.6 DYNAMIC INTERFACIAL TENSION AND ITS EFFECTS

The calibration of a mechanical or electrical transducer or force sensor is not required in the pendant drop tensiometry, unlike other tensiometry methods. Deformation is responsible for displacement, and gravitational motion, which is largely unchanging, is used to measure interfacial tension. Table 6.1 presents a range of systems using the pendant drop fitting approaches.

Low molecular surfactants, with dynamic surface tension and adsorption kinetics, are the subject of an ongoing study in a wide range of processes, such as emulsification, printing, coating, and mineral flotation. The Young–Laplace equation cannot be applied to rapid dynamic interfacial processes, as the pendant drop technique requires an equilibrium state to be present. Therefore, this technique cannot take into account fluid flow and inertial effects at higher droplet formation rates. Predicting the interfacial stresses requires a time-consuming technique called pendant drop tensiometry. However, it is crucial to prevent drastic shifts in parameters over time.

Fluctuations in the rate of surface area, diffusivity, and humidity occur when using the pendant drop method, as it is thermodynamically controlled along with changes in the chemical potential of the contents at its curved interface. The liquids that have low mutual solubility with liquid–liquid interfacial tensions can be dealt with through this type of method. In a volatile liquid system, the droplet's evaporation is a serious issue that needs to be handled or foreseen. Therefore, it is very important to note the droplet measurement volume along with the interfacial tension.

TABLE 6.1
Evaluating a Range of Systems Using the Current Pendant Drop Fitting Approaches and Comparing the Results to those from the Established Literature

Droplet Phase	Continuous Phase	Interfacial Tension (mN/m[1])	References
Water	Air	72.75	Vargaftik et al. (1983)
Glycerol	Air	63	Weast (1988)
CTAB (10 mM)	Air	37	Adamczyk (1999)
Toluene	Air	28.52	Berg (2010)
n-Tetradecane	Air	26.56	Berg (2010)
Perfluorohexane	Water	55	Kabalnov et al. (1990)

6.6.1 Young–Dupre Equation

The interfacial force and energy remains the basis for understanding and calculating the interfacial interaction in nature. The absorption, agglomeration, and surface fouling are all interrelated to the interfacial interaction. This can be exclusively measured by the CA. At the interface (solid–liquid), the correlation of CA and energy parameters is determined by the Young–Dupre equation.

$$(1+\cos\theta)\gamma_l = 2\left(\sqrt{\gamma_s^{lw}\gamma_l^{lw}} + \sqrt{\gamma_s^{+}\gamma_l^{-}} + \sqrt{\gamma_s^{-}\gamma_l^{+}}\right) \tag{6.28}$$

where θ is the CA, the symbol γ denotes the surface energy of a liquid or a solid, whereas γ_{lw} is the Lifshitz–van der Waals and Lewis acid/base components of the surface energy.

The calculation of energy parameters from the above equation is done on the assumption that the surface is smooth and flat, but in reality it is always rough. The roughness changes the actual contact area to larger than the geometric projection, creating deviation from the calculated energy parameters, as shown in Figure 6.11.

6.7 FACTORS AFFECTING CA MEASUREMENTS

Numerous physical characteristics of the base affect how wettability changes surface roughness. The physical characteristics of the material significantly affect the measurement of the CA. This portion discusses the key elements, including heterogeneity, surface roughness, and particular size and shape.

6.7.1 Roughness Factor

The material's surface roughness, which is crucial in determining the physical CA, has a significant impact on the moistening, spreading, and soaking processes (Extrand, 2002). Additionally, the solid's surface texture factors have an impact on the CA analysis's hysteresis. However, it is unclear how to quantitatively analyze

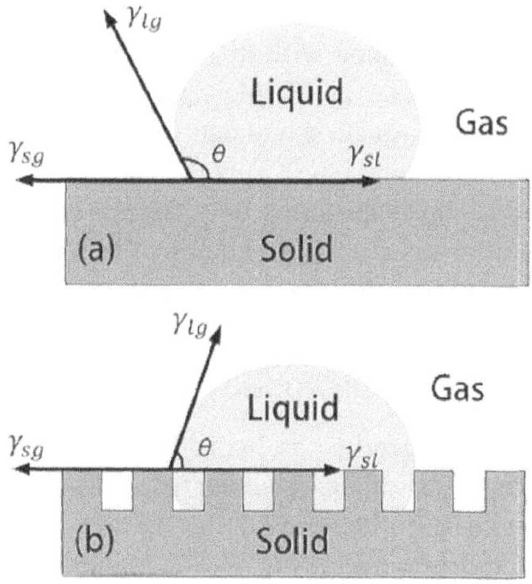

FIGURE 6.11 Schematic diagram of a liquid droplet on the smooth and rough planes. (Reproduced with permission from Ref. (Hu et al., 2024). Copyright Springer Nature, 2024.)

roughness and CA recurrence. The study by Oliver et al. identified several reasons why these ideas failed, some of which are listed below (Chau et al., 2009).

- Not many studies on surface roughness characterization.
- The inability of the system to differentiate between surface chemistry, surface roughness, heterogeneity, and anisotropy.

6.7.2 Heterogeneity Factor

Impurities on the surface substrate are connected predominantly because of the heterogeneity factor due to many reasons. Liquid CA readings on increasing and decreasing methylated quartz material studied by Fadeev and McCarthy (1999) said that it varies along with the change in the amount of added trimethylsilane groups, similar to that of the Cassie equation (Fadeev and McCarthy, 1999). Drelich et al. (2011) studied the wettability of the liquid on heterogeneous surfaces processed through self-assembled monolayers that consisted of hydrophilic and hydrophobic regions. According to the research, the Cassie equation needs to be altered to include the effects of the three equilibrium interfacial tensions on the SFE of the contact line. The modification is very important for the heterogeneity factor.

6.7.3 Particle Size/Shape Effect

Another significant parameter affecting the CA is the particle size and shape over the substrate surface, which affects bulk density, wettability, adhesion, and interaction with the fluids. The substrate features such as elongation, effective surface area, and surface roughness are explained using particle size and shape (Woo et al., 2015). The significance of these parameters on the properties of the surface can be very useful in predicting the results. The wettability properties of the surface by the effect of particle shape with different mill times are investigated by Kursun and Ulusoy (2006). The 2D SEM was implemented for identifying the milled particle shapes. The elongation and decrease in surface roughness increase the hydrophobicity, and the increase in surface roughness with a circular shape increases hydrophilicity. The shape and size of the material are given more importance as they influence the structural and chemical properties of the material that causes wettability changes.

6.7.4 Hydrophilicity and Hydrophobicity Analyses

The affinity for water is referred to as hydrophilicity, and hating as hydrophobicity. The wetting properties of the solid surface, which predominantly play an essential part, are primarily influenced by the oil sectors, which are engaged in processes like improved oil recovery, membrane technology, superhydrophilicity, and hydrophobicity (Ramachandran and Nosonovsky, 2016). Due to the unbalanced secondary force interaction, there is an affinity of liquid molecules on the surface as there is an interaction of the liquid drop on the surface of the material. The CA (θ^o) is a measurement resulting from the interaction between the solid/liquid, and also it is associated with the surface energy of solid materials. If the CA, $\theta < 90°$, it shows that the surface has a high affinity toward the liquid, and if the CA, $\theta > 90°$, it has low affinity and is generally termed as hydrophilic (Gourley et al., 1994). Figure 6.12 shows a variation in the CA with a hydrophobic and hydrophilic nature. The hydrophilic surface is defined as a surface having a high tendency to become wet or increase the hydration film on the surface. The CA, $\theta > 90°$ presents the least tendency of wetting the surface and exhibits the minimum desire toward the liquid. These surface characteristics are referred to as hydrophobic. A "bead"-like formation of liquid is seen on the surface due to the hydrophobicity, i.e., water-hating. If the CA, $\theta > 140°$, then the surface is called superhydrophobic (Barati et al., 2020).

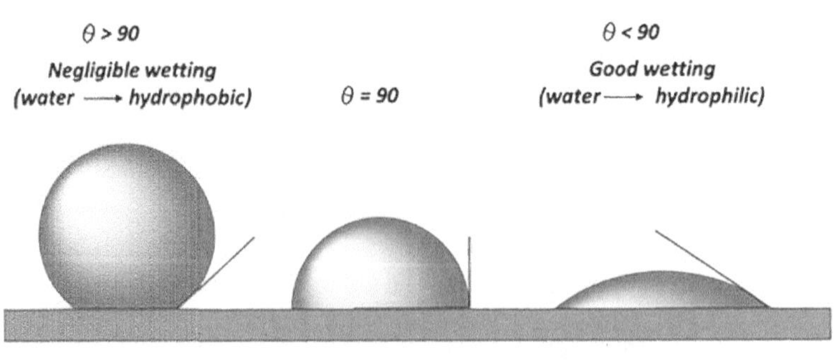

FIGURE 6.12 The CA on hydrophilicity and hydrophobicity analysis. (Reproduced with permission from Ref. (Hebbar et al., 2017). Copyright Elsevier, 2017.)

6.8 FACTORS AFFECTING SURFACE FREE ENERGY

6.8.1 The Kao Experiment

Researchers at Kao conducted experiments using various well-characterized fluorinated substrates, including both rough and flat surfaces (Onda et al., 1996). The investigation involved comparing CAs θ^* on the rough samples with those on flat materials, which should theoretically align with Young's angle θ. This assessment utilized different liquids to alter θ. Figure 6.13 illustrates a typical outcome where cos θ^* is plotted against cos θ, demonstrating alterations in wetting properties due to surface roughness. It is crucial to note that only one angle, θ^*, is considered, seemingly close to the advancing angle, disregarding hysteresis for each data point.

The x-axis of the plot does not cover the entire interval [−1, 1], with cos θ never falling below −0.3, equivalent to an angle θ of around 110°. This particular data point, acquired using water as a liquid, is noteworthy. CAs on flat surfaces seldom exceed approximately 120°, representing the maximum chemical hydrophobicity (Oliver et al., 1977). Despite the scarcity of data points on the hydrophobic side (cos $\theta < 0$), a remarkable effect is observed. Upon entering this domain, the apparent CA θ^* sharply rises to 170°, surpassing the chemical angle, indicating a superhydrophobic state induced by roughness.

In the hydrophilic domain (cos $\theta > 0$), cos θ^* initially increases linearly with cos θ, exhibiting a slope greater than unity (approximately 3). This behavior could be interpreted as a Wenzel regime (Equation 6.8). Microscopic images support this slope

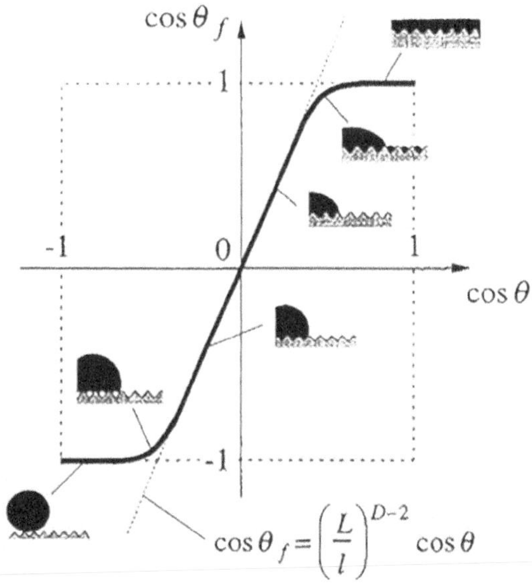

FIGURE 6.13 Cosine of the apparent contact angle θ_f on a textured surface, as a function of the cosine of the Young's angle θ measured on the same surface, yet flat; L and l are, respectively, the upper and the lower limit lengths of fractal behavior. (Reproduced with permission from Ref. (Onda et al., 1996). Copyright ACS Publication, 1996.)

(Shibuichi et al., 1996), revealing material roughness. Extracting such a complex and unseen quantity from a simple and economical experiment, using just a few drops to probe the surface, is noteworthy. However, this linear relationship deviates when the CA θ decreases below a critical value θ_c. Instead, a second linear regime emerges (with a slope smaller than unity), approaching $\theta^* = 0$ as $\theta = 0$. Essentially, a wettable solid remains wettable if rough. Section 6.8.4.2 delves into the explanation for this regime, attributing it to liquid penetration into micro-textures, where the liquid surrounds the drop used for measuring the CA. Unlike the Wenzel hypothesis, which assumes a dry solid beyond the drop, this regime, termed superhydrophilic, describes a scenario where the liquid lens sits on a mixture of solid and liquid, as elaborated in Section 6.8.4.

6.8.2 Work of Adhesion and Cohesion

Consider a cylindrical tube with a cross-sectional area of 1 cm², accommodating two immiscible liquids denoted as α and β, with an interface demarcating them (Aveyard and Haydon, 1973). The energy needed to part these layers and generate 2 cm² of liquid–air surface for each liquid is termed the "work of adhesion." This value is derived through the Dupre relationship (Dupré and Dupré, 1869), expressed as

$$W_{\alpha\beta} = \gamma_\alpha + \gamma_\beta - \gamma_{\alpha\beta} \tag{6.29}$$

Here, γ_α represents the measured interfacial tension between the two liquids, while γ_β and $\gamma_{\alpha\beta}$ signify the surface tensions at the liquid–air interfaces for the respective pure liquids.

In the scenario where the tube is filled solely with one liquid, the effort required to separate the column at any given cross-sectional plane for the formation of 2 cm² of a uniform liquid–air surface is termed the "work of cohesion" (Harkins and Cheng, 1921). This quantity can be determined using the Wan relationship, given by

$$W_{\alpha\alpha} = 2\gamma_\alpha \tag{6.30}$$

6.8.3 Natural Micro-Textures

Micro-textures also manifest on the surfaces of various plants and animals, as illustrated in Figure 6.14. Pliny the Elder, in his work Natural History, observed that water forms perfect spheres on a leaf with a woolly surface (Secundus, 2004). Dufour's review in 1833 (Dufour, 1833), Fogg's paper in 1944 (Fogg, 1944), and Cassie and Baxter's commentary on Fogg's work (Cassie and Baxter, 1945) are noteworthy examples. More recent and systematic investigations carried out over the past decade have yielded a notable assortment of micro-textures, exemplified in Figure 6.14.

Neinhuis and Barthlott identified approximately 200 plant species, such as asphodelus, drosera, eucalyptus, euphorbia, gingko biloba, iris, lotus, and tulipa, that possess such characteristics (Neinhuis and Barthlott, 1997; Mock et al., 2005). The surfaces of these plants typically share three key attributes: (1) they are covered with a wax layer or wax crystals, rendering them hydrophobic (with a Young CA exceeding 90°); (2) they feature textural elements like bumps, typically at a scale of 10 μm;

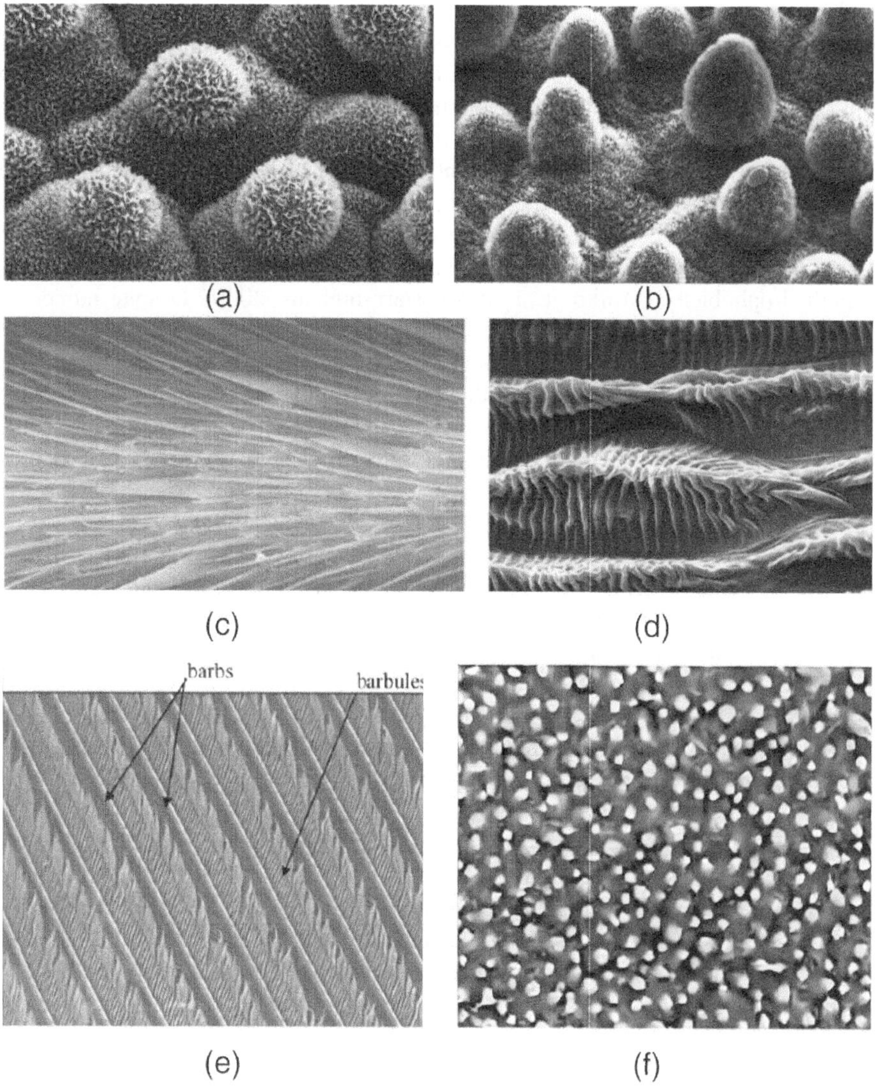

FIGURE 6.14 (a) Leaf of the so-called elephant's ear (*Colocasia esculenta*). (Reproduced with permission from Ref. (Barthlott and Neinhuis, 1997). Copyright Springer Nature, 1997.) (b) Lotus leaf. (Reproduced with permission from Ref. (Barthlott and Neinhuis, 1997). Copyright Springer Nature, 1997.) (c) Leg of a water strider. (Reproduced with permission from Ref. (Gao and Jiang 2004). Copyright Springer Nature, 2004.) (d) Petal surfaces of *Mutisia decurrens*. (Reproduced with permission from Ref. (Barthlott and Neinhuis, 1997). Copyright Springer Nature, 1997.) (e) SEM images of the pigeon feather structure. (Reproduced with permission from Ref. (Bormashenko, Bormashenko, et al., 2007). Copyright Elsevier, 2007.) (f) Low-magnification scanning electron microscope images of the surface structures on the lotus leaf. (Reproduced with permission from Ref. (Qu et al., 2010).)

and (3) a secondary texture, often comprising hairs and considerably smaller in size (around 1 μm in many instances), overlays the primary one. Neinhuis and Barthlott also replicated these surfaces using silicon to isolate the textural effects from the chemical coating. For instance, Figure 6.14a illustrates the upper surface of a *Colocasia esculenta* leaf, also known as the elephant ear, as observed through scanning electron microscopy (SEM). On plant leaves, discernible bumps at the 10–50 μm scale are commonplace (Figure 6.14a, b, f). Notably, hydrophobic plants, such as the lotus, and others exhibit papillae covered by nanostructures around 100 nm in size. The coexistence of dual roughness scales significantly contributes to achieving superhydrophobicity (Minko et al., 2003; Herminghaus, 2000). Despite numerous hypotheses, the hierarchical structure lacks a comprehensive understanding, though it is not essential for attaining high degrees of hydrophobicity (Gao and McCarthy, 2007). Such structures often yield both a high CA and a low CAH. For instance, the lotus leaf exhibits an advancing angle of approximately 160° and a receding angle exceeding 150°, conferring high mobility to water drops on these surfaces. In specific instances, like the rice leaf, papillae arrangement can be anisotropic, influencing wetting and adhesion along certain directions (Wang et al., 1998).

Feathers of various birds, including pigeons (Bormashenko, Bormashenko, et al., 2007) and ducks, as well as insects like cicadas, butterflies, and water striders (Lee et al., 2004), demonstrate hydrophobic and/or superhydrophobic properties. Insects' cuticles are shielded by a layer of epicuticular wax (typically 250 nm thick), preventing water intrusion and desiccation. Setae, present on the body or legs (Figure 6.14c), contribute to impressive superhydrophobic properties, enabling some animals to float or live underwater by spreading air on their bodies (Heckman, 1983; Noble-Nesbitt, 1963; Thorpe, 1950; Marmur, 2006). Setae often comprise tapered hairs with a length of 30 μm, a diameter of 1–10 μm, and an inclination angle of around 30° (Figure 6.14c). Similar to plants, a secondary texture—nanogrooves—exists, whose precise role remains under scrutiny. Other structures, such as the eye pattern of Culex pipiens (Figure 6.14d), exhibit well-ordered features at a remarkably small scale (posts of approximately 100 nm in size and height) (Gao et al., 2007). We further demonstrate that certain applications necessitate a reduction in pattern size.

6.8.3.1 Air Trapping

In the context of hydrophobic solids, the scenario diverges significantly from that observed in the case of hydrophilic solids. In instances where the solid surface exhibits sufficient roughness, the assumption that the liquid will conform to the solid, as posited in the Wenzel model (depicted in Figure 6.6), becomes untenable. Instead, the formation of air pockets beneath the liquid occurs, giving rise to the so-called Cassie or Fakir state (Shirtcliffe et al., 2006). This phenomenon materializes when the energetic cost associated with all liquid/vapor interfaces is less than the energy gained by the liquid refraining from adhering to the solid substrate.

To introduce quantifiability to this criterion, pillar-like textures are considered. Assuming flat liquid/air interfaces (justified by a condition of constant Laplace pressure in the liquid, particularly for defects significantly smaller than the drop size), the wet and liquid surface areas become proportional to $(r-\varphi S)$ and $(1-\varphi S)$, respectively. Therefore, air pockets are favored when the inequality (Alberti and DeSimone, 2005) is satisfied.

$$(r-\varphi_S)(\gamma_{SL}-\gamma_{SA})>(1-\varphi_S)\gamma \tag{6.31}$$

Utilizing the Young formula, this inequality can be reformulated as $\theta > \theta_c$, where

$$\cos\theta_c = -(1-\varphi_s)/(r-\varphi_s) \tag{6.32}$$

This criterion bears resemblance to the one established for the propagation of a liquid film within the texture. In this case, air replaces liquid, and the critical angle expected from Equation 6.31 aligns with π minus the critical angle below which hemi-wicking occurs (Equation 6.32). For highly rough solids ($r >> 1$), this criterion is consistently met. As a result, θ_c tends toward 90°, and θ exceeds this value due to the assumption of chemical hydrophobicity.

Materials adorned with long hairs exemplify this scenario, with the roughness factor ($r \approx 2\pi bh/p^2$) typically ranging from 5 to 10, as shown in Figure 6.14c. The leg of a small water-walking bug, Microvelia, remains uncontacted by the liquid (Bush et al., 2007).

The situation becomes less clear-cut when dealing with smaller pillar density φ_S, resulting in more modest roughness factors. The chemical CA θ typically falls within the range of 100°–110°, with a slightly negative cosine. This deviation from the conventional criterion for air trapping poses an ambiguity. However, despite a higher interfacial energy, Cassie states are frequently observed (Patankar, 2003). It appears that air present before introducing a drop can persist in a metastable state, as long as the drop refrains from initiating contact with the solid ground surface. The metastability of Cassie states is further explored in Section 6.8.3.3.

The Cassie regime garners particular interest due to its large CA and the additional benefit of a small CAH, attributed to the presence of an air cushion. Consequently, we term only this specific Cassie regime as superhydrophobic, showcasing remarkable properties like reduced adhesion, water repellence, and slip.

Considering the drop's placement on a combination of solid and air, an anticipated large apparent angle θ^* is expected. In the absence of solid, Young's equation (Equation 6.1, with the substitution of index S by A) predicts a "contact" angle of 180°, indicating no contact. Any deviation from this value provides insight into the proportion of solid in actual contact with the liquid.

The change in interfacial energy resulting from a displacement of the contact line by a distance dx is connected to the creation of new wet solid surface and liquid/vapor interfaces. The final balance equation is expressed as follows:

$$dE = \varphi_s(\gamma_{SL}-\gamma_{SA})\,dx + (1-\varphi_s)\gamma dx + \gamma dx\cos\theta^* \tag{6.33}$$

It is imperative to assume flat liquid/vapor interfaces. Considering curvature would alter the outcome, implying a potential (slight) increase in the angle as the drop size diminishes. Recent experiments by Rathgen et al. (2007), analyzing light diffracted through transparent Cassie materials, offer precise measurements of this curvature.

At equilibrium, E attains its minimum, determining the apparent angle θ^* (Bico et al., 1999):

$$\cos\theta^* = -1 + (1-\varphi_s)\cos\theta \tag{6.34}$$

The condition at each contact line, including the drop's edge and every liquid/vapor interface beneath it, is supplemented by the (local) Young condition. This condition is fulfilled through the existence of edges on the posts or, more broadly, the presence of substantial slopes on the rough material. $\theta > 90°$ indicates that sharp angles are conducive to meeting this condition. A prediction arises that stronger pinning occurs with smaller edge angles, as re-entrant designs enhance the robustness of Cassie states. It is observed that such designs can even induce air trapping in a hydrophilic scenario.

The prediction made by Equation 6.34 typically involves large angles. For $\theta \approx 110°$–$120°$ and φ_S ranging between 5% and 10%, apparent angles in the range of 160°–170° are obtained. Conversely, structures smaller than the wavelength of light lead to transparency. This prompts the intriguing question of determining the minimum size capable of generating water repellence, a question that remains unresolved. As φ_S decreases, θ^* increases. Yet, for a more intricate topology, φ_S should be a function of θ (Bico et al., 1999). Conversely, measuring θ^* in a Cassie situation should yield the solid fraction φ_S in contact with the liquid—an essential quantity for characterizing wetting and hydrodynamic slip, or any properties linked to solid/liquid contact (e.g., electrical or chemical). In the limit of small φ_S, it is noted that $\theta * = \pi - \varepsilon$ (with $\varepsilon << 1$), and Equation 6.34 can be reformulated as shown below highlighting the critical behavior in φ_S, underscoring the challenge of achieving a strictly nonwetting situation ($\varepsilon \approx 0$).

$$\varepsilon \approx 2\cos(\theta/2)\varphi_s^{1/2} \tag{6.35}$$

However, Gao and McCarthy (2006) approached, and potentially reached, this limit (within measurement uncertainties) by utilizing nonwoven assemblies of nanofibers, where both advancing and receding angles were 180°. Similar outcomes were reported for pulverulent hydrophobic solids achieved by compressing commercially available lubricant (Gao and McCarthy, 2007). In both instances, the texture is submicrometric and regular, with smooth rounded defects, suggesting very low hysteresis—a parameter we will delve into further.

6.8.3.2 Toward Nonsticking Water

The principal characteristic of a Cassie surface lies in its minimal CAH. Consequently, droplets exhibit exceptional mobility, resulting in unique phenomena such as bouncing, not typically observed on conventional materials (Richard and Quéré, 2000). While liquids tend to behave akin to Leidenfrost drops, where the underlying vapor minimizes friction, superhydrophobic materials generally display residual hysteresis (and thus adhesion), a value that remains subject to ongoing debate (Extrand, 2002).

In the discourse surrounding this matter, the potential for pinning the contact line on defects emerges as a pivotal factor. The shape of defects or the sharpness of edges becomes crucial, as highlighted in a classical experiment by Oner and McCarthy

(2000) wherein they manipulated post shapes. For the sake of this discussion, we narrow our focus to strong pinning on dilute defects, elucidated in 1984 by Joanny and de Gennes (Extrand, 2002) and Pomeau and Vannimenus (1985).

The model, as depicted in Figure 6.15, portrays a few defects where the line pins during liquid movement, subjected to a force F. The line encounters φ_S defects per unit area, translating to $\varphi_S\, dx$ defects per unit length for a displacement dx. As the line passes each defect, an energy Δ is accumulated and subsequently released in the liquid, dissipated by viscosity upon line depinning. Thus, the relationship is expressed as

$$F\, dx = \varphi_S dx \Delta \tag{6.36}$$

The conjecture is that Δ depends on defect shape, with complex contours generating higher Δ and, consequently, larger hysteresis, unlike small and rounded defects. For simplicity, we assume an equilibrium (Young) angle of 90° and consider defects as pillars (or disks) with radius b and mutual distance p (where $b << p$). Hence, $\varphi_S \sim b^2/p^2$.

Upon pinning on a defect of size b, the drop undergoes distortion. The tails form surfaces of zero curvature, expressed as $r = b \cosh(x/b)$, indicating that for $x \gg b$, $r \approx 1/2\ b\ e^{x/b}$. The maximum deformation (at $x = u$) occurs for the largest lateral deformation r, which is the typical distance p between two defects, resulting in

$$u \approx b\log\left(p/b\right) \tag{6.37}$$

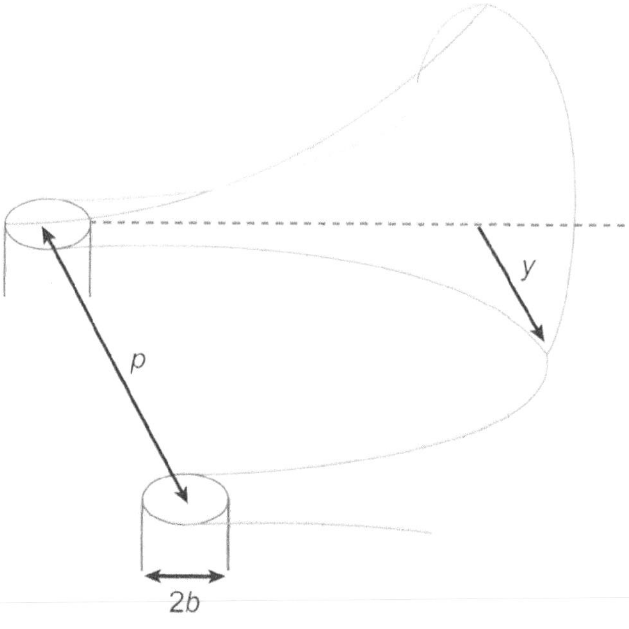

FIGURE 6.15 In a Cassie state, a drop is likely to pin on the edges of the defects as we displace it. The drop becomes distorted, and the energy stored in this deformation fixes the amplitude of the hysteresis. (Reproduced with permission from Ref. (Quéré, 2008). Copyright ACS Publication, 2009.)

The pinning force on a defect f is linked to b (the line pins on the contour of each defect): $f \approx \pi b\gamma$. This establishes a relationship between force and deformation:

$$f \approx \pi\gamma u / \log(p/b) \tag{6.38}$$

Equation 6.23 defines a linear spring of stiffness $K = \pi\gamma/\log(p/b)$ (Extrand, 2002). Hence, there is an energy $\Delta = 1/2Ku^2$ stored in the deformation. The force necessary to move the line can be written, per unit length, as

$$F = \gamma(\cos\theta_r - \cos\theta_a) \tag{6.39}$$

The combination of the provided equations leads to the expression:

$$(\cos\theta_r - \cos\theta_a) \sim \varphi_S \log(1/\varphi_S) \tag{6.40}$$

The CAH diminishes in the presence of defects' density. However, Equation 6.19 exhibits an intriguing behavior due to the logarithmic term: as φ_S decreases, the hysteresis decreases due to fewer defects (linear term), but the logarithmic term slowly diverges, resulting in appreciable residual hysteresis. This observation aligns with existing data, yet further quantitative validation and extension to more intricate patterns are imperative.

Hysteresis plays a crucial role in preventing drops from detaching from solids, irrespective of gravitational forces or airflow. Though calculating the sticking force comprehensively is challenging, a simplified approach aids in understanding how hysteresis influences this force (Quéré et al., 1998). Assuming the rear half of the drop adheres to the solid at an angle θ_r and the front half at an angle θ_a, the capillary sticking force $\pi\gamma(\cos\theta_r - \cos\theta_a)$ can be expressed, where l represents the quasicircular radius of the solid/liquid contact for $\Delta\theta << l$.

By assuming $l \approx R\varepsilon$, where ε is the difference between the mean angle and π and R is the drop radius, and employing Equations 6.35 and 6.40, the condition for a drop to move in a gravity field (e.g., on a vertical window pane) is given by

$$\varphi_s^{3/2} \log(1/\varphi_S) < R^2\kappa^2 \tag{6.41}$$

Here, the capillary length $\kappa^{-1} = (\gamma/\rho g)^{1/2}$ (2.7 mm for water) is introduced, and all numerical coefficients are neglected. Once again, the density of defects plays a pivotal role in determining wetting properties, where low densities are essential for suppressing adhesion. However, this limit also compromises the stability of the Cassie state, a consideration discussed subsequently.

6.8.3.3 Metastable Cassie States

As highlighted earlier, drops often exhibit a Cassie state despite possessing a lower Wenzel energy, rendering these metastable Cassie states delicate (Callies and Quéré, 2005). In Figure 6.16, we observe two millimetric water droplets on a microtextured substrate, characterized by a pillar density (φ_S) of 1% and a pillar height (h) of 12 µm. The initial drop settles without impact, presenting a Cassie state with visible

FIGURE 6.16 Millimetric water drops of the same volume on a superhydrophobic substrate covered with dilute pillars ($\varphi_S = 0.01$, $h = 2\,\mu m$). The drop on the right was thrown on the substrate, whereas the one on the left was carefully deposited. As observed, Cassie and Wenzel states (left and right, respectively) can coexist. (Reproduced with permission from Ref. (Y. Y. Yan et al., 2011). Copyright Elsevier, 2011.)

light passing underneath, while the subsequent drop, released from a few centimeters above, attains a Wenzel state upon impact due to the substrate's modest roughness ($r \approx 1.1$), resulting in a CA near Young's angle, slightly exceeding 90° in this scenario.

In cases where the Wenzel state represents the lower energy state, any disruption to a Cassie drop may induce its transition to the Wenzel state. Conversely, a Wenzel drop remains firmly entrenched in its stable configuration. Assessing the robustness of metastable Cassie states and elucidating the factors triggering impalement hold practical significance. Overcoming an energy barrier is imperative to reach the ground state. Evaluating this barrier entails considering the penetration of a Cassie drop. Assuming unchanged liquid/vapor interfaces during descent, the sole alteration in surface energy pertains to the (unfavorable) wetting of post walls, implying a positive energy per unit area

$$\Delta E = (\gamma_{SL} - \gamma_{SA})(r-1) = -\gamma(r-1) \quad (6.42)$$

Consequently, an energy barrier $\Delta E \approx 2\pi bh/p^2\gamma$ cos θ emerges, proportionate to pillar height h, serving as a natural parameter for regulating this parameter. Despite being typically insurmountable by thermal energy alone, external means such as pressure application, substrate vibration (Bormashenko, Pogreb, et al., 2007), or impact (Bartolo et al., 2006; Reyssat et al., 2006) can supply the requisite energy. Greater post height equates to heightened resistance against impalement. Once liquid infiltrates the texture, it remains firmly anchored, with the "printed" drop capable of conforming to the micropost network. Sbragaglia et al. (2007) delineated the Cassie/Wenzel transition dynamics as rapid, employing a zipping mechanism wherein one cavity row fills in approximately 10 μs across 100 μm before progressing to the subsequent row. The surface force involves wet surface formation and suspended liquid/air interface suppression, while the resisting force is predominantly viscous, aligning with the anticipated Washburn law for progression. For patterns with post heights (h) akin to pitch (p) of approximately 10 μm, the invasion time for a row of length x scales as $\eta x^2/\gamma h$, approximately 10 μs for water with $x = 100\,\mu m$.

The mechanism of nucleating solid/liquid contacts for initiating transition warrants attention. Interfaces above air pockets curve to match the drop's overall curvature (Figure 6.17). When the drop diminishes sufficiently, liquid can reach the underlying solid, facilitating propagation. Drop size significantly influences the wetting state, with smaller droplets more prone to the Wenzel state (Chow, 1983), evident in drop evaporation observations. As the drop size fluctuates continuously, investigators have identified a critical radius below which the drop abruptly shifts into the Wenzel regime (McHale et al., 2005).

In Figure 6.17, the curvature of the interface, denoted as δ/p^2 (for $h<p$), follows the notations. The relationship with the drop curvature $1/R$ leads to the determination of the depth of penetration (δ) of the interface within the texture: $\delta \sim p^2/R$. Notably, for smaller drops, δ increases. When δ approaches the pillar height (h), a solid/liquid contact may initiate on the substrate and propagate if the Cassie state is metastable. This implies a critical radius for a Cassie drop, scaling as (Reyssat et al., 2007)

$$R^* \sim p^2/h \tag{6.43}$$

It is essential to acknowledge that, according to the previous insights, depinning from the edge occurs if the drop radius is smaller than $p/|\cos\theta|$, which is a limiting condition for moderate chemical hydrophobicity. We assume that this second condition is screened by the first one, i.e., $p>h/|\cos\theta|$.

The Cassie state's robustness is enhanced when the critical radius R^* is small, preventing drops from transitioning into Wenzel drops. This can be achieved in two ways: by increasing h, using micro- or nanofibers to decorate solids (Reyssat et al., 2007), or by reducing both p and h. Miniaturizing the pattern size strengthens the resistance of the Cassie state, possibly explaining its existence in various natural materials.

Recent findings by Jiang and coauthors (Gao et al. 2007) reveal that the eyes of *C. pipiens* remain dry when exposed to tiny drops in foggy and moist environments. Although water condenses on most of the animal, the eyes remain dry, preserving their excellent vision. Microphotographs of the eye surface textures show remarkably

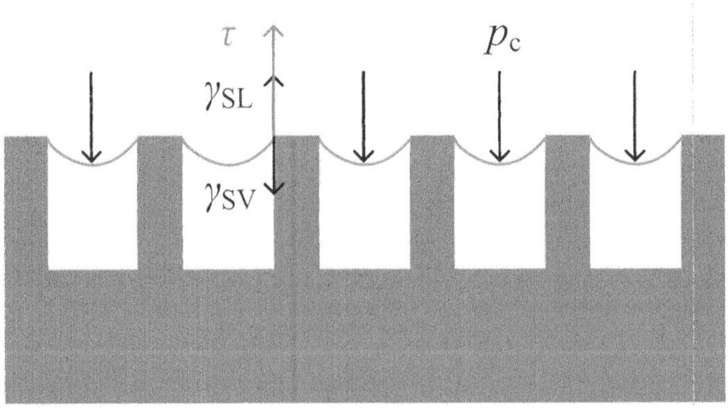

FIGURE 6.17 The liquid/vapor interface is curved, owing to the curvature of the drop, but the interface can also be curved if we apply pressure to the drop. (Reproduced with permission from Ref. (Cai et al., 2016). Copyright Springer Nature, 2016.)

small features, with $p \sim h \sim 100\,\text{nm}$. With these values, $R^* \sim 100\,\text{nm}$, making a drop at this scale is not only invisible but also quasi-instantaneously evaporating. In clouds, drops are polydisperse with a typical radius of 10 µm—small drops like these might resist the Wenzel transition on the nanopattern worn by the mosquito's eye.

Promising metastable Cassie states are observed on hydrophilic materials with specific designs. Oils with CAs around 40° can be suspended on special textures, resulting in an angle increase of approximately 100° (the superoleophobic effect) (Woodward et al., 2003). Herminghaus demonstrated that fakir wetting drops are achievable on overhangs or re-entrant angles, where a hydrophilic Young's condition can be satisfied. Fibers and defects with overhangs provide robust oleophobicity, resisting a Wenzel transition by applying Laplace pressure related to the size of the "holes" on the surface. Structures with pronounced re-entrant profiles reported by Tuteja et al. (2007) show that even a highly wetting liquid like octane can be suspended in a Cassie state with ACA and RCA as high as 160° and 140°, respectively.

6.8.4 Synthetic Micro-Textures

At a scale significantly smaller than the capillary length, patterns on a hydrophilic solid can induce superhydrophilicity. The Wenzel effect, focusing on surface roughness, plays a key role in enhancing hydrophilicity when the liquid accommodates the pattern, as illustrated in Figure 6.18. This enhancement coexists with the conventional partial wetting scenario, leaving the remaining solid area dry. However, these structures go beyond merely enhancing hydrophilicity; they can also serve as guides for the liquid within the array, resembling a wicking process. It is important to note that the observed phenomenon deviates from classical wicking and is aptly termed hemi-wicking. As the liquid film advances through the microstructures, it establishes an interface with air, potentially leaving a few dry islands in its wake.

Microstructures distributed in regular or irregular shapes in uniform or non-uniform patterns are seen in artificially created textured surfaces as shown in Figures 6.18 and 6.19. It is particularly used to cater to specific applications like

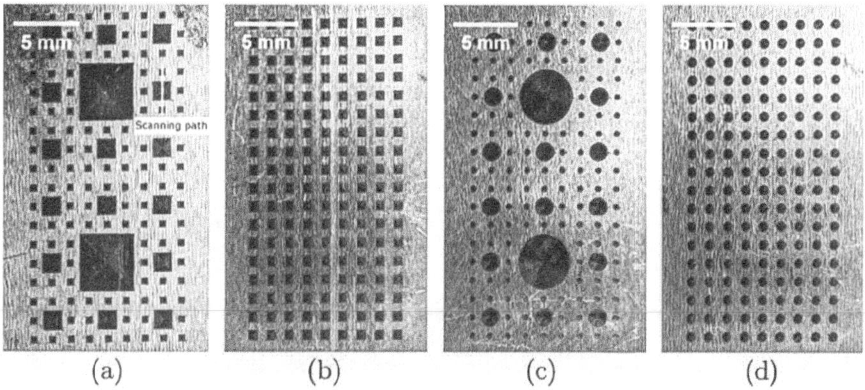

FIGURE 6.18 Textured surfaces with different patterns. (Reproduced with permission from Ref. (Wang et al., 2016). Copyright Elsevier, 2016.)

optics, surface energetics, lubrication, bioengineering, and thermodynamics, among a few (Jithin and Joshi, 2021). Many methods are used to create surface texturing. Some of the methods are chemical vapor deposition (CVD), laser beam texturing, electrical discharge texturing, direct laser texturing, etc.

Surface texturing is mainly needed to improve the use of the material in terms of biocompatibility, protein absorbance, and increase in the rate of osseointegration. Several studies show better performance in biocompatibility, particularly in Ti6Al4V implants (Sadeghi et al., 2019; Shibe and Chawla, 2014; Marla et al., 2018). In recent years, laser texturing is highly used, for example, picosecond laser was used to create different textures on Ti, stainless steel, Cu, Al_2O_3, and Al. The results showed that it creates a lower heat load and reduced heat-affected zone, and higher energy intensities can be achieved due to short pulse length and early evaporation of materials can be achieved through rapid heating.

Figure 6.18 shows a schematic representation of the arrangement of the circular and rectangular regular fractal surfaces. It is seen that the regularly spaced textures steeply reduce the friction coefficient in a mixed lubrication state. Figure 6.19 shows the laser texturing of Ti6Al4V and AA2024-T3 alloy surfaces with different patterns. Laser-textured surfaces provide better superhydrophobic surfaces compared to blank surfaces in terms of durability and robustness (Ahuir-Torres et al., 2018).

Figure 6.20 exhibits SEM images of an Al alloy with a CA of ~0° since there is a three-dimensional (3D) capillary effect (Gentleman and Ruud, 2010). Important parameters such as pulse-to-pulse overlap, pattern size, geometry, different structure depths, and roughness cause an effect on the wetting behavior of the surfaces after

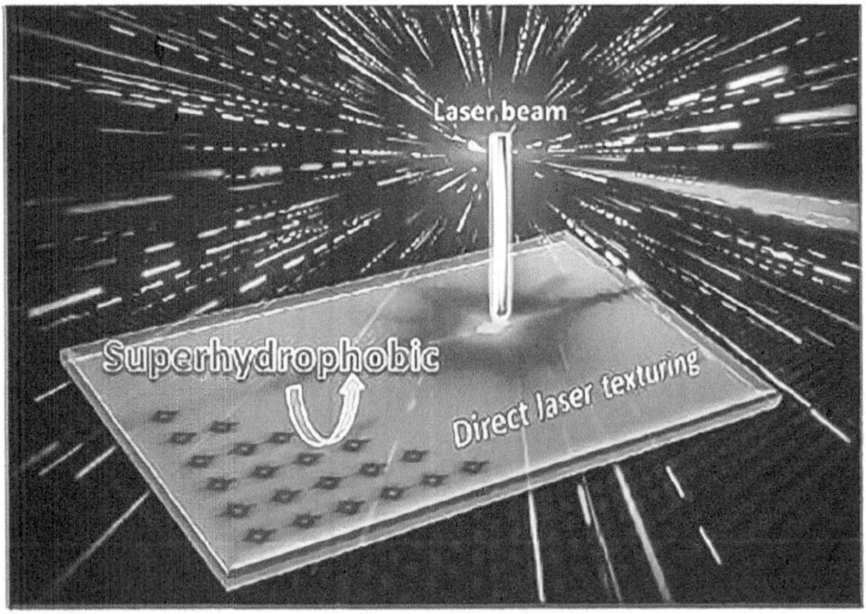

FIGURE 6.19 Schematic illustration of laser texturing of metal surfaces. (Reproduced with permission from Ref. (Tong and Xiong, 2022). Copyright Elsevier, 2022.)

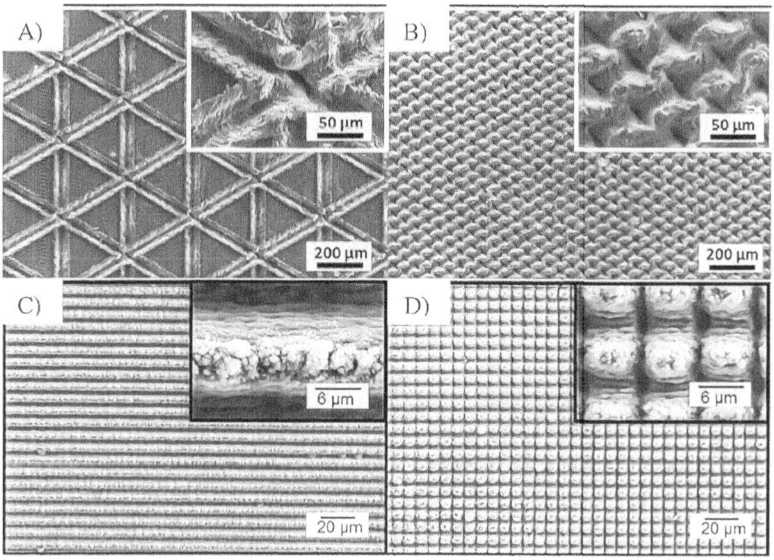

FIGURE 6.20 SEM images of laser-textured Al2024 surfaces with different processing parameters and different patterns. (Reproduced with permission from Ref. (Milles et al., 2019a, 2019b). Copyright Elsevier, 2019.)

laser texturing. Yu et al., used a line scanning method to mimic a superhydrophobic "lotus effect" surface and showed a CA of 180°. The main parameters in this research are an average laser power of 10 W, a scanning speed of 500 mm/s, a repetition rate of 20 kHz, and an incident angle of the laser of 90°.

Figures 6.21 and 6.22 are SEM images of Ti alloys by high-speed WEDM that have uniform height and a regular curved shape. Figure 6.21a and b presents a curved microarray and its side view. Figure 6.22 presents the microstructure of the planar

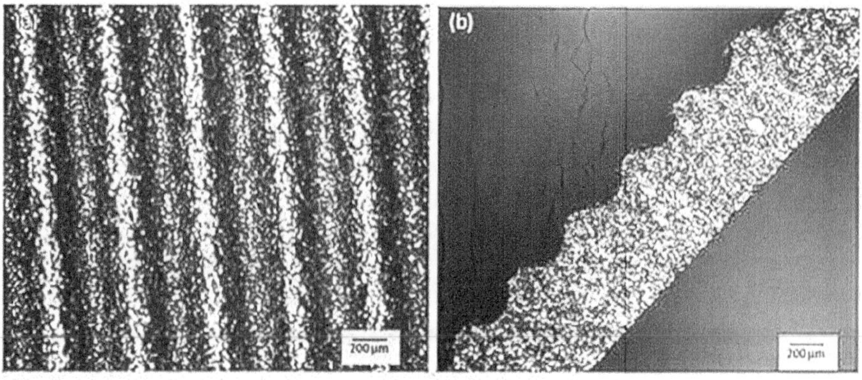

FIGURE 6.21 Optical image of (a) curved microarray and (b) side view of curved microarray. (Reproduced with permission from Ref. (Prakash et al. 2022). Copyright Springer Nature, 2021.)

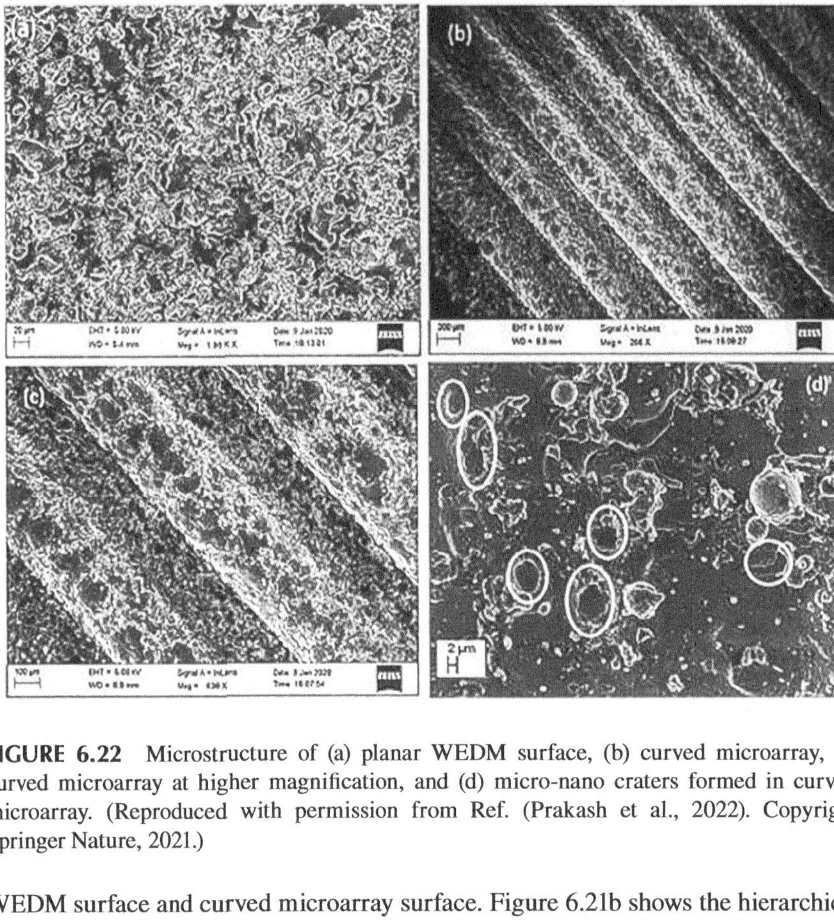

FIGURE 6.22 Microstructure of (a) planar WEDM surface, (b) curved microarray, (c) curved microarray at higher magnification, and (d) micro-nano craters formed in curved microarray. (Reproduced with permission from Ref. (Prakash et al., 2022). Copyright Springer Nature, 2021.)

WEDM surface and curved microarray surface. Figure 6.21b shows the hierarchical structure, which is uniformly fabricated. The high-frequency electric field generated between the wire and the workpiece is used to create patterns. The molten metal spatter and effect generate very small size craters on the surface of the workpiece and electrode. The spattered metal immediately becomes solid on the surface of the sample to form micro-nano craters with irregular structures and sizes. These patterns change the wettability and antibacterial property of the surface (Prakash et al., 2022).

The focal point in recent scholarly works has been the development of micro-textured surfaces characterized by specific wetting attributes, often leaning toward superhydrophobic characteristics. Various methodologies have been explored to fabricate such materials, encompassing both sophisticated and rudimentary techniques. For instance, exposing a glass surface to flame results in the deposition of soot, inducing a noticeable darkening of the glass. Subsequent application of water on this treated substrate manifests a distinctive behavior, resembling that of a pliable solid, with water droplets exhibiting rolling and bouncing dynamics upon the surface.

Broader techniques such as template synthesis, phase separation, diverse etching processes, crystallization, and electrospinning of microfibers have been proposed for the assembly of more intricate materials (Nakajima et al., 2001). Consequently, a plethora of surface textures has been achieved, ranging from highly disordered or fractal to meticulously ordered and well-defined structures, as illustrated in Figure 6.23. The majority of these surfaces exhibit unique wetting characteristics, prompting the exploration of optimal surface features for specific applications. The determination of the "best" surface is contingent upon the desired properties, a consideration that is subsequently deliberated upon in the following discourse.

6.8.4.1 Grooves

Emphasized earlier is the intrinsic striation of many solids due to grooves (Figure 6.23e–h), which can be intentionally etched for specific applications like directional wetting. Consideration is given to a rectangular groove with a width denoted as "w" and a depth of "δ," while the intricacies of distinct menisci are disregarded. To witness the spontaneous infiltration of the groove, the solid must decrease its energy through wetting ($\gamma_{SL} < \gamma_{SA}$). However, this alone is insufficient, as a liquid/vapor interface forms at the top. The anticipated flatness of the interface minimizes the corresponding surface area. Consequently, for a liquid advancement within the groove by a differential distance dx, the alterations in surface energies are expressed as (disregarding gravitational effects):

$$dE = (\gamma_{SL} - \gamma_{SA})(2\delta + w)dx + \gamma w dx \tag{6.44}$$

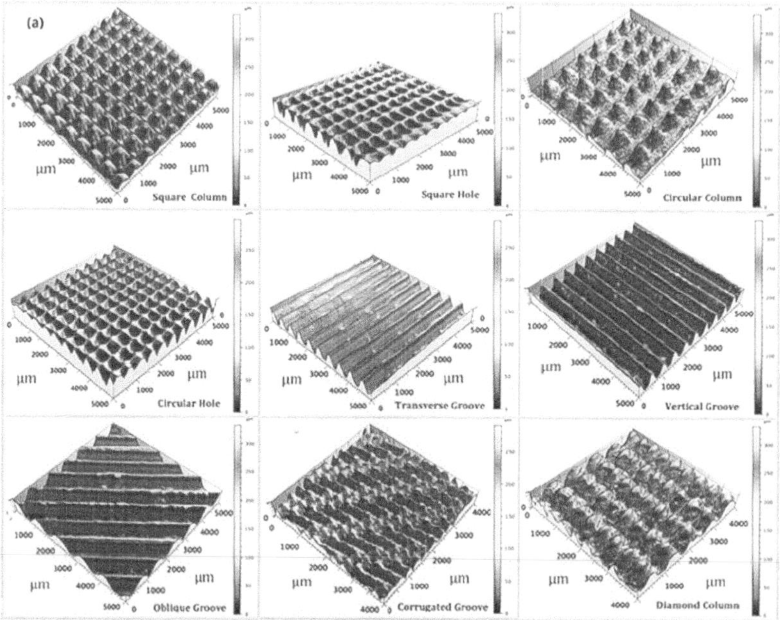

FIGURE 6.23 Different examples of synthetic microtextured surfaces. The three different surfaces (spikes, shallow cavities, and stripes) are crenelated on a micrometer scale. (Reproduced with permission from Ref. (He et al., 2021). Copyright Elsevier, 2021.)

Utilizing the Young equation, it is deduced that liquid progression is favorable ($dE < 0$) when the condition $\theta < \theta_c$ is met, where

$$\cos\theta_c = w/2\delta + w \tag{6.45}$$

Regardless of the values assigned to w and δ, this quantity, dependent solely on the aspect ratio δ/w, establishes a numerical range between 0 and 1. Within this range, the angle θ_c can be explicitly determined. In scenarios where the groove is narrow and/or deep (small w and/or large δ), the criterion for spontaneous penetration aligns with the classical porous medium ($\theta_c = 90°$). In a more general case ($\delta \sim w$), θ_c varies between 0° and 90°, implying that impregnating a groove is more challenging than a 3D porous medium. This challenge intensifies when the groove is shallow, as δ/w approaches 0, causing θ_c to converge to the criterion of complete wetting on a flat solid.

6.8.4.2 Assembly of Pillars

The discussion surrounding surfaces adorned with micro-posts, as illustrated in Figure 6.24, presents various arguments. The surface characteristics are defined by the pillar density, denoted as φ_S, and the level of roughness, denoted as r. In the case of an ethanol drop residing on a forest of pillars (Ishino et al., 2007), the drop assumes a lens shape, influencing the colors produced by the organized array of micro-posts. Notably, the drop is encircled by a film of ethanol, indicating the occurrence of hemi-wicking. In specific scenarios, the film conforms to the micropattern, adopting a square shape on a square array of micro-posts (Courbin et al., 2007). The impregnating front's propagation is portrayed in Figure 6.10, where the solid is coated by the liquid over a surface area proportional to $r - \varphi_S$. Simultaneously, the flat liquid/vapor interface evolves over a surface area proportional to $1 - \varphi_S$.

To analyze the energy variations per unit length perpendicular to the figure during the progression of a film by a distance dx (assuming it surpasses the defect scale), the expression is given by

$$dE = (\gamma_{SL} - \gamma_{SA})(r - \varphi_S)dx + \gamma_{LV}(1 - \varphi_S)dx \tag{6.46}$$

For favorable progression, it is crucial that the Young angle is smaller than a critical value θ_c, dependent solely on the solid's design (Bico et al., 2001):

$$\cos\theta_c = (1 - \varphi_s)/(r - \varphi_s) \tag{6.47}$$

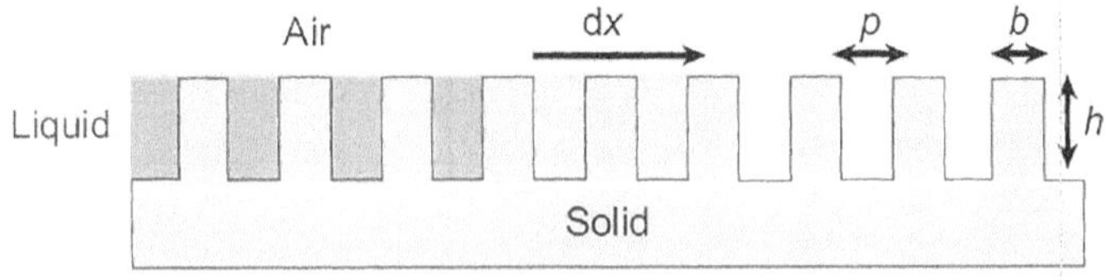

FIGURE 6.24 Liquid film (in dark grey) propagating on a solid, within a forest of micro-posts of height h, mutual distance p, and radius b. The condition for the progression is deduced from the variation of surface energy associated with it. (Reproduced with permission from Ref. (Quéré 2008).

The invasion of liquid on a micro-structured solid is thereby influenced by the geometry of the structures. In cases of dilute defects (small φ_S), cos θ_c is approximated as $1/r$. Consequently, the rougher the substrate, the larger θ_c, indicating a higher likelihood of hemi-wicking. For substrates comprising disconnected defects, like posts, the liquid front needs activation to traverse the jumps outlined in Figure 6.24. Wettability of liquids facilitates this process through the formation of menisci around each post, allowing the liquid to progress to the next row. Alternatively, the contact line may remain pinned in a metastable Wenzel state, necessitating an external energy source (e.g., vibrations) to initiate contact with the next rows of pillars. Equilibrium scenarios may even be envisioned where the drop coexists with a wet ring of finite extension, resembling a fried egg. For instance, if the energy barrier is surmounted due to Laplace pressure, the progression halts once the drop spreads sufficiently, reducing its Laplace pressure and impeding further motion.

The second regime, occurring for $\theta<\theta_c$, can be elucidated as a consequence of hemi-wicking. In this experimental scenario, the solid substrate exhibits pronounced roughness, characterized by a fractal structure. While the exact value of the parameter φ_S remains unknown, an anticipation is made that it is smaller than 1. Consequently, cos θ_c is expected to be on the order of $1/r$. The initiation of the second regime is in close proximity to the abscissa where the Wenzel regime (with a slope of r) intersects the line $\cos\theta^* = 1$, specifically at $\cos\theta \approx 1/r$. The apparent angle θ^* exhibits minimal dependence on θ, a qualitative understanding of which can be derived from the composite nature of the surface. This composite surface predominantly comprises liquid, interspersed with a few solid islands. The angle θ^*, while approaching 0 (the value in the absence of solid islands), cannot reach this theoretical minimum due to the presence of islands where the angle is $\theta>0$. The number of islands is expected to be a function of θ, posing challenges for the formulation of a general theory.

Considering a displacement of the contact line by a distance dx, the solid undergoes wetting on a surface fraction φ_S, and liquid interfaces are eliminated on a complementary fraction of $1-\varphi_S$. As the displacement leads to an increase in the liquid/vapor interface of the droplet, the overall change in surface energies per unit length of the line is expressed as

$$dE = \varphi_S\left(\gamma_{SL} - \gamma_{SA}\right)dx - \left(1-\varphi_S\right)\gamma dx + \gamma dx\cos\theta^* \quad (6.48)$$

The minimum of E provides the apparent angle θ^*, as originally demonstrated by Cassie (Cassie and Baxter, 1944; Li and Shan, 2012):

$$\cos\theta^* = 1 - \left(1-\varphi_s\right)\cos\theta \quad (6.49)$$

For small φ_S, the angle θ^* exhibits minimal variation with θ, consistent with the observations in Figure 6.13 for $\theta<\theta_c$. The relationship appears linear when the cosines of both angles are plotted, and the slope offers a means to determine φ_S. However, for a disordered surface, the actual behavior is anticipated to deviate from linearity. In the case of a disordered surface, the proportion of emerged islands is expected to be a function of θ, introducing complexity to the variation $\theta^*(\theta)$. The smaller the θ, the smaller the φ_S, contributing to the intricate nature of the actual relationship.

6.8.4.3 Dimples

Manufacturing surfaces with controlled micro-geometry features presents an effective strategy for enhancing the tribological performance of sliding interfaces. Incorporating artificial micro-dimples or micro-grooves onto surfaces has been shown to benefit tribological pairs across various conditions, ranging from boundary to hydrodynamic states (Kovalchenko et al., 2004). Numerous investigations have demonstrated substantial enhancements in friction coefficient, wear resistance, and load-carrying capacity of mechanical components subjected to tribological stresses by introducing appropriate surface textures. The notable improvement can be primarily attributed to the role of dimples in textured surfaces as reservoirs for lubricating oil, ensuring adequate lubrication even under boundary conditions. Moreover, each dimple provides a protective pocket for embedding wear particles, mitigating severe surface wear. Furthermore, dimples function as micro-hydrodynamic bearings during full or mixed lubrication, generating additional hydrodynamic pressure to enhance load-carrying capacity. Etsion et al. (1999) conducted comprehensive studies on these parameters, both theoretically and experimentally, highlighting the criticality of the dimple depth to diameter ratio.

Recent investigations by Raeymaekers et al. (2007) and Yan et al. (2010) have underscored the significant influence of the dimple area ratio, representing the ratio of dimples to surface areas, on the tribological properties of surface texture. Kligerman et al. (2005) explored the impact of texture design parameters on piston rings, determining optimal values for texture depth and density to minimize friction force. Wang et al. (2009) investigated the effect of dimple size on friction under line contact conditions, revealing that patterns with smaller dimple diameters exhibit greater friction reduction compared to those with larger diameters. In a study conducted by Yu et al., dimples arranged in a matrix pattern were examined, with the lubricant flow direction perpendicular to the major axis of elliptical dimples or the sides of square dimples. Friction tests were conducted using a reciprocating friction test rig to assess the influence of different dimple shapes on the tribological performance of sliding surfaces. The upper specimens' surfaces were textured with circular, square, and elliptical dimples, respectively.

Based on the findings, it is evident that elliptical dimples exhibit superior friction reduction effects compared to both square and circular dimples under varying test loads. This superiority can be attributed to the larger converging wedge present in elliptical dimples, facilitating the formation of a lubricant film between surfaces, thus preventing asperity contact more effectively. Conversely, square dimples rank next in terms of friction reduction, followed by circular dimples. The minimal difference in friction reduction between square and circular dimples is likely due to their similar dimple sizes and shapes, especially when the dimple area remains constant, leading to comparable effects on lubricant film generation. These observations highlight the importance of dimple shape and size in optimizing friction reduction in surface engineering applications.

6.8.5 Surface Chemistry

Under surface laser treatment (direct laser texturing), the titanium surface exhibits the following observations in the aspect of surface chemistry. There is an increased carbon content because of the accumulation of carbon. It is also predicted that the generation of new hydrophobic functional groups, such as -CH_3 or graphite carbon, produced by laser ablation on the surface of Ti alloys creates a hydrophobic surface and decomposition of carbon dioxide in air during laser processing (Bizi-Bandoki et al., 2013). There was a decrease in Ti content, indicating the observation of TiO_2. But TiO_2 does not produce a hydrophobic layer as such; rather, the presence of carbon on the surface provides an additional effect on the hydrophobic behavior of the surface (Rukosuyev et al., 2014). Therefore, the wettability of the surface is changed by the chemical composition after surface laser processing. The influence of surface chemical composition on wettability will continue from the beginning of laser processing to the stabilization of the surface. The surface morphology on wettability is formed directly after laser treatment and will not change with time. Also, the change of surface wettability over time is because of the chemical composition of the Ti surface. Indeed, surface morphology and surface chemistry are considered the most important factors in the case of laser-induced modification.

Huang et al. (2022) studied the superhydrophobic surface processing through chemical immersion treatment (CIT) of metals and discussed the detailed areas of surface chemistry. CIT includes an ethanol solution of 0.25 wt% chlorosilane reagent [$CF_3(CF_2)_9(CH_2)_2SiCl_3$]. -$CF_3$ and -CF_2 groups were identified in the XPS analysis. There is an increase in the surface hydrophobicity due to the presence of fluorocarbon functional groups, therefore reducing the surface energy. The fluorine species distribute randomly over the surface, showing fluorocarbon functional group adsorption over the entire surface. Laser texturing induces oxidation and hydroxylation (Gregorčič et al., 2017), which induces a chemical reaction between CIT and surface oxides. The oxides and hydroxides formed on the surfaces act as potential sites for chemical reactions, resulting in different wetting regimes. The adsorption of fluorine will reduce the adhesion of water and enhance hydrophobicity, providing a lotus effect.

Avik Samanta et al. (2020) studied the design of chemical surface treatment for the laser-textured alloy to achieve different wetting behavior. The processing was achieved by the combination of a nanosecond pulse texturing in air and CIT processes. The mechanism of chemical molecules on the processed surface is presented and explained in Figure 6.25. Here, the reactive group is $SiCl_3$

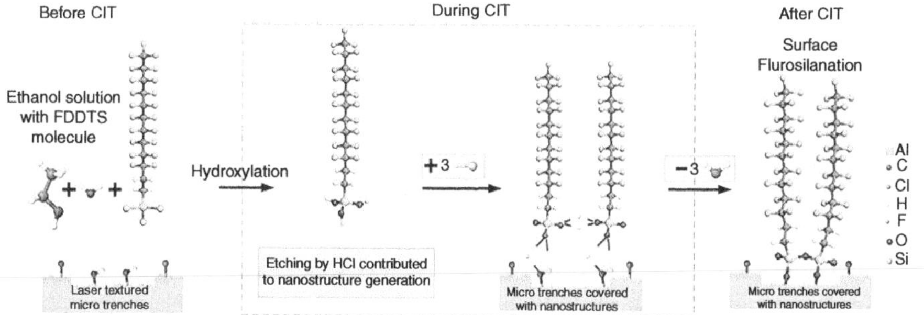

FIGURE 6.25 Schematic representation of the mechanism of attachment of chemical molecules on the laser-textured surface during the CIT process. (Reproduced with permission from Ref. (Samanta et al. 2020). Copyright ACS Publications, 2020.)

with $-(CH)_2(CF)_2CF_3$ group in which Cl is replaced with –OH by the hydroxylation process, and HCl is generated during the chemical reaction of fluorosilane molecules. The chemical attachment on the surface leads to an increase in alkalinity (fluorination or chlorination) and results in very low dispersive and nondispersive parts of the solid SFE. Therefore, the treated surface can become hydrophobic both to oil and to water. Also, the superoleophobic was achieved on treated surfaces where the θ for all the liquids (low surface tension) is more than 150°. The functional groups present on the surface lead to alkalination due to the CIT process. Surface alkalinity led to a reduction in SFE (Chen et al., 2016; Chang et al., 2010) of the textured solid surface, resulting in superoleophobicity.

Similar to this, there are superoleophilicity and superhydrophilicity. The surface texture was processed using 3-cyanopropyltricholosilane (CPTS) consisting of a highly polar functional group like a triple covalent bond of carbon with highly electronegative nitrogen (C≡N). Here, the nitrogen pulls the electrons toward itself and induces polarity. In the presence of water, the O (oxygen) pulls the electron toward itself, making the 2 H more positive. The slightly positive hydrogen atom in water reacts with a negative nitrogen atom and forms a hydrogen bond. The treated surface contains hydrogen bonding and dipole–dipole attraction, which contributed toward superhydrophilicity. Dual-scale structures with water-affinitive chemistry improve superhydrophilicity by allowing fluids to pass through higher-scale grooves, and the capillary pressure was maintained at lower-scale structures. On these surfaces, the water droplets immediately spread at a fast rate. If the surface chemistry does not have long-chain hydrocarbon or fluorocarbon, the dispersive component of the SFE is relatively high. Therefore, not only is surface polarity on the treated surface, but it also has a strong capillary effect due to the presence of grooves (Raphaël, 1989). Therefore, water and diiodomethane molecules can move fastly through the surface hierarchical structures because of the capillary effect shown in Figure 6.26. Superhydrophilicity is additionally shown to be the capability of hydrogen bond formation with the polar nitrile group.

6.8.6 Type and Properties of Interacting Liquids

Different liquids have distinct chemical properties, such as polarity and molecular structure, which can influence their interaction with the material surface. Polar liquids, like water, tend to have higher surface tensions and can result in higher SFE compared to nonpolar liquids (Ming Li et al., 2022a; Vialetto et al., 2022; Han et al., 2022). Yusof et al. (2022) studied material in the presence of phosphate-buffered saline solution and understood the dissolution of borates. The assessment revealed that a drastic reduction of pH value is seen due to the hydrolysis of B_2O_3 from glass, inducing variation of boron acid derivatives. Similarly, Suvin et al. (Edachery, Ravi, et al., 2022) studied using low-toxicity, environmentally friendly, and biodegradable green cutting fluid, namely "Cocos Nucifera," on AA5052, Ti6Al4V, and EN31. The studies compared commercial cutting fluids and green cutting fluids and showed that green cutting fluid increases wetting by 36% in AA5052, 43% in Ti6Al4V, and 43% in EN31 compared to commercial fluids (Figure 6.27). The fluids predominantly depend on surface roughness. In all three samples studied, the fluid wettability changed differently for different roughness. Also, the results showcased that there is

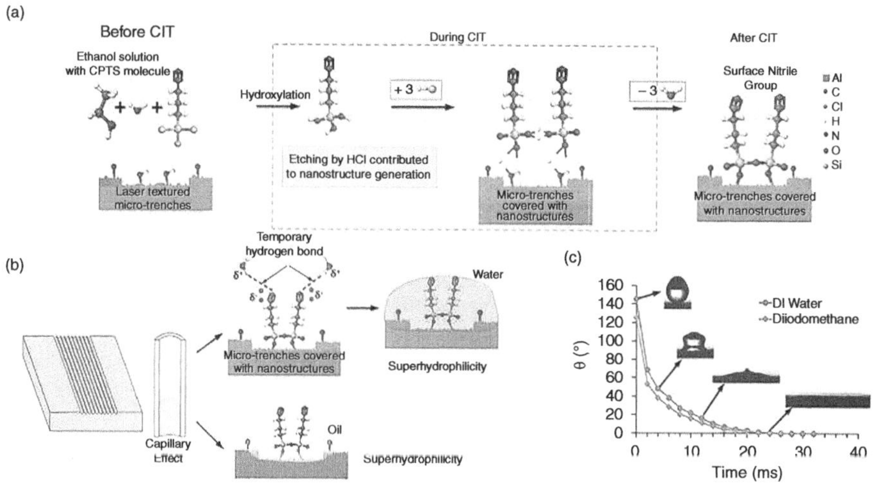

FIGURE 6.26 Scheme of a superhydrophobic and superoleophobic AA6061 surface. (a) Schematic representation of the mechanism of attachment of CPTS molecules on the laser-textured surface during the CIT process. (b) Combination of temporary hydrogen bond formation and capillary effect leading to superhydrophilicity and superoleophilicity. (c) Spreading of water and diiodomethane droplets on the treated surface. (Reproduced with permission from Ref. (Samanta et al., 2020). Copyright ACS Publications, 2020.)

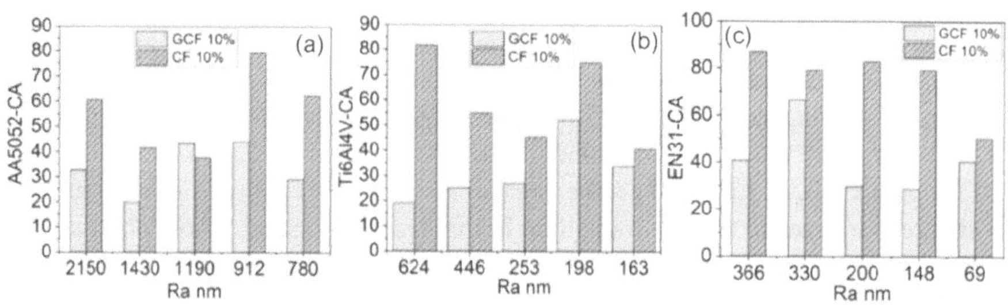

FIGURE 6.27 Comparison between GCF10% vs CF10% on (a) AA5052, (b) Ti6Al4V, and (c) EN31. (Reproduced with permission from Ref. (Edachery, Ravi, et al., 2022). Copyright Elsevier, 2022.)

a higher wetting characteristic at a 10% concentration of green cutting fluid, providing a sustainable and highly effective substitution for commercial ones (Figure 6.28).

Figure 6.28 shows CA measurements of green cutting fluids and commercial fluids on different materials. The rise in the concentration of the green fluid lubricants shows increased wettability as compared to the commercial fluid. Evidently, GCF benefits not only the environment but also provides a better alternative to commercial fluids to help machining processes.

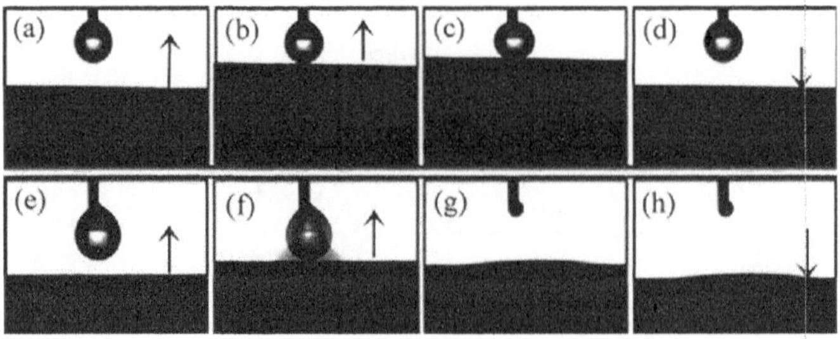

FIGURE 6.28 Approach, contact, deformation, and departure process of a 5 μL water droplet suspended on a syringe to the surfaces: (a–d) the as-prepared superhydrophobic surfaces; (e–h) the air-plasma-treated surfaces. The arrows represent the moving direction of the substrate. (Reproduced with permission from Ref. (Zhu et al., 2010). Copyright ACS Publications, 2010.)

6.9 APPROACHES FOR IMPROVING WETTABILITY (HYDROPHILICITY AND SUPERHYDROPHILICITY)

To improve wettability and surface energy, several surface modification approaches are used. The discussion of which are included herein.

6.9.1 Surface Treatments

Surface treatment is a surface modification technique that can be used to improve the wettability and surface energy of materials. It involves exposing a material surface to a low-temperature plasma, which generates reactive species that can interact with the surface (Bakhshzadmahmoudi et al., 2022; Carrino et al., 2002; Shin et al., 2022). Zhu et al. (2010) studied the rapid formation of superhydrophobic surfaces with a response to wettability transition. In this research, superhydrophobic surfaces were created using a straightforward solution-immersion process, followed by stearic acid chemical modification. The minimum processing time required to fabricate a superhydrophobic surface was 1.5 minutes. By alternating between air-plasma treatment and stearic acid coating, a reversible transition between superhydrophobic and superhydrophilic states was achieved. The complete wettability transition occurred within a short timeframe of 2 minutes.

The samples exhibited a rapid reversible transition in surface wettability by alternating between air-plasma treatment and stearic acid coating, allowing for quick changes between superhydrophobicity and superhydrophilicity. To illustrate this, we took the sample surface obtained by immersing it in NaOH and $(NH_4)_2S_2O_8$ aqueous solution for 0.5 minute as a model system. Before plasma treatment, it was observed that a water droplet (5 μL) suspended on a syringe had difficulty being pulled down to the surface in all cases, despite significant deformation shown in Figure 6.28. However, on air-plasma treatment, the superhydrophobic surface is converted to a superhydrophilic one. When a water droplet was dropped onto the surface, it immediately spread out within 0.3 second. Subsequently, when the plasma-treated surface was immersed in the stearic acid solution, the superhydrophobicity of the surface was restored, resulting in a CA of 159°. The entire wettability transition process was accomplished within just 2 minutes. This rapid wettability transition is crucial to process the application of surface in small devices, like transport fluids, water-proof coatings for moisture-sensitive devices, and rapid water motion. The good reversibility of the surface wettability was observed in Figure 6.29.

The wettability and SFE change of diamond films, triggered by different methods of surface treatment such as hydrogenation and oxidation, have been studied by Ostrovskaya et al. (2002). The results showcased an oxidation film surface that provides desorption of the hydrogen and thermal destruction of the surface. The wetting angle gives informative characteristics of the structure and state of the carbon surface film materials (roughness and chemical nature of surface functional groups).

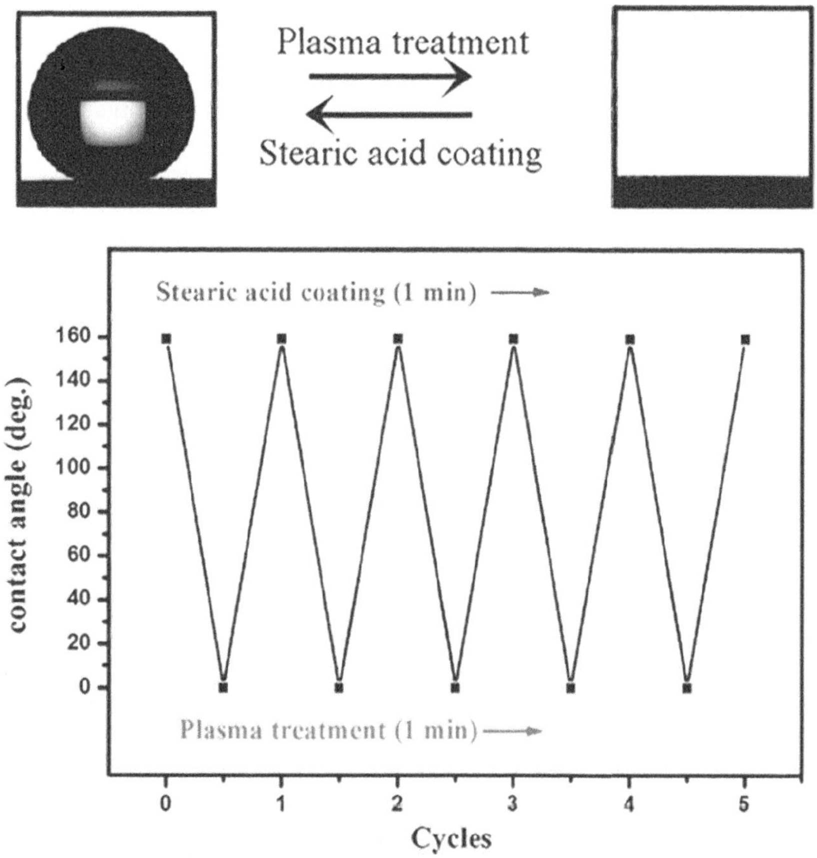

FIGURE 6.29 Reversible wettability transition can be rapidly realized through plasma treatment and stearic acid coating. (Reproduced with permission from Ref. (Zhu et al., 2010). Copyright ACS Publications, 2010.)

6.9.2 Plasma-Sprayed Crystalline Coating

In recent times, laser beam machining, chemical etching, sol–gel processing, nano-imprinting, emulsions, etc. have been widely used to change the hydrophobicity on the metal surface (Li et al., 2016; Bayer et al., 2010; Li et al., 2018; Liao et al., 2014). But all these methods are different in creating the hydrophobic surface in a single step. Bijalwan et al. (2022) studied to understand the wetting characteristics of hydrophilic high phosphorous pig iron metallic powder. Crystalline Fe-based powder is used to prepare the hydrophobic coating. The coating is done using the plasma spray technique assisted with a shroud for inert gas flow using N_2 and Ar gases. The primary gases N_2 and Ar are used because of their role in manipulating the surface characteristics, particularly wettability. During coating, the wetting characteristics are changed from hydrophilic to hydrophobic just by changing the type of encompassing inert gas and the gas flow rate. A hydrophobic surface could be fabricated using the same intrinsically hydrophilic starting material and without the use of any surface modifier that can affect its mechanical and thermal robustness. The results showed that the coating was fabricated on the metal surface by plasma spraying. The coatings were fabricated with crystalline as well as amorphous Fe-based powder. The variation of the coating wettability, from hydrophilic to hydrophobic behavior, was carried out in a single-step approach. The fabrication of nano/micropillar hierarchical structures on the coating surface provided hydrophobicity in the coating. Apart from this, roughening the top surface of the hydrophobic coatings on emery paper, their hydrophobic characteristics were also retained. Therefore, this is the best coating in a single-step approach to tuning the wettability of the coating from hydrophilic to hydrophobic behavior. Figure 6.30 presents the works of Anjaneyulu Udduttula et al. (2022), where the group synthesized nanostrontium phosphosilicate bio-ceramic with mimicking features as natural apatite ($Ca_5(PO_4)_3OH$:HAP) in the form of physical and chemical properties (Udduttula et al., 2022).

6.9.3 Plasma Immersion Ion Implantation

In plasma immersion ion implantation, the total immersion of the target material in plasma during ion implantation ensures uniform ion bombardment, even on devices with irregular geometry (Anders, 2000). Ion implantation can introduce energetic ions into materials to create lattice defects, and the coating hardness can be highly improved (Chason et al., 1997). Her-Hsiung Huang et al. (2021) studied the nitrogen plasma immersion ion implantation treatment of Ti6Al7Nb alloy, as shown in Figure 6.31. The work presented an in vitro analysis of the surface coatings like corrosion performances in simulated body fluids and biological responses, such as protein adsorption and bone cell growth. We also performed in vivo analysis in terms of initial osseointegration (i.e., bone-implant contact) two weeks after implantation in the femur of mini-pigs. The results also showed an increase in surface roughness, hydrophilicity, hardness, and corrosion resistance in simulated body fluids. The formation of TiN leads to the adsorption of fibronectin and the adhesion, proliferation, and mineralization of human bone marrow mesenchymal stem cells. The osseointegration was enhanced after implantation for two weeks in the femur of mini-pigs.

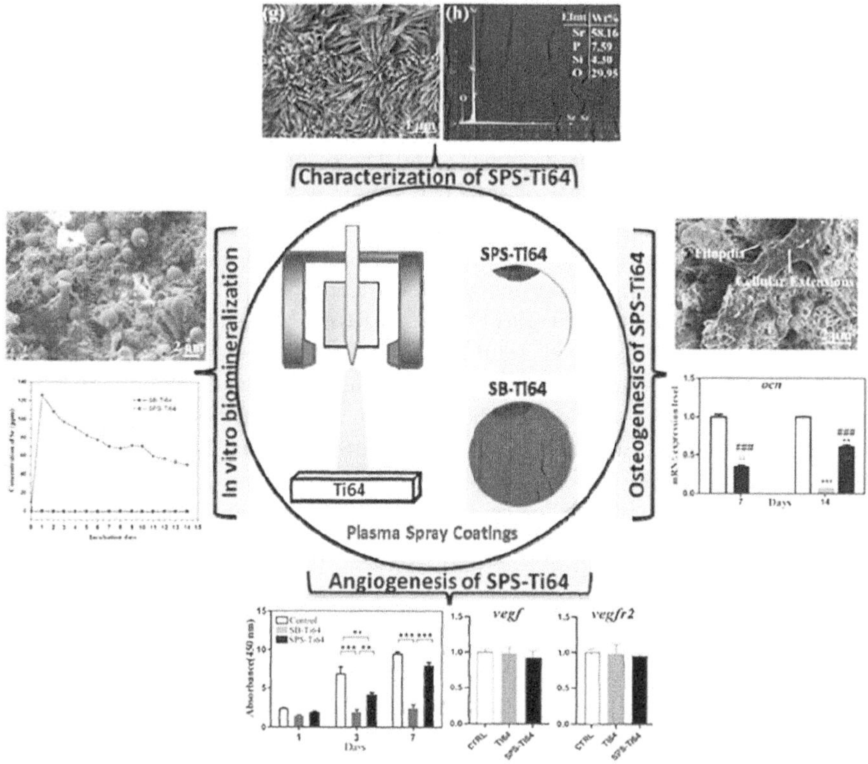

FIGURE 6.30 Graphical presentation for plasma-sprayed novel apatite-inspired $Sr_5(PO_4)_2SiO_4$ coating on Ti64 and its in vitro biological studies. (Reproduced with permission from Ref. (Udduttula et al., 2022). Copyright Elsevier, 2022.)

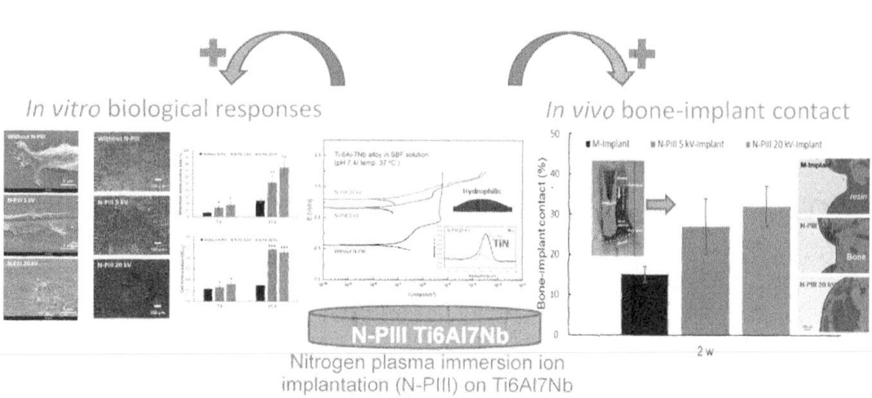

FIGURE 6.31 Graphical presentation of the plasma immersion ion implantation on Ti6Al7Nb. (Reproduced with permission from Ref. (H.-H. Huang et al., 2021). Copyright Elsevier, 2016.)

6.9.4 Selective Laser Melting (SLM)

Optimal wettability is a critical condition for effectively joining dissimilar metals. To enhance the surface properties of metals, the proposed method involves prefabricating surface textures. Chen et al. (2016) conducted a wetting study of Sn-Ag-Cu solder balls on a laser-textured copper plate. The results showed CA decreases by 31.9° when the surface texture on the copper plate was introduced with an appropriate size. The optimized surface texture and prediction of the wetting type were realized through theoretical calculations. Yiming Sun et al. (2023) studied the wettability and spreading behaviors of laser heating of AlSi5 on laser-textured DP980 steel. Surface morphologies and the diffusion behavior of the alloying elements were analyzed under the textured interface with different line spacings. The mechanism of surface texture in the wetting process was understood. The results presented that the wetting and spreading of Al on textured DP980 steel were different from those of untextured samples, and the wettability of Al drops on textured steel exhibited marked anisotropy. In the direction perpendicular to the groove texture, larger CAs and smaller spreading lengths were obtained. Therefore, in this work, $Fe_2(Al, Si)_5$, $Fe(Al, Si)_3$, and a little Fe_2Al_8Si are mainly formed on the surface at the wetting of AlSi5 on DP980 steel. The lower normalized free energy was obtained in the wetting behaviors on the textured surface, which provided the greater driving force for wetting and spreading of the droplets along the orthogonal directions, which is graphically shown in Figure 6.32.

Lin et al. (2022) discussed the atmospheric plasma spraying technique that was created using Ta_2O_5 coatings consisting of 0%, 3%, 5%, and 7% multiwalled carbon nanotubes (CNTs) on Ti6Al4V alloy. The results improved the indentation

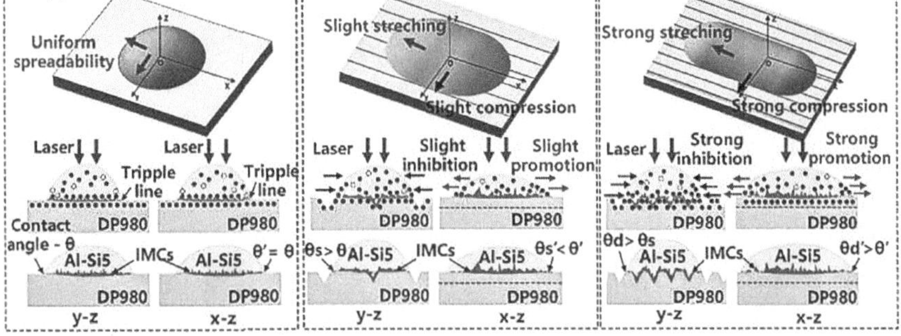

FIGURE 6.32 Schematic of wetting and spreading mechanism of $AlSi_5$/textured DP980 steel: (a) initial stage of heating; (b) wetting state of $AlSi_5$/DP980 steel: (a) initial stage of heating; (b) wetting state of AlSi5/DP980 steel; (c) wetting state of AlSi5/textured DP980 steel with sparse texture; (d) wetting state of $AlSi_5$/ textured DP980 steel with dense texture. (Reproduced with permission from Ref. (Y. Sun et al., 2023). Copyright Elsevier, 2023.)

fracture toughness of the resulting coatings by not degrading the biological properties. Mechanical methods such as machining, grinding, polishing, sandblasting, and friction stir processing (FSP) have been used to produce implants that were later implanted into rabbit tibiae. The results indicated a higher torque needed for the removal of the sandblasted titanium nail, indicating better bone fixation. The FSP method was adopted to obtain nanocomposite layers of TiO_2 and Ti6Al4V. This improves surface microhardness, corrosion resistance, and biocompatibility compared to the original samples. Therefore, the depth of the groove increases, and there is a decrease in corrosion resistance as there is an increase in the amount of TiO_2. The mechanism involved in physical and mechanical methods is relatively simple, but the ability to optimize the surface properties of Ti alloys is limited and is slightly inadequate in preparing complex samples. Yifeng Xiong et al. (2022) studied the crystal growth mechanism of Ti6Al4V alloy during SLM, which is seen in Figure 6.33, that showcases the effect of volume energy density, temperature, densification, and the coarsening of grains in the SLM process.

6.9.5 Chemical Vapor Deposition (CVD)

The CVD technique is seen as an easy and useful approach to producing high-quality solid coating, where the complex chemical reactions between reactant gas and substrate play an important role in the deposited coating (Fuentes-Fernandez et al., 2016; Gracio et al., 2010). The CVD method is far enough to deposit the coating on the substrate that even has some detrimental components for the chemical reaction, which will provide a promising approach to fabricating the desired coatings on any given material.

Yuanlai Xu et al. (2022) studied CVD on single-crystal Nickel–Cobalt–Manganese transition (NCM) nanoparticles by Al_2O_3. The surface of the NCM substrates was chemically bonded by Al_2O_3, and a trace of Al was found to be doped on the surface. The synergetic effects of the surface coating and doping improved adhesion between the coating layer and substrates. The results showed that the thickness of the surface coating can be optimized by CVD time. The chemical bonding with the substrate and the surface Al_2O_3 coating could act as a barrier that controls the electrolyte components from reaching the electrode surface. Figure 6.34 shows the studies conducted

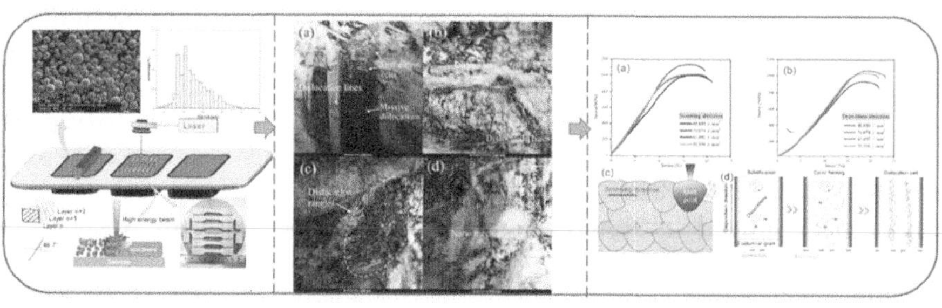

FIGURE 6.33 Graphical presentation of SLM-processed Ti6Al4V parts developed with different energy densities. (Reproduced with permission from Ref. (Xiong et al., 2022). Copyright Elsevier, 2022.)

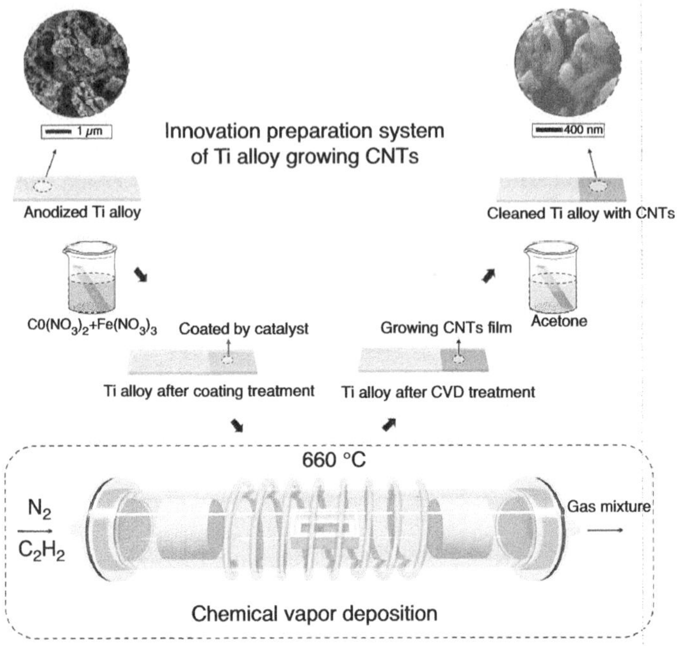

FIGURE 6.34 Schematic diagram of chemical vapor deposition of CF bundles to grow CNTs in situ on the anodized substrate surface. (Reproduced with permission from Ref. (Cheng et al. 2023). Copyright Elsevier, 2023.)

by the anodizing treatment used to modify Ti alloy through the CVD process. The studies concluded that the CVD process can be used to prepare vertically distributed CNTs. Also, the combination of anodizing, CVD, and reactive powder concrete to Ti substrate raises the shear strength of the material.

6.9.6 Sol–Gel Method

This chemical technique is employed to synthesize ceramic or glass materials at low temperatures, which is achieved by hydrolyzing and condensing metal alkoxide precursors (Brinker and Scherer, 2013). The advantage of this method is that it minimizes thermal stress in the material due to the low heating temperature. Also, it is widely utilized for producing protective layers such as hydroxyapatite and bioglass on metallic implant surfaces and improving adhesion and cell proliferation (Hsieh et al., 2002). Therefore, the sol–gel technique is cost-effective and enables the coating of intricately shaped elements (Asri et al., 2016). Nonetheless, the process provides fragile inorganic components that can be mitigated by incorporating organically modified precursors. This approach facilitates the production of flexible hybrid organic–inorganic materials, offering the flexibility to tailor and modify their composition, structure, and thickness to suit specific applications (Romero-Gavilán et al., 2020; Milošev et al., 2022).

In order to improve the osseointegration of Ti implants, Anne Ochsenbein et al. (2008) worked to understand the physical and chemical structure of the substrates, topography, and surface roughness along with the porous surface film. The group also studied the mechanism of cell–substrate interactions, and the cellular and molecular activity of osteoblastic cells grown on these different novel oxide layers was compared with that of untreated ones. Various oxide coatings on CP-Ti substrates using the sol–gel process have been investigated to reduce corrosion and improve biocompatibility. The results showed sol–gel-derived oxide coatings on the Ti surface with better cell interactions on the metal. Better chemical purity and different topography of the different oxide coatings are seen through their physicochemical character. The oxide layers formed showed different cell reactions. The SiO_2–TiO_2 coating even showed a significantly higher proliferation rate than that of CP-Ti with only a natural oxide surface layer. Similarly, TiO_2-ZrO_2 composite coatings were developed by the sol–gel process, showing excellent results as shown in Figure 6.35.

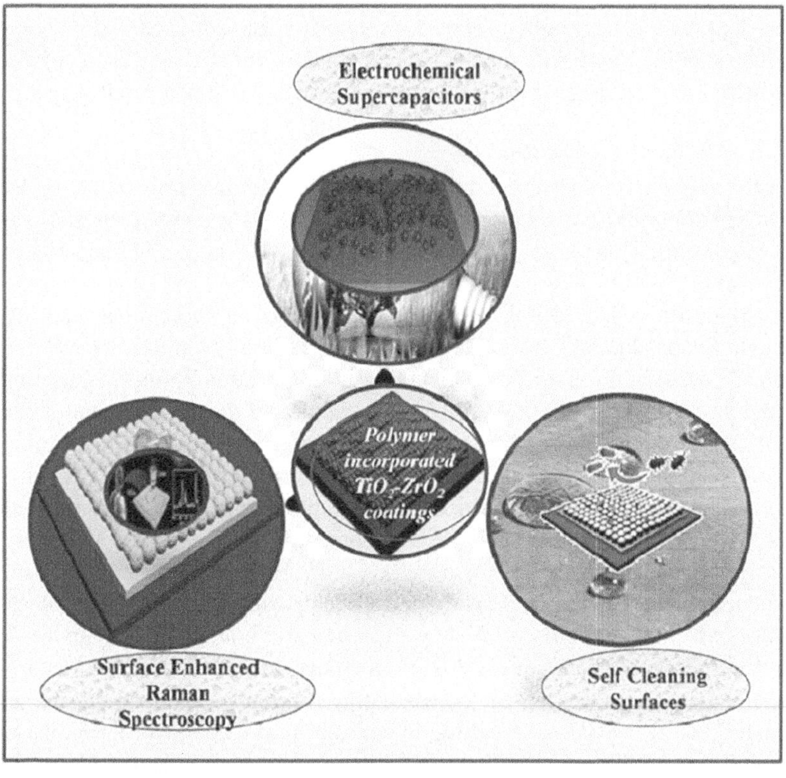

FIGURE 6.35 Pictorial representation showing different applications of the sol–gel process. (Reproduced with permission from Ref. (Mathew Simon et al., 2021). Copyright Elsevier, 2021.)

6.9.7 Electrochemical Treatment

Electrochemical techniques are one of the surface modification techniques that are simple and economical. Ang Gao et al. (2018) discussed the electrochemical surface engineering of Ti-based alloys for biomedical applications. The research mainly focused on three major electrochemical techniques is discussed below.

6.9.7.1 Low Voltage Anodization

The biocompatibility and corrosion resistance of Ti and its alloys are exposed to the presence of a surface oxide layer (Leyens and Peters, 2006). But, the stability of this layer, particularly a few nanometers thick, cannot be used for long-term applications, under load-bearing conditions (Brunette et al., 2001). While thermal oxidation can thicken and stabilize the surface oxide layer, heat treatments conducted at 500°C or higher can compromise the bulk mechanical properties, such as strength and fatigue resistance, and physical characteristics like the shape memory effect of NiTi alloys. Anodization, an uncomplicated and efficient electrochemical technique, offers a means to modify the surface properties of Ti and its alloys (Roy et al., 2011). By adjusting processing parameters such as the composition of the electrolyte, applied potential, temperature, and duration, it is possible to generate surface oxide layers with desirable properties while retaining excellent bulk attributes. Also, under suitable conditions, anodization can produce oxide layers with unique nanotopography, capable of influencing cell and tissue behavior or serving as carriers for targeted drugs and antibacterial agents.

6.9.7.2 Microarc Oxidation

When the applied voltages exceed the dielectric breakdown limit of the oxide layer formed in traditional anodization, it leads to increased gas evolution and frequent spark discharges. This particular form of anodization is commonly known as microarc oxidation (MAO), plasma electrolytic oxidation (PEO), or anodic spark deposition (Yerokhin et al., 1999). MAO has emerged as an economical and efficient technique for producing firmly adherent ceramic coatings on valve metal (current passes in one direction) surfaces, including titanium, aluminum, magnesium, and their alloys. Over the past decade, MAO has facilitated the development of ceramic coatings with remarkable bonding strength to the substrate, controllable morphology, and composition, as well as excellent wear and corrosion resistance (Walsh et al., 2009). In the case of dental and orthopedic implants made from titanium and its alloys, MAO treatment enhances their biocompatibility, improves bioactivity, and imparts antibacterial properties (Wang et al., 2015; He et al., 2017).

A continuous applied current is the most widely used idea to control the process (Walsh et al., 2009). Figure 6.36 shows a schematic voltage–time response curve in the PEO process (Cheng et al., 2015). The plots show three regions that are seen in the figure. After initiating the power supply, similar to anodization, the voltage increases linearly with time, resulting in constant thin oxide layer formation on the substrate (stage i). The thin oxide layer is seen in metal/oxide and oxide/electrolyte interfaces due to the field-assisted migration of O^{2-} anions and Ti^{4+} cations. But, the anodized oxide coating is not completely uniform. Flaws, defects, areas of concentrated stress, and nonuniform thickness in the coating are randomly spread on the

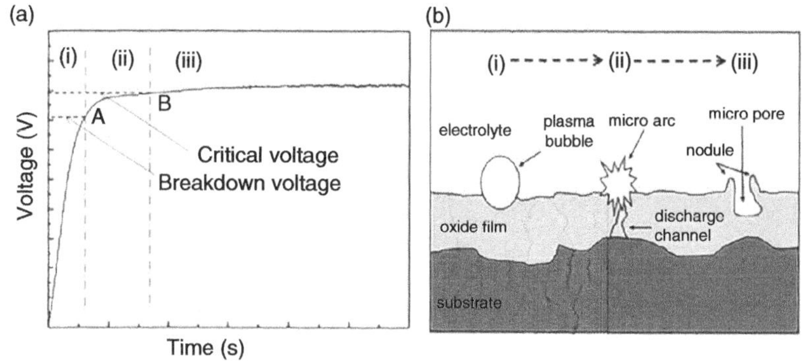

FIGURE 6.36 (a) Schematic of a typical voltage–time response curve containing three stages during the MAO process. (b) Schematic illustration of the formation of porous oxide film by plasma bubble implosion. (Reproduced with permission from Ref. (K. R. Shin et al., 2011). Copyright Elsevier, 2011.)

surface. If the applied voltage increases, the potential drop at above-mentioned weak spots will exceed the dielectric breakdown limit of the oxide layer. A spark discharge takes place and the applied voltage is called the breakdown voltage (point A) in Figure 6.36 (Guo et al., 2006). The voltage then continually increases with a smaller slope (stage ii), and a large number of small-size discharges are randomly distributed and rapidly moving over the entire anode surface (Matykina et al., 2007). With the increase in processing time, the spark size also increases with reduced density. The voltage, in point B, is called "critical voltage," beyond which a steady-state sparking is established on the anode surface and the cell voltage reaches a relatively stable value (stage iii) (Hussein et al., 2010). However, the continuous formation and breakdown of the oxide film cause the cell voltage to fluctuate.

6.9.7.3 Electrodeposition

It is a cost-effective and easy technique for the preparation of protective or other functional layers. The use of the electric field makes electrodeposition particularly suited for uniform film production on substrates of complicated shape, impregnation of porous substrates, and deposition on selected areas of the substrates. It offers rigid control on coating thickness, composition, and uniformity. The electrodeposition process can be conducted either by anodic or cathodic methods. However, anodic deposition has limited utility regarding possible materials to be deposited by this method (Zhitomirsky, 2002). There are two types of electrodeposition techniques, which are (1) cathodic electrolytic deposition (ELD) and (2) electrophoretic deposition (EPD), as shown in Figure 6.37. ELD is based on the electrochemical reduction of metal ions or produces colloidal particles by cathodic reactions for subsequent deposition, whereas EPD is achieved via the motion of charged particles toward an electrode under an applied electric field.

Methods such as CVD, electrochemical treatment, sol–gel, and biochemical modification are a few examples of chemical modification methods. Generally, a new substance layer is formed on the surface of the substrate through chemical, electrochemical, or biochemical reactions to improve the performance of the material. The CVD method was employed to deposit Ta films on Ti6Al4V porous scaffolds

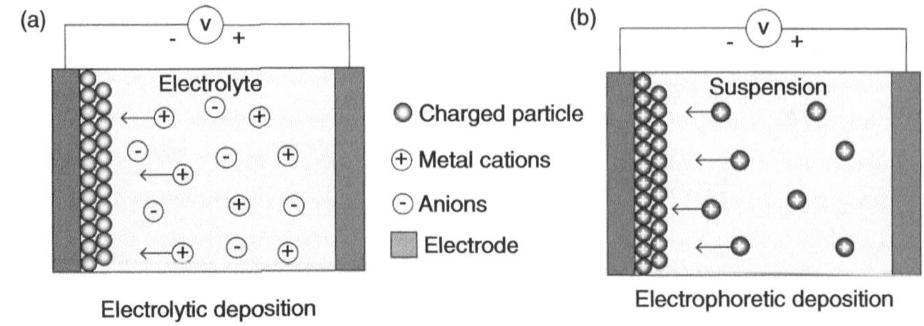

FIGURE 6.37 Schematic illustration of the two types of electrodeposition processes: (a) electrolytic deposition and (b) electrophoretic deposition. (Reproduced with permission from Ref. (Santos et al., 2015). Copyright INTECH, 2015.)

to improve its adhesion, growth, and cell proliferation. In the sol–gel process, the chemical reaction occurs in the solution phase. The mechanism involved is complex in the chemical modification methods.

Wu et al. (2006), Yang et al. (2009), and Cui et al. (2012) studied surface pretreatments utilized in surface grafting and FRP substrates. Through surface grafting, polymer chains or functional groups are covalently attached to the material surface, altering its surface chemistry and properties. This technique allows for precise control over the composition and structure of the grafted molecules, enabling tailored modifications to enhance wetting characteristics. By introducing hydrophilic functional groups, such as hydroxyl or carboxyl groups, the surface energy can be increased, promoting wetting by aqueous solutions. Li et al. (2018) studied Sn sensitization on polyethylene to obtain silver films for the reduction in bacterial process, where they achieved increased adhesive strength and wettability. Rajesh et al. (2019) and Cauchy et al. (2016) researched Sn sensitization on carbon fiber-reinforced epoxy resin composites to develop a conductive silver film. The resulting coating was not continuously uniform, suggesting further research to obtain uniform coatings. Conversely, grafting hydrophobic groups can reduce surface energy, resulting in a water-repellent surface. Surface grafting offers versatility, as it can be applied to various substrates, including polymers, metals, and ceramics. It also provides excellent stability, ensuring long-lasting improvements in surface properties.

Surface coating is an effective approach for improving the wettability and surface energy of materials (Tan et al., 2022). By applying a thin layer of coating on the surface material, the properties of the material can be modified to enhance its interaction with liquids. Various types of coatings can be utilized, such as polymer coatings, metal coatings, or ceramic coatings (Vishwakarma et al., 2022).

Döring et al. (2019) investigated coated endoprosthetic implants by applying an ion beam of high energy to see the surface and layer defects in nonloaded areas of already used coating systems. The results showed tribological behaviors and reduced roughness due to the postpolishing step. The use of coatings in a hard/soft bearing seems to be unfavorable due to the high friction coefficient. The wettability increases resulting in a reduced amount of droplets formation. The plasma-based thermal coatings have been utilized on AISI1045 sheets, and the improvement of strength, hardness, and wear resistance along with the material composition was studied. Similarly, Zhang et al. (2023) described the usage of CA to show the surface wettability of Co-Cr-Mo alloy with different textures. Also, the smaller the CA, the greater the wettability of the solid surface, and the better its tribological performance. Also, the laser-textured surface presents a relatively high CA due to the formation of needle-like protrusions inside the texture by the thermal effect of the laser. The ion beam etching can increase the contact area between the lubricating fluid and the surface of the substrate by removing the protruding melts on the texture edge, thereby reducing the CA and forming a better lubricating film. Figure 6.38 presents the contact state of the drop on the alloy Co-Cr-Mo before and after ion beam etching.

The cell growth and damage repair of the biocompatibility of Ti alloys is very beneficial using the surface coating. Zhou et al. investigated the nano-hydroxyapatite layer by in vitro biomineralization. The results showed that the single-layered silk fibroin (SF) was due to the nano-scale roughness of the hydroxyapatite deposition layer after biomineralization. But the improvement of osseointegration by a hybrid layer composed of hydroxyapatite and SF was more obvious. Few researchers have worked on the preparation of pure silica zeolite silicate coating without harmful ions (Al and V) on Ti6Al4V by secondary growth method, and the cells grown on this coating were evenly distributed, forming a large number of microfilaments that lead to proper adhesion and effective spreading. Zhang et al. synthesized calcium phosphate bioceramic coating doped with La_2O_3 on the surface of Ti6Al4V alloy. It showed that the coating with 0.6 wt% La_2O_3 showed improvement in cell proliferation and adhesion. The La^+ ions from calcium phosphate coating prevent the formation of osteoclasts.

These coatings can be designed to possess specific surface characteristics as well, such as hydrophilicity or hydrophobicity, to optimize wettability. Hydrophilic coatings, for example, can increase surface energy and promote wetting by allowing liquids to spread uniformly across the surface. On the other hand, hydrophobic coatings can reduce surface energy and repel liquids, resulting in a self-cleaning or water-resistant surface. The selection of appropriate coating materials and deposition techniques allows for precise control over the surface properties, enabling tailored modifications for diverse applications ranging from automotive coatings to medical devices. Surface coatings are a versatile method that offers flexibility in improving wettability and surface energy while providing additional benefits like corrosion resistance, wear protection, and aesthetic enhancement. Figure 6.39 showcases the detailed corrosion and adsorption mechanism of the doping studies of Gd ions into the poly 3-hydroxybutyrat e-3-hydrpxyvalerate coating that could improve the corrosion resistance of Ti-6Al-4V. The PHBV (polyhydroxy butyrate-CO-β-hydroxy valerate) improves the adhesion and

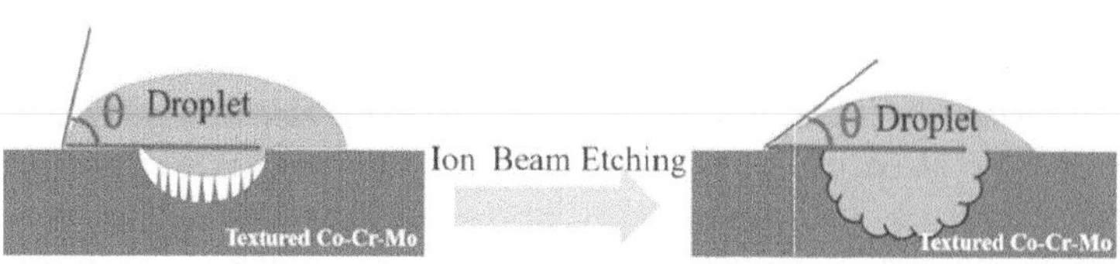

FIGURE 6.38 Graphical presentation of the mechanism through ion beam etching. (Reproduced with permission from Ref. (K. Zhang et al., 2023). Copyright Elsevier, 2023.)

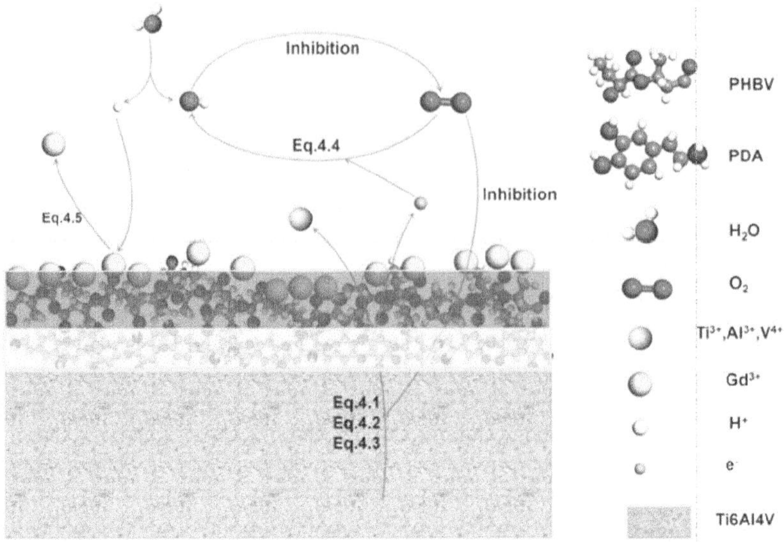

FIGURE 6.39 Graphical presentation of a mechanism involving Gd^{3+} modified PHBV coatings. (Reproduced with permission from Ref. (Qin et al., 2022). Copyright Springer Nature, 2021.)

proliferation of human gingival fibroblasts, and the addition of Gd ions into PHBV coating could further improve the cell proliferation by decreasing the surface roughness of PHBV coating and inhibiting the release of metal ions.

6.10 APPROACHES FOR IMPROVING HYDROPHOBICITY AND SUPERHYDROPHOBIC NATURE OF THE SURFACE

6.10.1 Electrochemical Deposition

A versatile electrochemical deposition technique has been employed to create the nanostructured surfaces necessary for achieving superhydrophobicity. He et al. (2010) recently introduced an electrochemical anodization method for synthesizing various ZnO nanostructures on a large scale using a mixed hydrofluoric acid and methanol electrolyte. By controlling the concentration of the electrolyte and reaction times, including less than one hour, different ZnO nanostructures such as nanodots, nanowires, and nanoflowers were successfully obtained (Figure 6.40).

He et al. (2010) also investigated the alteration in surface wetting properties by applying an external potential to ZnO nanostructure films. This change is attributed to the creation of surface defective sites on ZnO films due to the electric field effect. Another study by Huang et al. (2010) utilized a one-step deposition process to create superhydrophobic copper surfaces. In their method, a direct voltage (DC) was applied between two copper plates immersed in a dilute ethanolic stearic acid solution. This process led to the transformation of the surface of the anodic copper electrode into a superhydrophobic surface due to the reaction between copper and the stearic acid solution. The resulting

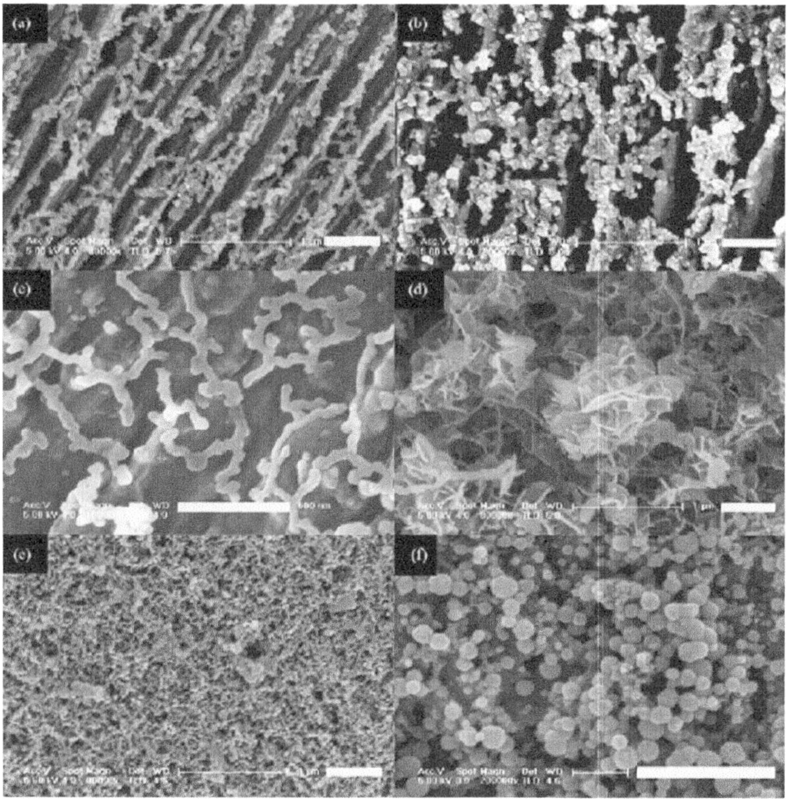

FIGURE 6.40 SEM images of ZnO nanostructures grown on the Zn foil with different reaction times in a mixed solution of methanol and water (50 vol% methanol, 1 mass% HF): (a) Sample A, 30 seconds; (b) Sample B, 1 minute; (c) Sample C, 5 minutes; (d) Sample D, 10 minutes; (e) Sample E, 15 minutes; (f) Sample F, 30 minutes. The length of the inserted scale bars is 500 nm. (Reproduced with permission from Ref. (He et al., 2010). Copyright Elsevier, 2010.)

surface was covered with flower-like low surface energy copper stearate films, exhibiting a water contact angle (WCA) of $153° \pm 2°$ with roll-off properties.

Meng et al. (2008) achieved the fabrication of super-amphiphobic surfaces, repelling both water and oil, on common engineering metals and their alloys (zinc, aluminum, iron, nickel, Zn–Fe alloy, and brass) by exploiting an electrochemical reaction in perfluorocarboxylic acid solutions. By controlling the chain length, concentration of perfluorocarboxylic acid, and process time, textured rough structures on various substrates were attained. SEM images of the prepared films over zinc substrates revealed petal-like nanosheet structures spreading uniformly, with a thickness ranging from about 30 to several hundred nanometers and micrometer-scale length and height. The prepared surfaces exhibited superamphiphobicity due to the synergistic effect of their unique surface compositions and microscopic structures.

An electrochemical route to directly fabricate hierarchical spherical cupreous microstructures on an indium tin oxide substrate was also proposed (Wang et al., 2008). After further chemisorptions of a self-assembled monolayer of n-dodecanethiol, the as-prepared compact surface became superhydrophobic with a CA as high as 152°. A superhydrophobic 3D porous copper film with a typical integration of micro or nanostructures by utilizing hydrogen bubbles as the dynamic template for metal electrodeposition was prepared (Y. Li et al., 2007). The pore diameter and wall thickness of the porous copper films were successfully tailored by adjusting the concentration of the electrodeposition electrolyte since the films were electrodeposited and grew within the interstitial spaces between the hydrogen bubbles.

6.10.2 Electrospinning Method

Electrospinning stands out as a straightforward technique for producing continuous fibers across nanometer to submicron scales. This method involves applying an electrical charge between an extrusion nozzle and a grounded collection plate. Recently, Islam et al. developed a superhydrophobic membrane using Pullulan (PULL), a natural polymer, through electrospinning of fluorinated silane-functionalized PULL. Their findings revealed that electrospinning solutions of pure PULL at 9 and 12 wt% concentrations resulted in CAs below 150°. However, when coupled with fluorinated silane, the CAs reached up to 155° and 151° for 9 and 12 wt% PULL membranes, respectively.

Kang et al. synthesized electrospun polystyrene (PS) fibers on a silicon wafer, exhibiting superhydrophobic characteristics due to the unique surface structure akin to lotus leaves' hierarchical arrangement (Figure 6.41) (Islam et al., 2010). Furthermore, Park et al. demonstrated the creation of solvent-resistant, mechanically robust, superhydrophobic nanofibrous mats by electrospinning poly(vinylidene fluoride) with tetraethyl orthosilicate (TEOS) inorganic silane material (Kang et al., 2008). The incorporation of TEOS sol into the nanofibrous matrix enhanced both solvent resistance and mechanical strength of the mats, yielding a constant WCA

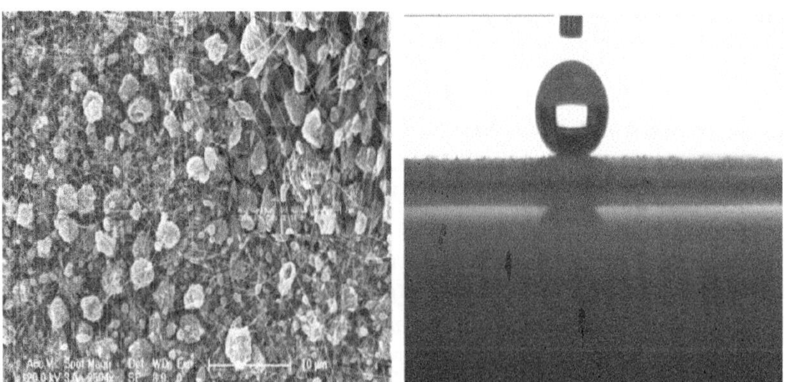

FIGURE 6.41 (a) Scanning electron microscope image of electrospun film prepared from 4 wt% PS/DMF solution. (b) Behavior of a water droplet on its surface. CA = 154.65°. (Reproduced with permission from Ref. (Zhan et al., 2010). Copyright Elsevier, 2010.)

of 156° regardless of TEOS content. Similarly, Zhan et al. achieved superhydrophobic PS surfaces with mechanical integrity using an advanced electrospinning technique (Zhan et al., 2010). By combining bead-on-string fibers and microfibers, they created a composite electrospun mat that enhanced surface roughness, thereby improving both hydrophobicity and mechanical properties. Control over WCA and mechanical strength was achieved by adjusting the mass ratio of bead-on-string fibers to micro-sized fibers.

6.10.3 Wet Chemical Reaction

For the widespread adoption of large-scale, multi-material, and hierarchical superhydrophobic functionality in commercial and industrial settings, a straightforward wet chemical coating method proves to be optimal. In their study, Qi et al. (2009) presented a simple technique for creating a superhydrophobic and antireflective structured silicon surface across a wafer scale using chemical etching. Through the utilization of KOH etching and silver catalytic etching, they generated pyramidal hierarchical structures on a crystalline silicon wafer (Figure 6.42), which displayed robust antireflection and superhydrophobic properties following fluorination. Post-fluoroalkylsilane treatment, the structured surface exhibited a CA of 169° and a sliding angle of less than 3°. The duration of silver-assisted etching played a pivotal role in determining the superhydrophobic characteristics, with optimal results achieved within a timeframe of less than 30 seconds to prevent structural damage and degradation of superhydrophobicity.

Similarly, monoalkyl phosphonic acid was utilized to establish a stable and durable superhydrophobic surface on nickel substrates via a wet chemical process. The chemical reaction between Ni and monoalkyl phosphonic acid gradually formed flowery microstructures, creating a continuous slipcover (Mei Li et al., 2007). Pan et al. (2010) developed a superhydrophobic rough structure on a copper wafer through the HNO_3 etching technique, aided by cetyltrimethyl ammonium bromide (CTAB) and ultrasonication. Subsequent modification with 1H,1H,2H,2H-perfluorodecyltriethoxysilane conferred stable superhydrophobicity upon the copper wafer.

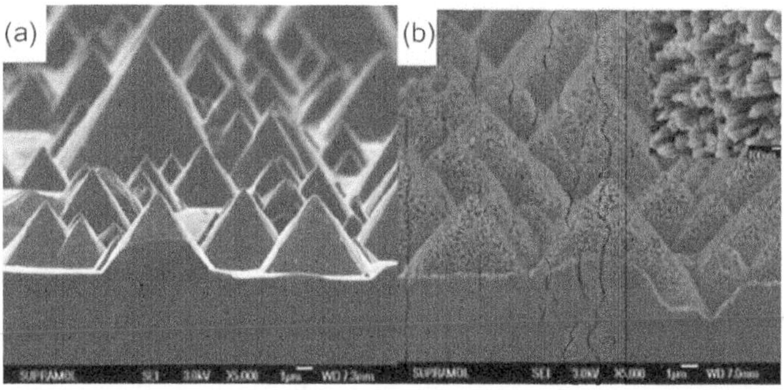

FIGURE 6.42 Cross-sectional SEM photographs of (a) silicon pyramids created with KOH etching and (b) hierarchical structures generated with Ag-assisted etching. (Reproduced with permission from Ref. (Qi et al., 2009). Copyright ACS Publication, 2009.)

Upon SEM analysis, dense and spherical micropits were observed on the copper wafer after etching with 5 M nitric acid in the presence of 1.2 mM CTAB under ultrasonication. The synergistic action of CTAB and ultrasonication facilitated the formation of these micropits, which endowed the surfaces with superhydrophobic properties, suitable not only for water but also for corrosive liquids. These findings are pivotal for expanding the utility of copper as superhydrophobic engineering materials across various industrial applications and environmental conditions.

Furthermore, superhydrophobic surfaces (WCA: $158.4° \pm 2°$, sliding angle: $2.7° \pm 1°$) were successfully prepared on macroporous silicon films with moderate porosity (37.8%) and excellent mechanical properties (Young's modulus, shear modulus, and collapse strength are 64.2, 24.1, and 0.32 GPa, respectively) through electrochemical etching under galvanostatic conditions combined with subsequent octadecyltrichlorosilane (OTS) modification (Wang et al., 2011). SEM analysis and fundamental calculations highlighted the significance of 3D microstructures in conferring mechanical stability and surface superhydrophobicity to macroporous silicon films. These findings hold promise for practical applications of superhydrophobic surfaces, particularly in domains with mechanical requirements such as microfluidics. Lastly, Wang et al. (2010) engineered a layer of flower-like superhydrophobic film on pure Mg surface via chemical etching in H_2SO_4, H_2O_2, and subsequent immersion in stearic acid ($CH_3(CH_2)16COOH$) ethanol solution. The resulting superhydrophobic surface exhibited a static WCA of 154° with a sliding angle of about 3°. The flower-like structure and bonding of $CH_3(CH_2)16COO$- on the Mg surface contributed to its superior water-repellent properties. Electrochemical impedance spectroscopy revealed a fourfold increase in transfer resistance of the superhydrophobic surface compared to bare Mg after one-hour immersion in 0.1 mol/L NaCl solution.

6.10.4 Hydrothermal Synthesis

Hydrothermal synthesis represents a valuable technique for producing nanomaterials with distinctive morphologies. In a study by Gu et al. (2010), superhydrophobic WO_3 nanostructure films were successfully fabricated on alumina plates and tungsten foils using a straightforward hydrothermal approach. By managing the adsorption/desorption process of n-dodecanethiol associated with the Ag deposits on WO_3 nanostructure surfaces, rapid and reversible changes in wettability, transitioning between superhydrophilicity and superhydrophobicity, were achieved. Remarkably, the transition from superhydrophilicity to superhydrophobicity required only 25 minutes, with the reverse change completed in just 30 seconds. The swift reversible alterations in wettability of WO_3 surfaces hold promise for applications in smart devices.

In another investigation, Xu and Cai (2008) utilized a hydrothermal method to establish superhydrophobic surfaces on cotton fabrics. Initially, ZnO nanocrystals were synthesized and applied to cotton fibers, followed by the fabrication of oriented ZnO nanorod arrays on the fibers to introduce nanoscale roughness. Subsequently, the fabrics underwent modification with dodecyltrimethoxysilane to achieve superhydrophobic surfaces. This method exhibited excellent reproducibility and utilized cost-effective laboratory equipment typical of conventional textile processing. Wu and Xia (2010) developed a surface featuring micro- and nano-scale ZnO structures

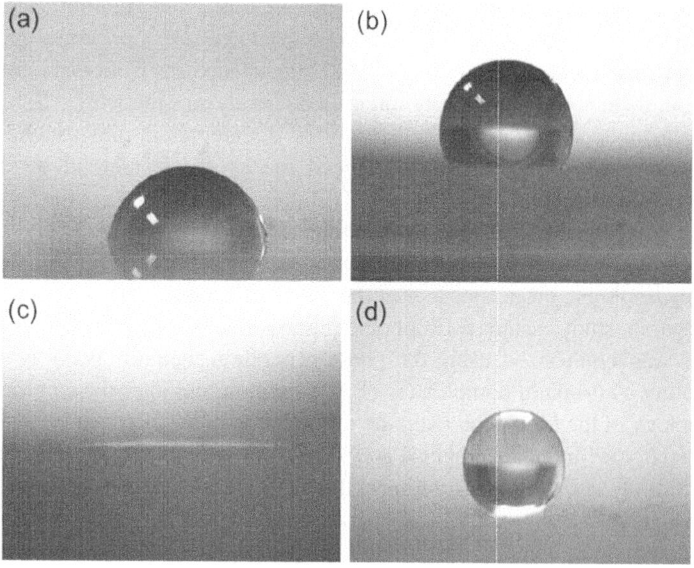

FIGURE 6.43 (a) Water droplet contact angle on a plane ZnO surface. (b) The contact angle on a Teflon-modified plane ZnO surface. (c) Water droplet spreading on a superhydrophilic "rose" like ZnO surface and (d) a water bead standing on a Teflon-modified "rose" like ZnO surface. (Reproduced with permission from Ref. (Wu et al., 2010). Copyright Elsevier, 2010.)

through an alkaline hydrothermal method. The WCA on this doubly rough surface was initially low, approximately 0°, but increased to around 168° after spin coating with Teflon (Figure 6.43). Microstructural analysis revealed a "rose" like ZnO crystal surface structure, contributing to both superhydrophilicity and superhydrophobicity. Furthermore, the prepared surface displayed notable chemical stability even after 15 days of immersion in water. This method offers simple control, cost-effectiveness, and a broad spectrum of potential applications, including self-cleaning superhydrophobic coatings on large substrates.

In a study by Song et al. (2010), a direct in situ hydrothermal synthesis method was introduced for constructing 3D flowerlike micro/nanoflakes on an iron plate surface. The micro-/nanostructure of the surface could be readily adjusted by varying the reaction temperature and precursor concentration. Following chemical modification with vinyltirethoxysilane, the resulting surface covered by flower-like micro/nanoflakes exhibited superhydrophobic characteristics.

6.10.5 Phase Separation Method

The phase separation method, known for its simplicity and cost-effectiveness, is widely employed in the fabrication of porous polymer membranes. In a study by Khoo and Tseng (2008), superhydrophobic 3D methyltrichlorosilane (MTS) nanoarchitectures were successfully crafted on commercially available glass and SiO_2 substrates. The diverse morphologies of these structures were achieved by manipulating various synthesis variables within the MTS reagent phase separation method.

Factors such as reagent concentration, reaction time, and humidity significantly influenced the morphology of the MTS nano-architecture. Concentration played a pivotal role, leading to the evolution of MTS nanostructures from discrete and short fibers to the formation of randomly integrated fibrous quasinetworks. Subsequently, with increasing concentration, the structures disintegrated into rougher discrete fibers. Song et al. (2010) further enhanced the synthesis method by improving 3,6-O-ditertbutyldimethyl silyl chitosan. Utilizing a simple and cost-effective phase separation method, durable superhydrophobic films were prepared. The roughness of these films displayed a hierarchical organization at three levels, demonstrating extreme hydrophobic properties across the entire pH range.

In a separate study, a superhydrophobic methylsilicone resin film with luminescent properties was synthesized using the phase separation method (Wang et al., 2009). Incorporating a europium complex (Eu(BA)3Phen) did not alter the wetting behavior or morphology of the film. Moreover, the introduction of luminescent properties to the superhydrophobic film holds practical applications, such as enhancing water repellence in light-emitting diodes and facilitating self-cleaning of road signs used at night.

Wei et al. (2010) conducted a synthesis of styrene and 2,2,3,4,4,4-hexafluoro-butyl methacrylate copolymers through bulk polymerization. Subsequently, superhydrophobic copolymer films were prepared using the phase separation technique. Dissolving the copolymer in tetrahydrofuran and subsequently adding ethanol induced phase separation. The microstructures of the polymer films, as depicted in Figure 6.44, were controlled by the degree of phase separation, enhanced appropriately by the concentration of ethanol.

6.10.6 Self-Assembly and Layer-by-Layer Methods

Methods like self-assembly and layer-by-layer (LBL) deposition present cost-effective approaches for crafting superhydrophobic surfaces. Xu and Wang (2010) illustrated the creation of optically transparent and superhydrophobic coatings through the self-assembly of dual-sized silica particles from a mixed dispersion. By adjusting the size and ratio of these particles without relying on organic/inorganic templates, a desirable micro-/nano-hierarchical structure for superhydrophobicity was achieved. The entire assembly process occurred under mild conditions within just half an hour, offering advantages for large-scale coating. Another approach by Song et al. (2009) involved fabricating superhydrophobic surfaces through the application of an OTS self-assembled monolayer (SAM) on silicon micro/nanotextured surfaces produced by the aluminum-induced crystallization (AIC) of amorphous Si (a-Si) technique. This innovative method significantly enhanced the hydrophobicity of the silicon surface, resulting in superhydrophobic surfaces with a WCA of 155° and a sliding angle of less than 1°. The utilization of well-established technologies, such as OTS SAMs, chemistry and AIC of a-Si, holds great potential for applications in the MEMS industry.

Yang et al. (2009) demonstrated a convenient and effective method for preparing superhydrophobic films using mulberry-like $CaCO_3/SiO_2$ composite particles. Originating from violent stirring and surface modification, these particles, along with the self-assembly of polydimethylsiloxane, provided a simple procedure with potential applications for fabricating large-area superhydrophobic surfaces for practical

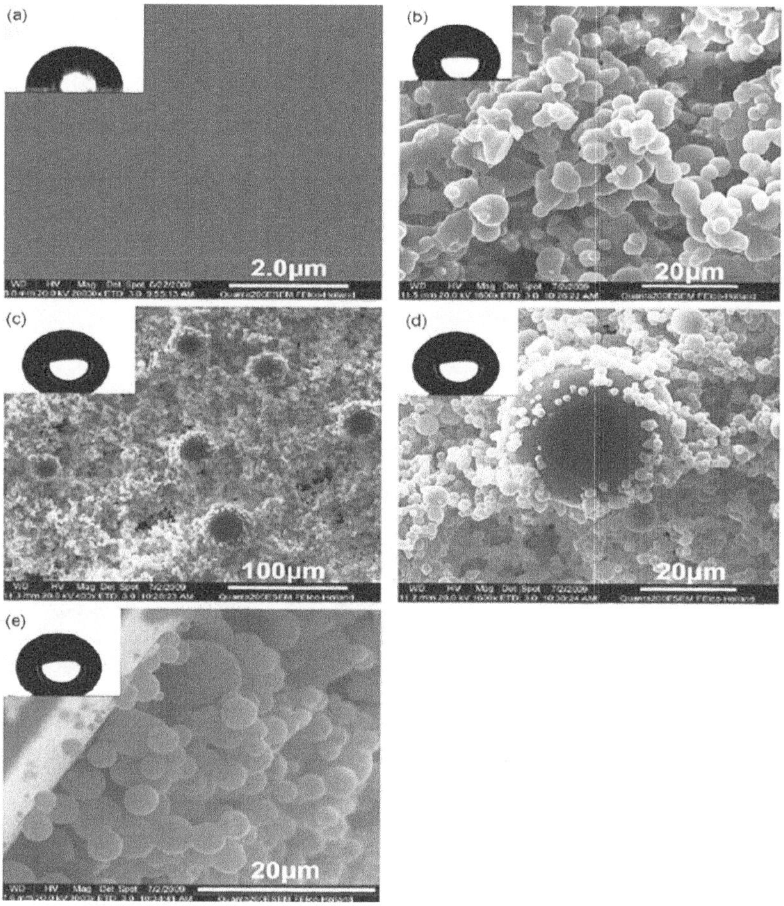

FIGURE 6.44 Environmental Scanning Electron Microscope (ESEM) images of copolymer films obtained with different ethanol content in copolymer solution: (a) 0%; (b) 30%; (c) 50%; (d) is the higher magnification ESEM image of (c); (e) is the section image of (c). Insets are the water contact angles on the copolymer films. (Reproduced with permission from Ref. (Wei et al., 2010). Copyright Elsevier, 2010.)

self-cleaning applications. On the other hand, the LBL deposition technique relies on electrostatic charge interactions between different layers, such as polyanion and polycation. Zhao et al. (2010) achieved the transformation of highly hydrophilic cotton fabrics into superhydrophobic ones through electrostatic LBL assembly of polyelectrolyte/silica nanoparticle multilayers on cotton fibers, followed by a fluoroalkylsilane treatment (Figure 6.45). The surface morphology of the silica nanoparticle-coated fibers, influencing the hydrophobicity, was tailored by controlling the number of multilayers.

Examining the cotton fabrics assembled with five multilayers or more revealed slippery superhydrophobicity with a CAH lower than 10°. The buoyancy of the superhydrophobic fabric was demonstrated using a miniature boat made with the fabric, exhibiting remarkable loading capacity and durability after at least 30 machine

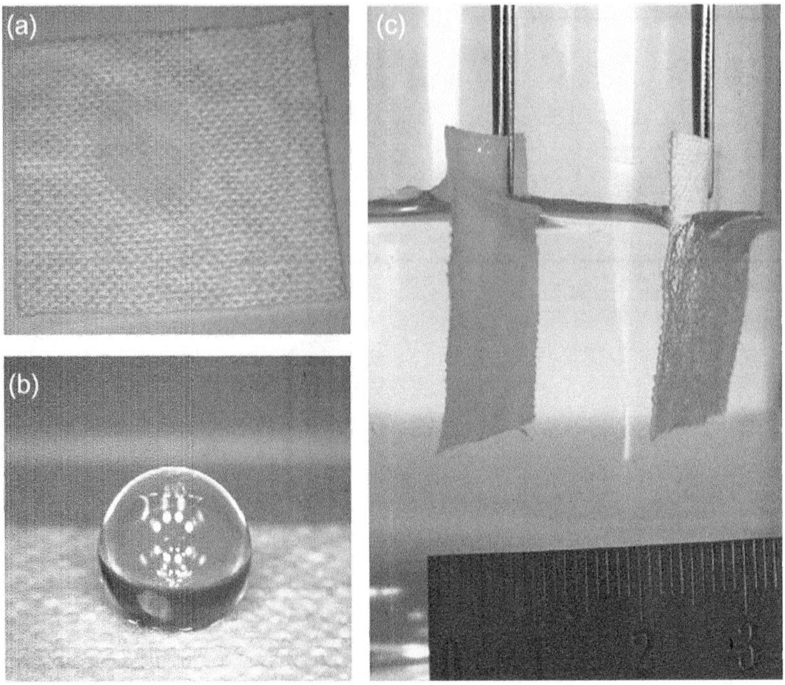

FIGURE 6.45 Photographs of water drops on (a) untreated cotton fabric and (b) superhydrophobic cotton fabric assembled with $(PAH/SiO_2)_5$ multilayers. (c) Untreated (left) and superhydrophobic (right) cotton fabrics being immersed in water. (Reproduced with permission from Ref. (Zhao et al., 2010). Copyright Elsevier, 2010.)

washing cycles. Jindasuwan et al. (2009) employed a LBL deposition approach to dip-coat a water-repellent film onto a glass substrate. Inspired by the lotus leaf's unique surface features, they achieved this by depositing polyelectrolyte layers, followed by SiO_2 particles to modify surface roughness, and a semifluorinated silane to reduce surface energy. Adjusting the SiO_2 content allowed for the modification of the film's surface roughness and, consequently, its hydrophobicity. In a different study, Amigoni et al. (2009) constructed hybrid organic/inorganic surfaces by alternating layers of amino-functionalized silica nanoparticles and epoxy-functionalized silica nanoparticles. The last layer of amino-functionalized silica particles was hydrophobized by grafting a newly designed highly fluorinated aldehyde, resulting in stable, highly superhydrophobic surfaces. The hydrophobicity increased with the number of layers, showcasing the potential of this method.

6.10.7 Solution Immersion Method

The utilization of a solution immersion method is an efficient and economical procedure, making it a time-saving approach that requires inexpensive reagents. This technique is straightforward to implement, devoid of any necessity for special equipment or intricate procedures. In a study by Kong et al. (2008), a distinctive

FIGURE 6.46 SEM images of (a) Cu_2O micro-crystal and (b) slice-like $Cu_2(OH)_3NO_3$ crystal. (Reproduced with permission from Ref. (Kong et al., 2008). Copyright Elsevier, 2008.)

solution-immersion method was introduced for crafting superhydrophobic surfaces on copper substrates. A slice-like $Cu_2(OH)_3NO_3$ crystal was meticulously developed on the copper foil surface (Figure 6.46) through successive immersion in an aqueous solution containing sodium hydroxide and cupric nitrate. The superhydrophobic structure was achieved by modifying the slice-like $Cu_2(OH)_3NO_3$ crystal with 1H,1H,2H,2H-perfluorodecyltriethoxysilane (FAS-17).

In a separate investigation, Liu et al. (2008) employed a room temperature solution-immersion process to fabricate a novel hierarchical CuO architecture on Cu foils. This architecture comprised densely packed nanoplates and nanoribbons. The resulting film exhibited excellent superhydrophobicity after a simple modification with commercially available wax, a cost-effective low-surface-energy material. Another approach, as demonstrated by Liu et al. (2009), involved the creation of stable superhydrophobic films on zinc substrates using a straightforward immersion technique. This technique entailed immersing the substrates into a methanol solution of hydrolyzed 1H,1H,2H,2H-perfluorooctyltrichlorosilane [CF3(CF2)5(CH2)2SiCl3, PFTS] for 5 days at room temperature, followed by a brief annealing at 130°C in air for 1 hour. The resulting superhydrophobic film served as an effective corrosion-resistant coating for the zinc interface during prolonged exposure to an aqueous solution of sodium chloride (3% NaCl) for nearly one month. Li et al. (2008) adopted a commercially available waterproof reagent, potassium methyl siliconate (PMS), to develop superhydrophobic surfaces on cellulose-based materials via a solution-immersion method. The silanol, formed through the reaction of a PMS aqueous solution with CO_2, was assembled on the cellulose molecule surface through hydrogen bond interactions. Subsequently, superhydrophobic coatings were formed via a subsequent polycondensation reaction, resulting in the covalent attachment of a polymethylsilsesquioxane layer to the surfaces of the cellulose fibers. This process yielded uniform nanoscale roughness protuberances with inherent microscale roughness characteristics, leading to a superhydrophobic character with satisfactory durability.

6.11 DYNAMICS OF FLUIDS

Pearl drops or liquid marbles exhibit remarkable dynamic properties due to their high CA and low hysteresis. This unique combination results in enhanced liquid mobility, surpassing typical levels. Unlike conventional scenarios where viscous dissipation occurs predominantly near contact lines, pearl drops experience minimal dissipation in this regard. A familiar example illustrating this heightened mobility is the behavior of liquid nitrogen drops on surfaces. While one might attribute the mobility to the low viscosity of liquid nitrogen, similar dynamics can be expected for water drops falling through air, where friction primarily arises from air flow around the drop. In subsequent subsections, we delve into the rapid motions of these drops. However, it is pertinent to note that viscous motions can also occur under specific conditions, particularly when the liquid composing the pearl possesses exceptionally high viscosity. We elaborate on these phenomena in the ensuing sections.

6.11.1 Creeping Motions

In this section, our focus is on liquids characterized by high viscosity, where viscosity serves as the predominant source of friction, even in the absence of contact lines. Initially, we differentiate between scenarios involving significantly large drops, flattened under the influence of gravity (resembling puddles). Subsequently, we delve into the more intricate scenario of droplets. For the sake of clarity and ease of analysis, we assume that gravity acts as the driving force behind the motion. Specifically, we examine situations where the liquid is deposited on an incline with a tilting angle denoted as α. Our objective is to characterize the stationary velocity, V, of the descent in these conditions.

6.11.2 Puddles

In the context of liquid deposition on a plate, the formation of a puddle, characterized by a drop flattened under the influence of gravity, occurs when dealing with drops exceeding the capillary length, κ^{-1}. The thickness of these puddles, in the absence of wetting (as κ^{-1} tends to zero), converges toward $2\kappa^{-1}$. When dealing with viscous liquids denoted by η, the terminal velocity is determined by the equilibrium between gravitational force (the plate inclined at an angle $\alpha << 1$) and viscous friction. At a scaling level and expressed per unit volume, this equilibrium is formulated as

$$\eta \frac{V}{\kappa^{-2}} \sim \rho g \alpha \tag{6.50}$$

Considering velocity gradients across a thickness on the order of κ^{-1}. By incorporating the capillary length definition, the relation (Lu et al., 2022) is derived as

$$V_o \sim \frac{\gamma}{\eta} \alpha \tag{6.51}$$

It is noteworthy that this velocity is independent of the quantity ρ, which governs both the driving and friction forces. The impact of ρg is manifested in the puddle thickness—higher ρg results in a thinner puddle and consequently increased viscous resistance. These findings extend to scenarios involving very large puddles under partial wetting conditions, considering the CA in the puddle thickness, leading to $V \sim V_0 \sin^2(\theta/2)$. However, for puddles of intermediate size, typically in the centimeter range, the contribution of viscous dissipation near the contact line becomes significant, causing the velocity to be notably smaller than predicted by the aforementioned expression.

6.11.3 Creeping Droplets

Drops smaller than the capillary length adopt a shape influenced by surface tension. Initially, let us examine drops in partial wetting scenarios. When hysteresis is substantial, these drops tend to remain stationary—they need to be sufficiently large for gravity to surpass the adhesive force caused by the disparity between advancing and receding angles. In cases where hysteresis is negligible (Figure 6.47), the drop will move, and its descent velocity results from the equilibrium between gravity and viscosity.

Consider the scenario of a drop without CAH, characterized by a small angle θ with its substrate. Due to the modest value of θ, the drop appears quite flat, leading to slow motion (since most of the liquid remains close to the solid substrate). We can assume that the dynamic CA remains similar to the static value, θ. Various methods exist for assessing the viscous resistance near a moving liquid wedge. Here, we adopt

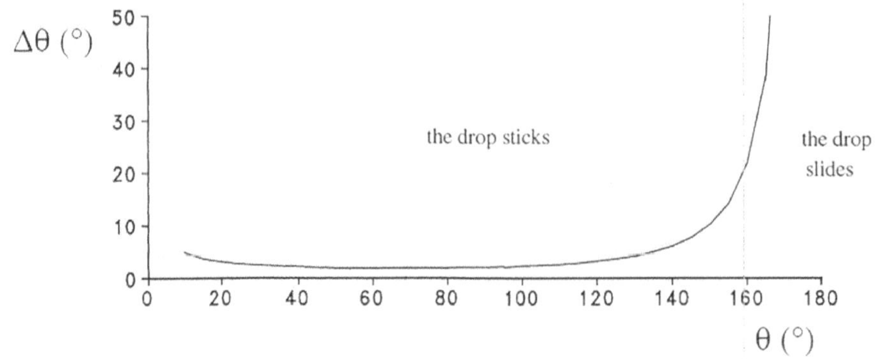

FIGURE 6.47 The quantity $\Delta\theta_c(\theta)$ is plotted in a $(\Delta\theta, \theta)$ diagram, denoting $\Delta\theta$ as the contact angle hysteresis and θ as the mean contact angle. This line, drawn for $R_0 = 0.5$ mm and $\alpha = 90°$, divides the space into two regions: in the main upper one, the drop sticks, while it slides below the line, and on its right. (Reproduced with permission from Ref. (Quéré, 2005). Copyright IOP Publishing, 1934.)

the hydrodynamic perspective (Kligerman et al., 2005; Majhy et al., 2020): a liquid wedge exhibits a thickness $\delta = \theta y$ near the contact line, where $y = 0$ denotes the contact line position. Integrating $\eta V/\delta$ over the drop surface results in a logarithmic divergence of the viscous force. This divergence is managed by introducing two cut-off lengths (typically the drop size and a molecular dimension): the logarithmic value approximates 10–15, a significant figure that cannot be disregarded when estimating drop velocity orders of magnitude.

At a scaling level, the viscous force can be expressed as $\eta Vl/\theta$, where l represents the solid/liquid contact. Expanding l at a small angle, we find $l \sim R_0/\theta^{1/3}$, and balancing this force with gravity ($\rho g R_0^3\ \alpha$, in the case of small slopes), yields

$$V \sim \alpha \frac{\rho g R o^2}{\eta} \theta^{2/3} \tag{6.52}$$

This velocity approaches zero as the drop flattens due to wetting reasons ($\theta \to 0$). Even for substantial CAs, its value remains modest. For instance, with $\theta = 1$ rd, $R_0 = 1$ mm, $\eta = 1$ Pa s, and $\alpha = 0.1$ rd, considering the logarithmic term, the velocity is approximately 100 μm/s[1]. For $R_0 = 100$ μm, this velocity reduces to 1 μm/s[1]. Similar to Stokes' law of sedimentation, V decreases rapidly as the size of the moving body decreases, as experimentally verified by Podgorski et al. (2001). (It is worth noting that with the presence of CAH opposing motion, the velocity will be lower than predicted by Equation 6.39 (Dussan, 1985).) The subsequent subsection will explore how these trends are significantly altered in the zero-wetting limit.

6.11.4 Rolling Droplets

In the context of drops approaching the limit of zero wetting, their shape tends to become nearly spherical, as illustrated in Figure 6.49. A "contact" with the substrate, defined as the zone below the drop that is parallel to the substrate, can be established

FIGURE 6.48 Millimetric water drop deposited on a plate at 250°C. A vapor film (visible in the picture) intercalates between the liquid and the solid, and thus induces a nonwetting situation. The bar indicates 1 mm. (Reproduced with permission from Ref. (Quéré, 2005). Copyright IOP Publishing, 1934.)

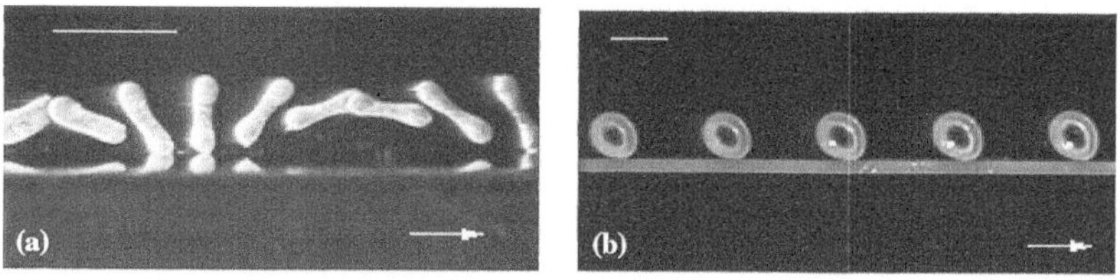

FIGURE 6.49 Different shapes of a liquid marble in the high velocity regime. Both the camera and the plane are tilted (by an angle of 35°), and the arrows indicate the direction of the motion. The white line represents 1 cm. (a) The drop ($R_0 = 1.3$ mm) deforms into a peanut (interval between two pictures is 9 ms); (b) the drop ($R_0 = 2.5$ mm) deforms into a doughnut (interval between two pictures is 23 ms). (Reproduced with permission from Ref. (Quéré, 2005). Copyright IOP Publishing, 1934.)

with a certain size denoted by l. However, as the drying process progresses ($\varepsilon \to 0$) approaching 0, the size l diminishes proportionally to εR_0 (denoting ε as $\pi - \theta$). Nevertheless, despite this reduction, a contact, as defined earlier, remains observable (Figure 6.48), a phenomenon attributed by Mahadevan and Pomeau to the drop's weight. The gravitational force tends to lower the center of mass, thereby increasing the contact's size rapidly, a phenomenon recognized as the Hertz geometry, depicting a sphere slightly penetrating a plane.

Conversely, surface tension counteracts the deformation caused by this contact formation, ultimately leading to an equilibrium between gravity and surface tension to determine l. Given the quasispherical nature of the drop with a radius, R_0, the Laplace pressure within it amounts to $2\gamma/R_0$. Transitioning across the flat interface beneath the drop, an equivalent pressure must be encountered, akin to the weight of the drop $4\pi\rho g R_0^3/3$ divided by the contact's surface area πl^2. From this analysis, it follows that the contact size scales as (Newton et al., 2006)

$$l \sim R_0^{\,2} / \kappa^{-1} \tag{6.53}$$

This scaling law diverges significantly from typical wetting contacts, where the contact size linearly correlates with the drop size R_0. The implications of this discrepancy on drop dynamics will be discussed subsequently.

The determination of the CA of a nonwetting liquid marble presents relatively straightforward measurement opportunities, particularly when placed on a transparent substrate like glass, facilitating direct observation. Viscous liquid marbles conform to the scaling predicted by Equation 6.53 (Newton et al., 2006). Moreover, this scaling holds true for bubbles situated atop a container filled with a wetting liquid, inducing nonwetting behavior in the bubble. Numerical derivation of Equation 6.53 through the solution of the Young–Laplace equation for a nonwetting liquid enables precise analysis of its numerical coefficient (Newton et al., 2006). Such computational resolutions aid in discerning between wetting and weight contacts, especially for large, albeit non-180°, CAs, where the contact size is expected to adhere to the larger of the values computed from both equations.

An intriguing scenario arises when droplets are set in motion. Mahadevan and Pomeau first highlighted that the mode of motion minimizing viscous dissipation under small Reynolds numbers involves a solid-like rotation of the drops, hence justifying the term "liquid marbles" for these nonwetting drops. Richard and Quéré subsequently confirmed this concept experimentally. The fit overlaid onto the data indicates a trajectory closely resembling a cycloid, indicative of solid-like rotation.

6.12 QUICK MOTIONS

6.12.1 Inertial Deformations

In the context of the experiments described herein, altering the velocity, often achieved by tilting the substrate, is expected to result in deformations of droplets. These deformations can be either viscous or inertial in nature, leading to a breakdown of the Mahadevan–Pomeau picture. For a creeping droplet in partial wetting scenarios, where drop velocity tends to be modest, inertial deformations are usually negligible. In such cases, the primary effect involves viscous deformation at the rear of the droplet, causing elongation (potentially forming a distinctive cuspidal shape) and leaving a lengthy liquid trail behind (Podgorski et al., 2001). This extended liquid trail can be problematic in applications where maintaining a dry solid surface after the passage of the droplet is desirable.

In situations characterized by zero wetting, the conventional effect observed in partial wetting scenarios is absent due to the lack of contact lines and solid rotation, minimizing viscous dissipation. Consequently, the terminal velocity of a viscous liquid marble descending a steeply inclined plate is primarily determined by friction with the air. This results in a significantly higher velocity than usual, leading to inertial deformations. Figure 6.49 illustrates how a quasispherical liquid marble can deform when moving at a much higher velocity than previously assumed or observed (Newton et al., 2006). These high-velocity regimes are easily explored by placing a millimetric viscous marble on a tilted plate with an inclination angle (α) of 36°, contrasting with less than 8° as in the previous section.

Upon observation, it is apparent that the droplet loses its quasispherical shape in such conditions. A detailed motion analysis reveals a complex combination of slip and rotation. The rotational speed (ω) cannot be straightforwardly determined from the droplet size and translational velocity (Aussillous and Quéré, 2004). Nevertheless, ω is significant enough to induce a strong centrifugal force, causing the globule to transform into shapes resembling a peanut or a doughnut. These deformations align with the natural behavior of a deformable rotating body, as proposed by early researchers such as Poincaré (Ji and Wang, 2022) and Rayleigh (1914) and comprehensively described by Brown and Scriven (n.d.-a).

Despite the high velocity, these droplets maintain their integrity during motion, exhibiting no leakage. This intriguing property, preserving the content of the droplet throughout its motion, is attributed to the absence of viscous dissipation at the rear of the droplet. Unlike scenarios where a liquid trail is deposited due to specific viscous behavior, in these conditions, viscous effects are generally negligible. This is especially true considering the possible presence of a dynamic air film between the substrate and the droplet, akin to a dynamic version of a Leidenfrost film. The properties of this film require further investigation. Consequently, the expected terminal velocity of the droplet is anticipated to result from a balance between weight (ρgR0^3sin α) and air friction (with density ρa), $\rho a V^2 R_0^2$. For simplicity, it is assumed that the droplet surface area remains on the order of R_0^2 despite deformation. This yields a typical descent velocity on the order of $\sqrt{\rho g R_0 \sin \alpha / \rho a}$, typically a few meters per second, emphasizing once again the remarkable mobility of droplets in nonwetting regimes.

6.12.2 Shape of Revolving Drops

The morphology of a rotating drop under significant deformation can be delineated with greater precision, dating back to inquiries by astrophysicists exploring the configurations of revolving celestial entities. Newton hypothesized that the Earth's form should approximate an ellipsoid, with flattened poles and a bulging equator due to centrifugal forces. Subsequently, Laplace and Rayleigh extended this inquiry to higher rotational velocities, suggesting that the ellipsoid may morph into a disc, or even assume a toroidal shape—although Rayleigh expressed reservations regarding the stability of the latter form, possibly due to the Rayleigh instability, which tends to fragment elongated liquid threads into smaller droplets. Applying basic scaling principles aids in comprehending the characteristic dimensions of these configurations. For a disc with thickness h and radius R, the Laplace pressure gradient striving to revert to a spherical form scales as γ/h^2. This is counterbalanced by the centrifugal force ($R^2h \sim R_0^3$), leading to a relationship for the disc radius in terms of the rotation speed:

$$R \sim R_0\Omega^{2/3} \tag{6.54}$$

Here, $\Omega s = (\rho\omega^2R_0^3/\gamma)^{1/2}$ denotes the rotation number introduced by Brown and Scriven. The regime of significant deformation under consideration implies Ωs of approximately unity, suggesting observations with relatively large drops. Analogous reasoning applies to configurations such as peanuts, tori, and other geometries. The expressions for the Laplace restoring force and volume conservation vary for each shape, yielding distinct relationships between the size R of the structure and the parameter Ωs. Nonetheless, as depicted in Figure 6.49, diverse shapes can coexist under similar experimental conditions, prompting inquiries into their relative stability.

Chandrasekhar initially demonstrated that a peanut configuration possesses lower surface energy compared to a torus. However, the latter, albeit metastable, configuration can be observed when a substrate is present beneath the drop. Eliminating the substrate induces instability in the torus, causing it to pinch and eventually transition into a peanut shape. The precise mechanism underlying the stabilization of the torus remains unresolved.

6.12.3 Pure Sliding

The earlier sections of this study primarily addressed the behavior of viscous nonstick droplets. However, the investigation into liquids with low viscosity, such as water, has been relatively limited despite its practical significance. As previously discussed, the droplet's terminal velocity is anticipated to be influenced by air friction surrounding its movement, although distinctions exist. One notable variation pertains to the initial phase. In the case of a water droplet on a superhydrophobic surface, a straightforward relationship governs the droplet's position: $1/2\ gt^2 \sin\alpha$, indicating a pure sliding motion. This pattern is observable over short distances, typically within the scale of a leaf.

Subsequently, air friction constrains the droplet's speed, triggering rotational motion within the droplet (the delay in initiating rotation may stem from inertia or viscosity, contingent upon the values of ρ and η). Consequently, the droplet's deformation during the terminal velocity phase diverges notably from the aforementioned scenario. Initially, the droplet elongates akin to a classical droplet, with a trailing effect likely developing at its rear. However, the trailing liquid bends upwards, approximately perpendicular to the substrate, preventing liquid deposition at the droplet's rear. Although many aspects of this phenomenon require further elucidation, it is noteworthy that this characteristic effectively maintains the substrate dry after the droplet's passage.

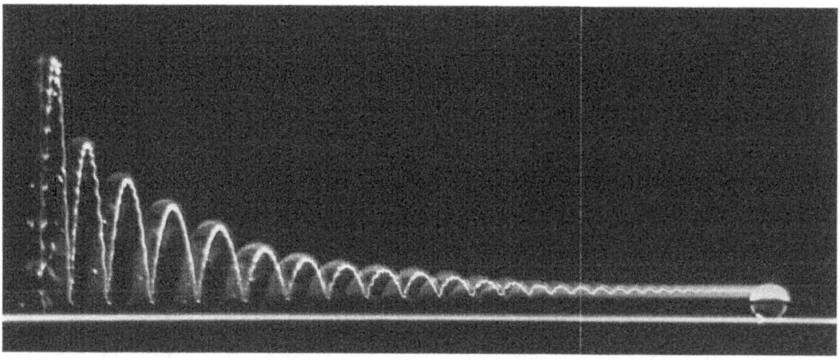

FIGURE 6.50 Trajectory (as seen in a long-time exposure photograph) of a millimetric water drop hitting a very slightly inclined superhydrophobic plate. Many rebounds are observed. Note also the vibration of the line of light after each take-off, demonstrating the vibration of the drop. (Reproduced with permission from Ref. (Quéré, 2005). Copyright IOP Publishing, 1934.)

6.12.4 Bouncing Drops

6.12.4.1 Water Repellency

When encountering solid surfaces, nonwetting drops exhibit a distinctive behavior: they rebound upon impact, as depicted in Figure 6.50. This phenomenon was initially documented by Hartley and Brunskill (n.d.-b) in their study on water droplets hitting a nonwettable foliage. Such behavior poses challenges for pesticide applications: the repellent nature of the leaf causes ineffective treatment, leading to potential soil pollution and disease spread due to drop dispersion. Nevertheless, this characteristic proves advantageous for superhydrophobic materials, including waterproof fabrics, concrete, and hydrophobic glass, which maintain dryness even during rainfall, earning them the designation of water-repellent.

6.12.4.2 Elasticity of Shock

During impact, the kinetic energy of the droplet can undergo storage through surface deformation. In standard scenarios, the shock induces deformation, followed by either retraction (in partial wetting) or spreading (in complete wetting). These movements result in significant dissipation of initial kinetic energy, primarily due to moving contact lines. In cases of nonadhesive droplets, the absence of contact lines occurs due to nonwetting conditions (as in liquid marbles or the Leidenfrost effect) or due to the presence of a dynamic air film between the solid and the liquid upon impact. This reduction in viscous dissipation facilitates bouncing, especially for drops with CAs exceeding 150°–160° and with sufficient impact velocity.

Remarkably, shocks exhibit considerable elasticity on superhydrophobic surfaces, with observed restitution coefficients reaching 0.91 (Richard and Quéré, 2000). Leidenfrost drops even approach unity in their restitution coefficients (Biance et al., 2006). Higher impact velocities correlate with reduced elasticity, attributed to inertial effects. In such shocks, kinetic energy transforms into internal vibrational energy, evidenced by drop vibrations postimpact, which dissipate due to viscosity. Higher velocities lead to greater drop deformations and consequently lower elasticity

(Biance et al., 2006). Additionally, significant deformations during impact may result in drop fragmentation, generating satellites, albeit with negligible mass compared to the main droplet, especially at Weber numbers around 10.

Viscosity plays a minor role in the shock, suggesting bouncing as primarily a capillary-inertial phenomenon akin to a spring forced into vibration by a shock. Unlike solid shocks that induce local deformations, the entire drop deforms, acting as a spring with stiffness denoted by γ. The response time of such a spring scale as $(M/\gamma)^{1/2}$, where M represents the spring mass. For drops, $M \sim \rho R^3$, implying a contact time scaling as $R^{3/2}$ and independent of velocity. Experimental observations confirm these scalings for impact velocities typically exceeding 10 cm s^{-1} for millimeter-sized drops (Richard et al., 2002). At very low impact velocities, the contact time increases as velocity decreases, attributed to the balance between potential, kinetic, and surface tension energies, necessitating consideration of all these parameters (Okumura et al., 2003).

6.13 SPECIAL PROPERTIES

6.13.1 Anisotropy

Various textures, both natural (as depicted in Figure 6.22) and synthetic (illustrated in Figure 6.23), exhibit isotropy. However, there is an intriguing possibility in engineering directional structures, such as arrays featuring parallel grooves or microwrinkles, to induce anisotropic wetting, notably within the Cassie regime. This results from differential pinning of contact lines, leading to distinct contact angles and hysteresis along and perpendicular to the grooves. Preferential axial motion characterizes such designs, which prove advantageous in situations necessitating liquid guidance.

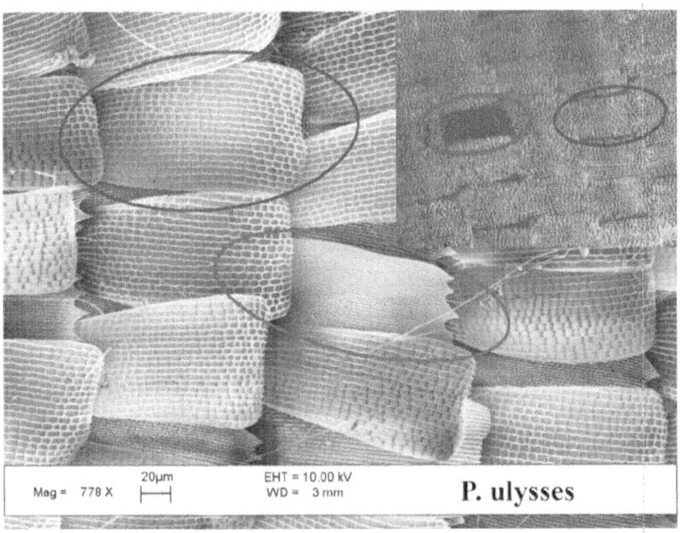

FIGURE 6.51 The wings of *Papilio ulysses*. The way the tiles are displayed together with the detail of the texture confer anisotropy to the texture. (Reproduced with permission from Ref. (Tam et al., 2013). Copyright Optica Publications, 2013.)

Examples of such directional patterns are found in nature such as the scales on the wings of the *Papilio ulysses* butterfly (Figure 6.51). The arrangement and micro-texture of these scales collectively contribute to the material's directionality. Water striders, exemplified by their inclined hairs in Figure 6.22c, exploit another form of anisotropy. These insects strike the surface perpendicularly to the grooves, thereby generating a significant contact force, before rotating their legs by 90° to align them with the direction of motion for skating. This motion arises from alternating pinning and gliding events.

6.13.2 Wettability Switches

Switching the wettability of textured surfaces can be achieved when the CA surpasses 90° (Figure 6.13). This phenomenon offers opportunities to create switches between superhydrophobic and superhydrophilic states. Various physicochemical factors influence solid wettability, such as light exposure for photocatalytic textures (Zhang et al., 2004) and heat for temperature-sensitive coatings (Sun et al., 2004), which can trigger this transition. Feng and Jiang (2006) provide a detailed review on this topic.

Additionally, electric fields, as demonstrated by Lippman, impact wettability. Applying voltage across a droplet reduces its CA, especially pronounced on textured surfaces. Even a modest voltage (around 10 V) can cause a drop with an angle of approximately 160° to spread significantly (Krupenkin et al., 2004). However, achieving reversible switching becomes challenging as the liquid becomes irreversibly pinned in the superhydrophilic state upon intimate contact with the rough solid, contrary to expectations for a switch. Krupenkin et al. (2007) suggested employing a short, intense pulse of current through a thin conductive layer on the sample, which evaporates the liquid near the surface, thereby restoring a Cassie-suspended state.

Presently, there are no clear instances of a spontaneous transformation from a Wenzel state (even potentially metastable) to a Cassie state. This situation is problematic, especially during vapor condensation, leading naturally to a Wenzel state, often transitioning toward mixed and ambiguous Cassie/Wenzel states (Narhe and Beysens, 2004). Achieving authentic antidew materials requires further research and development efforts.

6.13.3 Giant Slip

Experimental findings confirm the capacity of superhydrophobic materials to induce slippage in water flow, particularly when these materials are in the Cassie state. The extent of this phenomenon is quantified by the slip length, which represents the extrapolated distance at which the liquid velocity ceases (Figure 6.52). Initially proposed by Navier, this length primarily pertains to the molecular scale but can extend to approximately 10 nm on flat hydrophobic surfaces, as evidenced by experiments utilizing a surface force apparatus (Cottin-Bizonne et al., 2005). The presence of surface roughness can significantly augment this hydrophobic behavior (Cécile Cottin-Bizonne et al., 2003).

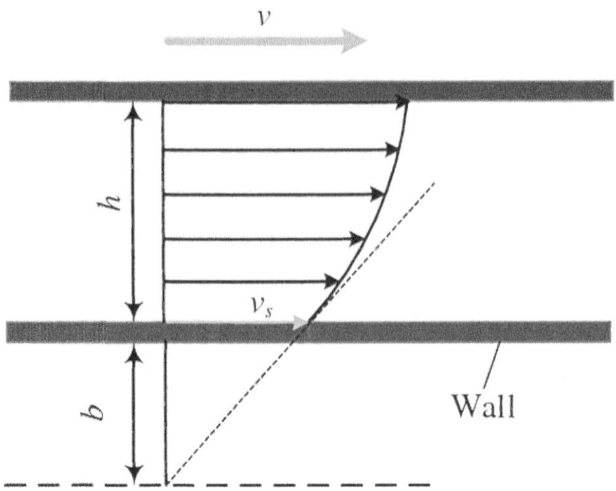

FIGURE 6.52 The slip length b is the distance inside the solid for which the velocity profile of a flowing liquid vanishes. (Reproduced with permission from Ref. (Liu et al., 2021). Copyright Springer Nature, 2021.)

Ou et al. (2004) observed micrometric slip lengths through pressure measurements needed to propel a specific water flux along a microgroove-patterned square channel. In the context of Poiseuille flow, the flux correlates with $W^4\nabla p/\eta$, where ∇p represents the pressure gradient and W signifies the width and depth of the channel. Conversely, for extensive slip ($\lambda > W$), the flux scales as $\lambda W^3\nabla p/\eta$, where slip at the wall diminishes the pressure gradient by a factor of λ/W. Ou et al. estimated a pressure reduction of approximately 40%, suggesting slip lengths on the order of 10 µm (Roach et al., 2007).

Joseph et al. (2006) employed microparticle imaging velocimetry to directly measure slip lengths for water flowing over hydrophobic CNTs. Slip length varies significantly depending on the hydrophobic nature. In the Wenzel state (induced by liquid compression), no measurable slip is observed, aligning with Richardson's assertion that roughness negates potential slippage if the liquid conforms to it (Richardson, 1973). Conversely, micrometric slip lengths are evident in the Cassie state and increase linearly with post distance, as observed in experiments conducted under constant φS. The magnitude of this slip can be substantially attenuated if liquid/air interfaces are curved due to, for instance, pressure exertion on the liquid (Steinberger et al., 2007).

The anticipation of an extensive slip in a Cassie situation is logical: liquid glides on air due to the significant viscosity contrast between water and air (typically by a factor of 100). However, a portion of the liquid contacts the tops of the posts, limiting total surface slip. This argument can be quantified further, as demonstrated in a recent analysis by Ybert et al. (2007). For a flow with typical velocity V, the area "affected" by post presence should scale as b, the postradius (where $b << p$, the postarray pitch). The friction force per pillar, and hence per surface area p^2, should scale as ηVb, with η representing liquid viscosity. This yields a viscous stress $\sigma \sim \eta Vb/p^2$, which surpasses the stress arising from underlying air flow provided

$\eta Vb/p^2 > \eta_a V/h$, introducing air viscosity η_a and postheight h. Therefore, a geometric requirement $bh/p^2 > \eta_a/\eta$ can be met by adjusting post height h. With water, η_a/η is approximately 10^{-2}, and the criterion is satisfied with posts of dimensions $b = 1\,\mu m$, $h = 10\,\mu m$, and $p = 10\,\mu m$, where the factor bh/p^2 is 10^{-1}. The stress σ can also be expressed as $\eta V/\lambda$, yielding an effective slip length λ:

$$\lambda \sim p^2/b \sim p/\varphi_S^{1/2} \tag{6.55}$$

For a constant pillar density φ_S, λ varies linearly with pillar spacing p. For $b \ll p$ ($\varphi_S \ll 1$), extensive slip lengths are anticipated, ranging between a few p, i.e., 1–10 μm (as observed experimentally), significantly larger than slip lengths on flat hydrophobic surfaces. When air friction predominates pillar friction ($h < \eta_a p^2/\eta b$), the slip length becomes $\lambda \sim \eta h/\eta_a$, potentially substantial.

The CA θ^* is also influenced by density φ_S (Equation 6.34), thus correlating θ^* and λ. At low φ_S, $\theta^* = \pi - \varepsilon$. Incorporating Equation 6.43 into Equation 6.35 [where $\lambda \sim p/\varepsilon$, which directly links nonwetting quality to slip amplitude], it is practical to maximize slip and operate with low post densities. Nonetheless, in this scenario, liquid is prone to infiltrating the texture, leading to slip properties failure. Hence, an optimum must be sought, and the microstructure design warrants scrutiny. The design can also induce slip anisotropy; for instance, grooves typically exhibit larger slip along them than perpendicular to them. It can be proven that slip length (again approximately p, the distance between grooves) differs by a factor of two in both directions, reflecting the factor-of-two difference in viscous force on a slender object (like a cylinder) in both directions (Philip, 1972).

6.14 APPLICATION OF HYDROPHILIC/HYDROPHOBIC SURFACES AND SCOPE FOR FUTURE RESEARCH

6.14.1 Anti-Icing

The phenomenon of icing is widely existing in nature and is a menace in the field of aerospace, wind power generation, and other fields. Researchers conducted studies on anti-icing technology, and further research is needed to obtain an efficient, cost-effective, eco-friendly, and durable anti-icing surface (Lv et al., 2014). One of the important equipment in aircraft is the fans that are affected in a low-temperature environment. In such cases, the icy layer deteriorates the aerodynamic shape of the aircraft, therefore minimizing the speed of the fan, and increasing the weight, which may lead to an aircraft crash at the time of landing. The icing on the surface of fan blades tries to decrease the power generation efficiency as well. Many superhydrophobic surfaces of carbon fiber composites with easy processes provide stable, excellent hydrophobic, and anti-icing properties (Allahdini et al., 2022).

6.14.2 Antifouling

CNTs are used in antifouling because of their excellent hydrophobicity and high electrical conductivity among other properties. Biofouling is mainly used to prevent the unwanted growth and accumulation of microorganisms. This can occur at solid–liquid, liquid–gas, and solid–gas contact, leading to the formation of biofilms. Biofouling is the accumulation of unwanted material on surfaces. From living tissues, such as organ implants and human teeth, to semi-aquatic surfaces, biofilm occurs everywhere. Even if there is 99.99% contamination of microorganisms removed, 0.001% of them is enough to get the system contaminated (Jin et al., 2022). Further investigation is needed into antifouling techniques such as the creation of superhydrophobic antimicrobial coatings, the use of biological mimics in surface modification, and the combination of chemical processes. Research should also be done on other methods like engineered polymers for surface modification, small molecules for surface grafting, surface coatings, and cell repellent (Gao et al., 2022).

6.14.3 Self-Cleaning

Apart from many applications, one of the finest properties of a superhydrophobic or hydrophilic substance is self-cleaning, as not all coatings can produce such a surface. Surfaces with a WCA > 90° or WCA > 150° are considered hydrophobic and superhydrophobic surfaces, respectively. Also, by combining small-scale/nano-scale roughness with hydrophobicity, a superhydrophobic surface can be achieved. The application of this method is mainly used in coatings to reduce maintenance costs, increase prevention from snow adhesion, and self-cleaning solar panels for commercial as well as individual purposes. In automotive parts, the application is vast such as self-cleaning coatings on car bodies and mirrors, door and plastic roofs of the high-rise buildings, household taps, and further in sensors and aerospace (Y. Wu et al., 2022). Further research must focus on exploring a better mechanism for self-cleaning coating material. Such research should aim to enhance surfaces' ability to self-clean and their longevity, and create surfaces with particular extra functions in addition to self-cleaning.

6.14.4 Antifogging

It is the phenomenon of humid air condensed and transformed into discrete liquid droplets on solid materials having a less or equal temperature at the dew point. This causes a blurred view or an irregular pattern in the presence of scattering of incident light in all directions. Much interest is shown in superhydrophilic and superhydrophobic surfaces as they can curb the effects of fogging through a change in the fog droplet morphology. On these surfaces, the material tends to form a thin film of water. Therefore, the scattering of light is limited and the fogging is minimized. Future research must concentrate on the increase in the use of eco-friendly hydrophilic materials. New morphologies and different strategies are needed to bring innovative antifogging materials. More importantly, researchers must develop less time-consuming and eco-friendly antifogging materials, which should be biocompatible as well (Minjing Li et al., 2022b).

6.15 INFERENCES

This chapter discusses theory with the examination of particle interactions at the air/liquid interface that takes place during the stabilization/detachment process. Measuring the CA can give a numerical assessment of the liquid's attraction to the solid surface;

however, as the particle size, hydrophilicity/hydrophobicity, surface texture, and variability are all highly delicate and influential factors, obtaining the exact CA measurement and obtaining consistent results can be difficult. From Young's equation to Wenzel and Cassie–Baxter's equation were discussed for explaining the roughness of the solid surface. All other methods to calculate surface energies were discussed, such as Wu, OWRK, Neumann, and Zisman methods. A quick and effective way to comprehend the physicochemical interplay of the surface with the liquid is to determine the CA.

A new type of understanding of ACA and RCA and their uses in identifying solid surface energy, along with its parameters, are needed with corrections. At maximum adhesion force, the experimental condition is unstable to measure the CA of solid surfaces. Therefore, further research is needed in this area to identify the solid/liquid surfaces and their affinity toward fluids. To comprehend the depth of CAH, which provides the basis for interactions at interfaces/contact lines, both liquid/solid and liquid/liquid systems must be examined in detail (Edachery, Swamybabu, et al., 2022).

REFERENCES

Adamczyk, Z, G Para, and P Warszyński. 1999. "Influence of Ionic Strength on Surface Tension of Cetyltrimethylammonium Bromide." *Langmuir* 15 (24): 8383–87.

Adamson, AW, and AP Gast. 1967. *Physical Chemistry of Surfaces*, volume 150. Interscience Publishers.

Ahuir-Torres, JI, MA Arenas, W Perrie, and J De Damborenea. 2018. "Influence of Laser Parameters in Surface Texturing of Ti6Al4V and AA2024-T3 Alloys." *Optics and Lasers in Engineering* 103:100–9.

Alberti, G, and A DeSimone. 2005. "Wetting of Rough Surfaces: A Homogenization Approach." *Proceedings of the Royal Society A: Mathematical, Physical and Engineering Sciences* 461 (2053): 79–97.

Allahdini, A, R Jafari, and G Momen. 2022. "Transparent Non-Fluorinated Superhydrophobic Coating with Enhanced Anti-Icing Performance." *Progress in Organic Coatings* 165:106758.

Amigoni, S, ET de Givenchy, M Dufay, and F Guittard. 2009. "Covalent Layer-by-Layer Assembled Superhydrophobic Organic–Inorganic Hybrid Films." *Langmuir* 25 (18): 11073–77. https://doi.org/10.1021/la901369f.

Anders, A. 2000. *Handbook of Plasma Immersion Ion Implantation and Deposition*. Vol. 8. Wiley New York.

Asri, RIM, WSW Harun, MA Hassan, SAC Ghani, and Z Buyong. 2016. "A Review of Hydroxyapatite-Based Coating Techniques: Sol–Gel and Electrochemical Depositions on Biocompatible Metals." *Journal of the Mechanical Behavior of Biomedical Materials* 57:95–108.

Aussillous, P, and D Quéré. 2004. "Shapes of Rolling Liquid Drops." *Journal of Fluid Mechanics* 512:133–51. https://doi.org/10.1017/S0022112004009747.

Aveyard, R, and DA Haydon. 1973. *An Introduction to the Principles of Surface Chemistry*. Cambridge University Press.

Bakhshzadmahmoudi, M, S Jamali, and E Ahmadi. 2022. "Wettability Modification of Polystyrene Surface by Cold Atmospheric Pressure Plasma Jet." *Colloid and Polymer Science*, 300 (2):103–110.

Barati, Gh, M Aliofkhazraei, S Khorsand, S Sokhanvar, and A Kaboli. 2020. "Science and Engineering of Superhydrophobic Surfaces: Review of Corrosion Resistance, Chemical and Mechanical Stability." *Arabian Journal of Chemistry* 13 (1): 1763–1802. https://doi.org/10.1016/j.arabjc.2018.01.013.

Barthlott, W, and C Neinhuis. 1997. "Purity of the Sacred Lotus, or Escape from Contamination in Biological Surfaces." *Planta* 202:1–8.

Bartolo, D, F Bouamrirene, E Verneuil, A Buguin, P Silberzan, and S Moulinet. 2006. "Bouncing or Sticky Droplets: Impalement Transitions on Superhydrophobic Micropatterned Surfaces." *Europhysics Letters* 74 (2): 299.

Bayer, IS, A Steele, PJ Martorana, and E Loth. 2010. "Fabrication of Superhydrophobic Polyurethane/Organoclay Nano-Structured Composites from Cyclomethicone-in-Water Emulsions." *Applied Surface Science* 257 (3): 823–26.

Berg, JC. 2010. *An Introduction to Interfaces & Colloids: The Bridge to Nanoscience*. World Scientific.

Berg, JM, LG Tomas Eriksson, PM Claesson, and KGN Borve. 1994. "Three-Component Langmuir-Blodgett Films with a Controllable Degree of Polarity." *Langmuir* 10 (4): 1225–34.

Beyhaghi, S, S Geoffroy, M Prat, and KM Pillai. 2014. "Wicking and Evaporation of Liquids in Porous Wicks: A Simple Analytical Approach to Optimization of Wick Design." *AIChE Journal* 60 (5): 1930–40. https://doi.org/10.1002/aic.14353.

Biance, A-L, F Chevy, C Clanet, G Lagubeau, and D Quéré. 2006. "On the Elasticity of an Inertial Liquid Shock." *Journal of Fluid Mechanics* 554:47–66. https://doi.org/10.1017/S0022112006009189.

Bico, J, C Marzolin, and D Quéré. 1999. "Pearl Drops." *Europhysics Letters* 47 (2): 220.

Bico, J, C Tordeux, and D Quéré. 2001. "Rough Wetting." *Europhysics Letters* 55 (2): 214.

Bijalwan, P, KK Pandey, S Sharma, P Singh, A Dan, A Banerjee, AN Bhagat, and A Kumar Keshri. 2022. "Single-Step Approach to Tune the Wettability of Plasma Sprayed Crystalline and Amorphous Fe-Based Coating." *Surfaces and Interfaces* 30:101979. https://doi.org/10.1016/j.surfin.2022.101979.

Bizi-Bandoki, P, S Valette, E Audouard, and S Benayoun. 2013. "Time Dependency of the Hydrophilicity and Hydrophobicity of Metallic Alloys Subjected to Femtosecond Laser Irradiations." *Applied Surface Science* 273:399–407.

Bormashenko, E, R Pogreb, G Whyman, Y Bormashenko, and M Erlich. 2007. "Vibration-Induced Cassie-Wenzel Wetting Transition on Rough Surfaces." *Applied Physics Letters* 90 (20): 201917.

Bormashenko, E, Y Bormashenko, T Stein, G Whyman, and E Bormashenko. 2007. "Why Do Pigeon Feathers Repel Water? Hydrophobicity of Pennae, Cassie–Baxter Wetting Hypothesis and Cassie–Wenzel Capillarity-Induced Wetting Transition." *Journal of Colloid and Interface Science* 311 (1): 212–16.

Brinker, CJ, and GW Scherer. 2013. *Sol-Gel Science: The Physics and Chemistry of Sol-Gel Processing*. Academic Press.

Brown, RA, and LE Scriven. n.d.-a. *The Shape and Stability of Rotating Liquid Drops*. https://doi.org/10.1098/rspa.1980.0084.

Brunette, DM, P Tengvall, M Textor, and P Thomsen. 2001. *Titanium in Medicine: Material Science, Surface Science, Engineering, Biological Responses and Medical Applications*. Springer.

Brzoska, JB, N Shahidzadeh, and F Rondelez. 1992. "Evidence of a Transition Temperature for the Optimum Deposition of Grafted Monolayer Coatings." *Nature* 360 (6406): 719–21.

Bush, JWM, DL Hu, and M Prakash. 2007. "The Integument of Water-Walking Arthropods: Form and Function." *Advances in Insect Physiology* 34:117–92.

Butt, H-J, J Liu, K Koynov, B Straub, C Hinduja, I Roismann, R Berger, et al. 2022. "Contact Angle Hysteresis." *Current Opinion in Colloid & Interface Science* 59 (June):101574. https://doi.org/10.1016/j.cocis.2022.101574.

Cai, T-m, Z-h Jia, H-n Yang, and G Wang. 2016. "Investigation of Cassie-Wenzel Wetting Transitions on Microstructured Surfaces." *Colloid and Polymer Science* 294 (5): 833–40. https://doi.org/10.1007/s00396-016-3836-4.

Callies, M, and D Quéré. 2005. "On Water Repellency." *Soft Matter* 1 (1): 55–61.

Carrino, L, G Moroni, and W Polini. 2002. "Cold Plasma Treatment of Polypropylene Surface: A Study on Wettability and Adhesion." *Journal of Materials Processing Technology* 121 (2–3): 373–82.

Cassie, ABD, and S Baxter. 1944. "Wettability of Porous Surfaces." *Transactions of the Faraday Society* 40:546–51.

Cassie, ABD, and S Baxter. 1945. "Large Contact Angles of Plant and Animal Surfaces." *Nature* 155 (3923): 21–22.

Cauchy, X, M Gagne, JE Klemberg-Sapieha, F Sirois, and D Therriault. 2016. "Wet Chemical Metallization of Aerospace Composites as a Lightning Protection Strategy." *Engineering Solutions for Sustainability: Materials and Resources* II, 187–92.

Chang, F-M, S-L Cheng, S-J Hong, Y-J Sheng, and H-K Tsao. 2010. "Superhydrophilicity to Superhydrophobicity Transition of CuO Nanowire Films." *Applied Physics Letters* 96 (11): 114101.

Chason, E, ST Picraux, JM Poate, JO Borland, MI Current, TD de La Rubia, DJ Eaglesham, OW Holland, ME Law, and CW Magee. 1997. "Ion Beams in Silicon Processing and Characterization." *Journal of Applied Physics* 81 (10): 6513–61.

Chau, TT, WJ Bruckard, PTL Koh, and AV Nguyen. 2009. "A Review of Factors That Affect Contact Angle and Implications for Flotation Practice." *Advances in Colloid and Interface Science* 150 (2): 106–15.

Checco, A, P Guenoun, and J Daillant. 2003. "Nonlinear Dependence of the Contact Angle of Nanodroplets on Contact Line Curvature." *Physical Review Letters* 91 (18): 186101.

Chen, H, J Peng, L Fu, X Wang, and Y Xie. 2016. "Solder Wetting Behavior Enhancement via Laser-Textured Surface Microcosmic Topography." *Applied Surface Science* 368:208–15.

Chen, T, H Liu, H Yang, W Yan, W Zhu, and H Liu. 2016. "Biomimetic Fabrication of Robust Self-Assembly Superhydrophobic Surfaces with Corrosion Resistance Properties on Stainless Steel Substrate." *Rsc Advances* 6 (50): 43937–49.

Cheng, F, Y Xu, J Zhang, L Wang, H Zhang, Q Wan, W Li, L Wang, and Z Lv. 2023. "Growing Carbon Nanotubes In-Situ via Chemical Vapor Deposition and Resin Pre-Coating Treatment on Anodized Ti-6Al-4V Titanium Substrates for Stronger Adhesive Bonding with Carbon Fiber Composites." *Surface and Coatings Technology* 457:129296. https://doi.org/10.1016/j.surfcoat.2023.129296.

Cheng, Y, Z Peng, X Wu, J Cao, P Skeldon, and GE Thompson. 2015. "A Comparison of Plasma Electrolytic Oxidation of Ti-6Al-4V and Zircaloy-2 Alloys in a Silicate-Hexametaphosphate Electrolyte." *Electrochimica Acta* 165:301–13.

Chow, RT-P. 1983. "On the Ability of Drops or Bubbles to Stick to Non-Horizontal Surfaces of Solids." *Journal of Fluid Mechanics* 137:1–29.

Cottin-Bizonne, C, B Cross, A Steinberger, and E Charlaix. 2005. "Boundary Slip on Smooth Hydrophobic Surfaces: Intrinsic Effects and Possible Artifacts." *Physical Review Letters* 94 (5): 056102.

Cottin-Bizonne, C, J-L Barrat, L Bocquet, and E Charlaix. 2003. "Low-Friction Flows of Liquid at Nanopatterned Interfaces." *Nature Materials* 2 (4): 237–40.

Courbin, L, E Denieul, E Dressaire, M Roper, A Ajdari, and HA Stone. 2007. "Imbibition by Polygonal Spreading on Microdecorated Surfaces." *Nature Materials* 6 (9): 661–64.

Cui, G, S Qi, X Wang, G Tian, G Sun, W Liu, X Yan, D Wu, Z Wu, and L Zhang. 2012. "Interfacial Growth of Controllable Morphology of Silver Patterns on Plastic Substrates." *The Journal of Physical Chemistry B* 116 (40): 12349–56.

Döring, J, M Crackau, C Nestler, F Welzel, J Bertrand, and CH Lohmann. 2019. "Characteristics of Different Cathodic Arc Deposition Coatings on CoCrMo for Biomedical Applications." *Journal of the Mechanical Behavior of Biomedical Materials* 97:212–21. https://doi.org/10.1016/j.jmbbm.2019.04.026.

Drelich, J, E Chibowski, DD Meng, and K Terpilowski. 2011. "Hydrophilic and Superhydrophilic Surfaces and Materials." *Soft Matter* 7 (21): 9804–28.

Dufour, L. 1833. *Recherches Anatomiques et Physiologiques Sur Les Hémiptères: Accompagnées de Considérations Relatives à l'histoire Naturelle et à La Classification de Ces Insectes*. Vol. 4. Bachelier.

Dupré, A, and P Dupré. 1869. *Théorie Mécanique de La Chaleur*. Gauthier-Villars.

Durian, DJ, and C Franck. 1987. "Wetting Phenomena of Binary Liquid Mixtures on Chemically Altered Substrates." *Physical Review Letters* 59 (5): 555.

Dussan VEB. 1985. "On the Ability of Drops or Bubbles to Stick to Non-Horizontal Surfaces of Solids. Part 2. Small Drops or Bubbles Having Contact Angles of Arbitrary Size." *Journal of Fluid Mechanics* 151:1–20. https://doi.org/10.1017/S0022112085000842.

Edachery, V, S Ravi, AF Badiuddin, A Tomy, SV Kailas, and PS Suvin. 2022. "Wetting Behaviour of a Green Cutting Fluid (GCF); Influence of Surface Roughness and Surface Energy of AA5052, Ti6Al4V and EN31." *Materials Today: Proceedings* 62:7605–9. https://doi.org/10.1016/j.matpr.2022.04.835.

Edachery, V, V Swamybabu, D Adarsh, and SV Kailas. 2022. "Influence of Surface Roughness Frequencies and Roughness Parameters on Lubricant Wettability Transitions in Micro-Nano Scale Hierarchical Surfaces." *Tribology International* 165:107316.

Etsion, I, Y Kligerman, and G Halperin. 1999. "Analytical and Experimental Investigation of Laser-Textured Mechanical Seal Faces." *Tribology Transactions* 42 (3): 511–6. https://doi.org/10.1080/10402009908982248.

Extrand, CW. 2002. "Model for Contact Angles and Hysteresis on Rough and Ultraphobic Surfaces." *Langmuir* 18 (21): 7991–9.

Fadeev, AY, and TJ McCarthy. 1999. "Trialkylsilane Monolayers Covalently Attached to Silicon Surfaces: Wettability Studies Indicating That Molecular Topography Contributes to Contact Angle Hysteresis." *Langmuir* 15 (11): 3759–66.

Feng, XJ, and L Jiang. 2006. "Design and Creation of Superwetting/Antiwetting Surfaces." *Advanced Materials* 18 (23): 3063–78.

Fogg, GE. 1944. "Diurnal Fluctuation in a Physical Property of Leaf Cuticle." *Nature* 154 (3912): 515.

Fowkes, FM. 1987. "Role of Acid-Base Interfacial Bonding in Adhesion." *Journal of Adhesion Science and Technology* 1 (1): 7–27.

Fuentes-Fernandez, EMA, JJ Alcantar-Peña, G Lee, A Boulom, H Phan, B Smith, T Nguyen, S Sahoo, F Ruiz-Zepeda, and MJ Arellano-Jimenez. 2016. "Synthesis and Characterization of Microcrystalline Diamond to Ultrananocrystalline Diamond Films via Hot Filament Chemical Vapor Deposition for Scaling to Large Area Applications." *Thin Solid Films* 603:62–8.

Furmidge, CGL. 1962. "Studies at Phase Interfaces. I. The Sliding of Liquid Drops on Solid Surfaces and a Theory for Spray Retention." *Journal of Colloid Science* 17 (4): 309–24.

Gao, A, R Hang, L Bai, B Tang, and PK Chu. 2018. "Electrochemical Surface Engineering of Titanium-Based Alloys for Biomedical Application." *Electrochimica Acta* 271:699–718. https://doi.org/10.1016/j.electacta.2018.03.180.

Gao, L, and TJ McCarthy. 2006. "'Artificial Lotus Leaf' Prepared Using a 1945 Patent and a Commercial Textile." *Langmuir* 22 (14): 5998–6000.

Gao, L, and TJ McCarthy. 2007. "A Commercially Available Perfectly Hydrophobic Material (θA/θR= 180/180)." *Langmuir* 23 (18): 9125–7.

Gao, P, Y Wang, J Wang, F Wang, W Ma, Z Zhang, X Men, and Y Lu. 2022. "Rational Design of Durable Anti-Fouling Coatings with High Transparency, Hardness, and Flexibility." *ACS Applied Materials & Interfaces* 14 (25): 29156–66.

Gao, X, and L Jiang. 2004. "Water-Repellent Legs of Water Striders." *Nature* 432 (7013): 36.

Gao, X, X Yan, X Yao, L Xu, K Zhang, J Zhang, B Yang, and L Jiang. 2007. "The Dry-Style Antifogging Properties of Mosquito Compound Eyes and Artificial Analogues Prepared by Soft Lithography." *Advanced Materials* 19 (17): 2213–17.

Gentleman, MM, and JA Ruud. 2010. "Role of Hydroxyls in Oxide Wettability." *Langmuir* 26 (3): 1408–11.

Gindl, M, G Sinn, W Gindl, A Reiterer, and S Tschegg. 2001. "A Comparison of Different Methods to Calculate the Surface Free Energy of Wood Using Contact Angle Measurements." *Colloids and Surfaces A: Physicochemical and Engineering Aspects* 181 (1–3): 279–87.

Gourley, L, M Britten, SF Gauthier, and Y Pouliot. 1994. "Characterization of Adsorptive Fouling on Ultrafiltration Membranes by Peptides Mixtures Using Contact Angle Measurements." *Journal of Membrane Science* 97:283–89.

Gracio, JJ, QH Fan, and JC Madaleno. 2010. "Diamond Growth by Chemical Vapour Deposition." *Journal of Physics D: Applied Physics* 43 (37): 374017.

Gregorčič, P, B Šetina-Batič, and M Hočevar. 2017. "Controlling the Stainless Steel Surface Wettability by Nanosecond Direct Laser Texturing at High Fluences." *Applied Physics A* 123:1–8.

Gu, C, J Zhang, and J Tu. 2010. "A Strategy of Fast Reversible Wettability Changes of WO3 Surfaces between Superhydrophilicity and Superhydrophobicity." *Journal of Colloid and Interface Science* 352 (2): 573–79.

Guo, C, S Wang, H Liu, L Feng, Y Song, and L Jiang. 2008. "Wettability Alteration of Polymer Surfaces Produced by Scraping." *Journal of Adhesion Science and Technology* 22 (3–4): 395–402.

Guo, HF, MZ An, HB Huo, S Xu, and LJ Wu. 2006. "Microstructure Characteristic of Ceramic Coatings Fabricated on Magnesium Alloys by Micro-Arc Oxidation in Alkaline Silicate Solutions." *Applied Surface Science* 252 (22): 7911–16.

Han, R, A-M Zhang, S Tan, and S Li. 2022. "Interaction of Cavitation Bubbles with the Interface of Two Immiscible Fluids on Multiple Time Scales." *Journal of Fluid Mechanics* 932:A8.

Harkins, WD, and YC Cheng. 1921. "The Orientation of Molecules in Surfaces. VI. Cohesion, Adhesion, Tensile Strength, Tensile Energy, Negative Surface Energy, Interfacial Tension, and Molecular Attraction." *Journal of the American Chemical Society* 43 (1): 35–53.

Hartley, GS, and RT Brunskill. n.d.-b. *Surface Phenomena in Chemistry and Biology*. Pergamon Press.

He, S, M Zheng, L Yao, X Yuan, M Li, L Ma, and W Shen. 2010. "Preparation and Properties of ZnO Nanostructures by Electrochemical Anodization Method." *Applied Surface Science* 256 (8): 2557–62.

He, X, X Zhang, X Wang, and L Qin. 2017. "Review of Antibacterial Activity of Titanium-Based Implants' Surfaces Fabricated by Micro-Arc Oxidation." *Coatings* 7 (3): 45.

He, Z, X Lan, Q Hu, H Li, L Li, and J Mao. 2021. "Antifouling Strategies Based on Super-Phobic Polymer Materials." *Progress in Organic Coatings* 157 (August):106285. https://doi.org/10.1016/j.porgcoat.2021.106285.

Hebbar, RS, AM Isloor, and AF Ismail. 2017. "Chapter 12 – Contact Angle Measurements." In *Membrane Characterization*, edited by Nidal Hilal, Ahmad Fauzi Ismail, Takeshi Matsuura, and Darren Oatley-Radcliffe, 219–55. Elsevier. https://doi.org/10.1016/B978-0-444-63776-5.00012-7.

Heckman, CW. 1983. "Comparative Morphology of Arthropod Exterior Surfaces with the Capability of Binding a Film of Air Underwater." *Internationale Revue Der Gesamten Hydrobiologie Und Hydrographie* 68 (5): 715–36.

Herminghaus, S. 2000. "Roughness-Induced Non-Wetting." *Europhysics Letters* 52 (2): 165.

Hsieh, M-F, L-H Perng, and T-S Chin. 2002. "Hydroxyapatite Coating on Ti6Al4V Alloy Using a Sol–Gel Derived Precursor." *Materials Chemistry and Physics* 74 (3): 245–50. https://doi.org/10.1016/S0254-0584(01)00474-6.

Hu, Y, J Zhong, L He, S Liu, C Zhang, and W Fu. 2024. "Study the Regularity of Micro-Texture Parameters of Bone Milling Tool Surface for Wettability Improved and Bone Surface Damage Reduction." *The International Journal of Advanced Manufacturing Technology* 131 (9): 5343–52. https://doi.org/10.1007/s00170-024-13098-6.

Huang, H-H, C-F Liu, S Wang, C-S Chen, and J-H Chang. 2021. "Nitrogen Plasma Immersion Ion Implantation Treatment of Ti6Al7Nb Alloy for Bone-Implant Applications: Enhanced in Vitro Biological Responses and in Vivo Initial Bone-Implant Contact." *Surface and Coatings Technology* 405:126551. https://doi.org/10.1016/j.surfcoat.2020.126551.

Huang, W, B Nelson, S Tian, R Ordikhani-Seyedlar, RCY Auyeung, A Samanta, H Hu, S Shaw, C Lamuta, and H Ding. 2022. "Superhydrophobic Surface Processing for Metal 3D Printed Parts." *Applied Materials Today* 29:101630.

Huang, Y, DK Sarkar, and X-G Chen. 2010. "A One-Step Process to Engineer Superhydrophobic Copper Surfaces." *Materials Letters* 64 (24): 2722–24.

Huhtamäki, T, X Tian, JT Korhonen, and RHA Ras. 2018. "Surface-Wetting Characterization Using Contact-Angle Measurements." *Nature Protocols* 13 (7): 1521–38. https://doi.org/10.1038/s41596-018-0003-z.

Hussein, RO, X Nie, DO Northwood, A Yerokhin, and A Matthews. 2010. "Spectroscopic Study of Electrolytic Plasma and Discharging Behaviour during the Plasma Electrolytic Oxidation (PEO) Process." *Journal of Physics D: Applied Physics* 43 (10): 105203.

Ishino, C, M Reyssat, E Reyssat, K Okumura, and D Quere. 2007. "Wicking within Forests of Micropillars." *Europhysics Letters* 79 (5): 56005.

Islam, Md S, N Akter, and MR Karim. 2010. "Preparation of Superhydrophobic Membranes by Electrospinning of Fluorinated Silane Functionalized Pullulan." *Colloids and Surfaces A: Physicochemical and Engineering Aspects* 362 (1–3): 117–20.

Ji, L, and C Wang. 2022. "Poincaré's Works Leading to the Poincaré Conjecture." *Archive for History of Exact Sciences* 76 (3): 223–60. https://doi.org/10.1007/s00407-021-00283-2.

Jin, H, L Tian, W Bing, J Zhao, and L Ren. 2022. "Bioinspired Marine Antifouling Coatings: Status, Prospects, and Future." *Progress in Materials Science* 124:100889.

Jindasuwan, S, O Nimittrakoolchai, P Sujaridworakun, S Jinawath, and S Supothina. 2009. "Surface Characteristics of Water-Repellent Polyelectrolyte Multilayer Films Containing Various Silica Contents." *4th International Conference on Technological Advances of Thin Films and Surface Coatings* 517 (17): 5001–5. https://doi.org/10.1016/j.tsf.2009.03.116.

Jithin, S, and SS Joshi. 2021. "Surface Topography Generation and Simulation in Electrical Discharge Texturing: A Review." *Journal of Materials Processing Technology* 298:117297. https://doi.org/10.1016/j.jmatprotec.2021.117297.

Johnson, RE. 1964. "Contact Angle, Wettability, and Adhesion." *Advances in Chemistry Series* 43: 112.

Johnson Jr, RE, and RH Dettre. 1964. "Contact Angle Hysteresis: I. Study of an Idealized Rough Surface." *Contact Angle, Wettability, and Adhesion*, edited by Frederick M. Fowkes, 112–135. ACS Publications.

Joseph, P, C Cottin-Bizonne, J-M Benoit, C Ybert, C Journet, P Tabeling, and L Bocquet. 2006. "Slippage of Water Past Superhydrophobic Carbon Nanotube Forests in Microchannels." *Physical Review Letters* 97 (15): 156104.

Kabalnov, AS, KN Makarov, OVj Shcherbakova, and AN Nesmeyanov. 1990. "Solubility of Fluorocarbons in Water as a Key Parameter Determining Fluorocarbon Emulsion Stability." *Journal of Fluorine Chemistry* 50 (3): 271–84.

Kang, M, R Jung, H-S Kim, and H-J Jin. 2008. "Preparation of Superhydrophobic Polystyrene Membranes by Electrospinning." *Colloids and Surfaces A: Physicochemical and Engineering Aspects* 313:411–14.

Khoo, HS, and F-G Tseng. 2008. "Engineering the 3D Architecture and Hydrophobicity of Methyltrichlorosilane Nanostructures." *Nanotechnology* 19 (34): 345603.

Kligerman, Y, I Etsion, and A Shinkarenko. 2005. "Improving Tribological Performance of Piston Rings by Partial Surface Texturing." *Journal of Tribology* 127 (3): 632–38. https://doi.org/10.1115/1.1866171.

Kong, L, X Chen, G Yang, L Yu, and P Zhang. 2008. "Preparation and Characterization of Slice-like $Cu_2(OH)_3NO_3$ Superhydrophobic Structure on Copper Foil." *Applied Surface Science* 254 (22): 7255–58. https://doi.org/10.1016/j.apsusc.2008.05.317.

Kovalchenko, A, O Ajayi, A Erdemir, G Fenske, and I Etsion. 2004. "The Effect of Laser Texturing of Steel Surfaces and Speed-Load Parameters on the Transition of Lubrication Regime from Boundary to Hydrodynamic." *Tribology Transactions* 47 (2): 299–307. https://doi.org/10.1080/05698190490440902.

Krupenkin, TN, J Ashley Taylor, EN Wang, P Kolodner, M Hodes, and TR Salamon. 2007. "Reversible Wetting – Dewetting Transitions on Electrically Tunable Superhydrophobic Nanostructured Surfaces." *Langmuir* 23 (18): 9128–33.

Krupenkin, TN, J Ashley Taylor, TM Schneider, and S Yang. 2004. "From Rolling Ball to Complete Wetting: The Dynamic Tuning of Liquids on Nanostructured Surfaces." *Langmuir* 20 (10): 3824–27.

Kursun, H, and U Ulusoy. 2006. "Influence of Shape Characteristics of Talc Mineral on the Column Flotation Behavior." *International Journal of Mineral Processing* 78 (4): 262–68.

Lee, W, M-K Jin, W-C Yoo, and J-K Lee. 2004. "Nanostructuring of a Polymeric Substrate with Well-Defined Nanometer-Scale Topography and Tailored Surface Wettability." *Langmuir* 20 (18): 7665–69.

Leyens, C, and M Peters. 2006. *Titanium and Titanium Alloys: Fundamentals and Applications*. Wiley Online Library.

Li, J, S Zhao, F Du, Y Zhou, and H Yu. 2018. "One-Step Fabrication of Superhydrophobic Surfaces with Different Adhesion via Laser Processing." *Journal of Alloys and Compounds* 739:489–98.

Li, M, C Li, BRK Blackman, and S Eduardo. 2022a. "Mimicking Nature to Control Bio-Material Surface Wetting and Adhesion." *International Materials Reviews* 67 (6):658–81.

Li, M, J Xu, and Q Lu. 2007. "Creating Superhydrophobic Surfaces with Flowery Structures on Nickel Substrates through a Wet-Chemical-process Electronic Supplementary Information (ESI) Available: Water Contact Angles before and after Immersion in Organic Solvents, Hot Water and Conservation in Air for Four Days; XPS Spectra; Image of Diesel Permeating through the Modified Steel Mesh. See." *Journal of Materials Chemistry* 17 (45):4772–6.

Li, M, T Yang, Q Yang, Z Fang, H Bian, C Zhang, X Hou, and F Chen. 2022b. "Bioinspired Anti-fogging and Anti-fouling Artificial Compound Eyes." *Advanced Optical Materials* 10 (17): 2200861.

Li, Q, Y Yan, M Yu, B Song, S Shi, and Y Gong. 2016. "Synthesis of Polymeric Fluorinated Sol–Gel Precursor for Fabrication of Superhydrophobic Coating." *Applied Surface Science* 367:101–8.

Li, R, and Y Shan. 2012. "Contact Angle and Local Wetting at Contact Line." *Langmuir* 28 (44): 15624–28.

Li, S, S Zhang, and X Wang. 2008. "Fabrication of Superhydrophobic Cellulose-Based Materials through a Solution-Immersion Process." *Langmuir* 24 (10): 5585–90. https://doi.org/10.1021/la800157t.

Li, W, Y Chen, S Wu, J Zhang, H Wang, D Zeng, and C Xie. 2018. "Preparing High-Adhesion Silver Coating on APTMS Modified Polyethylene with Excellent Anti-Bacterial Performance." *Applied Surface Science* 436:117–24.

Li, Y, W-Z Jia, Y-Y Song, and X-H Xia. 2007. "Superhydrophobicity of 3D Porous Copper Films Prepared Using the Hydrogen Bubble Dynamic Template." *Chemistry of Materials* 19 (23): 5758–64.

Liao, R, Z Zuo, C Guo, Y Yuan, and A Zhuang. 2014. "Fabrication of Superhydrophobic Surface on Aluminum by Continuous Chemical Etching and Its Anti-Icing Property." *Applied Surface Science* 317:701–9.

Liber-Kneć, A, and S Łagan. 2021. "Surface Testing of Dental Biomaterials—Determination of Contact Angle and Surface Free Energy." *Materials* 14 (11): 2716.

Lin, WT, ZW Lin, TY Kuo, CS Chien, JW Huang, YL Chung, CP Chang, MZ Ibrahim, and HT Lee. 2022. "Mechanical and Biological Properties of Atmospheric Plasma-Sprayed Carbon Nanotube-Reinforced Tantalum Pentoxide Composite Coatings on Ti6Al4V Alloy." *Surface and Coatings Technology* 437:128356.

Liu, H, S Szunerits, W Xu, and R Boukherroub. 2009. "Preparation of Superhydrophobic Coatings on Zinc as Effective Corrosion Barriers." *ACS Applied Materials & Interfaces* 1 (6): 1150–53. https://doi.org/10.1021/am900100q.

Liu, J, X Huang, Y Li, Z Li, Q Chi, and G Li. 2008. "Formation of Hierarchical CuO Microcabbages as Stable Bionic Superhydrophobic Materials via a Room-Temperature Solution-Immersion Process." *Solid State Sciences* 10 (11): 1568–76. https://doi.org/10.1016/j.solidstatesciences.2008.02.005.

Liu, Y, X Feng, Y Li, J Lu, and Z Su. 2021. "Research on Boundary Slip of Hydrostatic Lead Screw under Different Driving Modes." *Scientific Reports* 11 (1): 22307. https://doi.org/10.1038/s41598-021-01524-8.

Lu, Y, D Liu, Y Cai, C Gao, Q Jia, and Y Zhou. 2022. "AFM Measurement of Roughness, Adhesive Force and Wettability in Various Rank Coal Samples from Qinshui and Junggar Basin, China." *Fuel* 317:123556.

Lv, J, Y Song, L Jiang, and J Wang. 2014. "Bio-Inspired Strategies for Anti-Icing." *ACS Nano* 8 (4): 3152–69.

Majhy, B, VP Singh, and AK Sen. 2020. "Understanding Wetting Dynamics and Stability of Aqueous Droplet over Superhydrophilic Spot Surrounded by Superhydrophobic Surface." *Journal of Colloid and Interface Science* 565:582–91.

Marangoni, C. 1871. "Ueber Die Ausbreitung Der Tropfen Einer Flüssigkeit Auf Der Oberfläche Einer Anderen." *Annalen Der Physik* 219 (7): 337–54.

Marla, D, SA Andersen, Y Zhang, JH Hattel, and J Spangenberg. 2018. "A Study of Laser Surface Modification of Polymers: A Comparison in Air and Water." *Journal of Manufacturing Processes* 32:432–37.

Marmur, A. 2006. "Underwater Superhydrophobicity: Theoretical Feasibility." *Langmuir* 22 (4): 1400–2.

Marmur, A, and D Valal. 2010. "Correlating Interfacial Tensions with Surface Tensions: A Gibbsian Approach." *Langmuir* 26 (8): 5568–75.

Mathew Simon, S, G George, MS Sajna, VP Prakashan, Jose TA Jose, P Vasudevan, AC Saritha, PR Biju, C Joseph, and NV Unnikrishnan. 2021. "Recent Advancements in Multifunctional Applications of Sol-Gel Derived Polymer Incorporated TiO_2-ZrO_2 Composite Coatings: A Comprehensive Review." *Applied Surface Science Advances* 6:100173. https://doi.org/10.1016/j.apsadv.2021.100173.

Matykina, E, A Berkani, P Skeldon, and GE Thompson. 2007. "Real-Time Imaging of Coating Growth during Plasma Electrolytic Oxidation of Titanium." *Electrochimica Acta* 53 (4): 1987–94.

McCarthy, T. 2006. "A Perfectly Hydrophobic Surface." *Journal of the American Chemical Society* 128 (28): 9052–53.

McHale, G, S Aqil, NJ Shirtcliffe, MI Newton, and H Yildirim Erbil. 2005. "Analysis of Droplet Evaporation on a Superhydrophobic Surface." *Langmuir* 21 (24): 11053–60.

Meng, H, S Wang, J Xi, Z Tang, and L Jiang. 2008. "Facile Means of Preparing Superamphiphobic Surfaces on Common Engineering Metals." *The Journal of Physical Chemistry C* 112 (30): 11454–58.

Milles, S, B Voisiat, M Nitschke, and AF Lasagni. 2019a. "Influence of Roughness Achieved by Periodic Structures on the Wettability of Aluminum Using Direct Laser Writing and Direct Laser Interference Patterning Technology." *Journal of Materials Processing Technology* 270 (August):142–51. https://doi.org/10.1016/j.jmatprotec.2019.02.023.

Milles, S, M Soldera, B Voisiat, and AF Lasagni. 2019b. "Fabrication of Superhydrophobic and Ice-Repellent Surfaces on Pure Aluminium Using Single and Multiscaled Periodic Textures." *Scientific Reports* 9 (1): 13944.

Milošev, I, D Hamulić, P Rodič, C Carrière, S Zanna, H Budasheva, D Korte, M Franko, D Mercier, and A Seyeux. 2022. "Siloxane Polyacrylic Sol-Gel Coatings with Alkly and Perfluoroalkyl Chains: Synthesis, Composition, Thermal Properties and Log-Term Corrosion Protection." *Applied Surface Science* 574:151578.

Minko, S, M Müller, M Motornov, M Nitschke, K Grundke, and M Stamm. 2003. "Two-Level Structured Self-Adaptive Surfaces with Reversibly Tunable Properties." *Journal of the American Chemical Society* 125 (13): 3896–900.

Mishchenko, L, B Hatton, V Bahadur, J Ashley Taylor, T Krupenkin, and J Aizenberg. 2010. "Design of Ice-Free Nanostructured Surfaces Based on Repulsion of Impacting Water Droplets." *ACS Nano* 4 (12): 7699–7707.

Mittal, KL. 2003. *Contact Angle, Wettability and Adhesion, Volume* 3. CRC Press.

Mock, U, R Förster, W Menz, and J Rühe. 2005. "Towards Ultrahydrophobic Surfaces: A Biomimetic Approach." *Journal of Physics: Condensed Matter* 17 (9): S639.

Nakajima, A, K Hashimoto, and T Watanabe. 2001. *Recent Studies on Super-Hydrophobic Films*. Springer.

Narhe, RD, and DA Beysens. 2004. "Nucleation and Growth on a Superhydrophobic Grooved Surface." *Physical Review Letters* 93 (7): 076103.

Neinhuis, C, and W Barthlott. 1997. "Characterization and Distribution of Water-Repellent, Self-Cleaning Plant Surfaces." *Annals of Botany* 79 (6): 667–77.

Newton, MI, DL Herbertson, SJ Elliott, NJ Shirtcliffe, and G McHale. 2006. "Electrowetting of Liquid Marbles." *Journal of Physics D: Applied Physics* 40 (1): 20. https://doi.org/10.1088/0022-3727/40/1/S04.

Noble-Nesbitt, J. 1963. "Transpiration in Podura Aquatica L. (Collembola, Isotomidae) and the Wetting Properties of Its Cuticle." *Journal of Experimental Biology* 40 (4): 681–700.

Nosonovsky, M, and B Bhushan. 2012. "Lotus versus Rose: Biomimetic Surface Effects." In *Green Tribology: Biomimetics, Energy Conservation and Sustainability*, edited by Michael Nosonovsky and Bharat Bhushan, 25–40. Springer.

Ochsenbein, A, F Chai, S Winter, M Traisnel, J Breme, and HF Hildebrand. 2008. "Osteoblast Responses to Different Oxide Coatings Produced by the Sol–Gel Process on Titanium Substrates." *Acta Biomaterialia* 4 (5): 1506–17. https://doi.org/10.1016/j.actbio.2008.03.012.

Okumura, K, Chevy, F, Richard, D, Quéré, D, and Clanet, C. 2003. "Water Spring: A Model for Bouncing Drops." *Europhysics Letters* 62 (2): 237. https://doi.org/10.1209/epl/i2003-00340-1.

Oliver, JF, C Huh, and SG Mason. 1977. "Resistance to Spreading of Liquids by Sharp Edges." *Journal of Colloid and Interface Science* 59 (3): 568–81.

Onda, T, S Shibuichi, N Satoh, and K Tsujii. 1996. "Super-Water-Repellent Fractal Surfaces." *Langmuir* 12 (9): 2125–27.

Öner, D, and TJ McCarthy. 2000. "Ultrahydrophobic Surfaces. Effects of Topography Length Scales on Wettability." *Langmuir* 16 (20): 7777–82.

Oss, CJ Van, RJ Good, and MK Chaudhury. 1988. "Additive and Nonadditive Surface Tension Components and the Interpretation of Contact Angles." *Langmuir* 4 (4): 884–91.

Ostrovskaya, L, V Perevertailo, V Ralchenko, A Dementjev, and O Loginova. 2002. "Wettability and Surface Energy of Oxidized and Hydrogen Plasma-Treated Diamond Films." *Diamond and Related Materials* 11 (3–6): 845–50.

Ou, J, B Perot, and JP Rothstein. 2004. "Laminar Drag Reduction in Microchannels Using Ultrahydrophobic Surfaces." *Physics of Fluids* 16 (12): 4635–43.

Owens, DK. 1970. "Some Thermodynamic Aspects of Polymer Adhesion." *Journal of Applied Polymer Science* 14 (7): 1725–30.

Owens, DK, and RC Wendt. 1969. "Estimation of the Surface Free Energy of Polymers." *Journal of Applied Polymer Science* 13 (8): 1741–47.

Pan, L, H Dong, and P Bi. 2010. "Facile Preparation of Superhydrophobic Copper Surface by HNO3 Etching Technique with the Assistance of CTAB and Ultrasonication." *Applied Surface Science* 257 (5): 1707–11.

Patankar, NA. 2003. "On the Modeling of Hydrophobic Contact Angles on Rough Surfaces." *Langmuir* 19 (4): 1249–53.

Patel, AJ, P Varilly, and D Chandler. 2010. "Fluctuations of Water near Extended Hydrophobic and Hydrophilic Surfaces." *The Journal of Physical Chemistry B* 114 (4): 1632–37.

Paty, M. 1993. "Laplace, Pierre Simon, 1749–1827." In *Dictionnaire des philosophes, deuxième éd. rev. et augm.*, edited by Huisman, Denis, vol. 2, 1677–19. Presses Universitaires de France.

Philip, JR. 1972. "Integral Properties of Flows Satisfying Mixed No-Slip and No-Shear Conditions." *Zeitschrift Für Angewandte Mathematik Und Physik ZAMP* 23 (6): 960–68.

Podgorski, T, J-M Flesselles, and L Limat. 2001. "Corners, Cusps, and Pearls in Running Drops." *Physical Review Letters* 87 (3): 036102. https://doi.org/10.1103/PhysRevLett.87.036102.

Pomeau, Y, and J Vannimenus. 1985. "Contact Angle on Heterogeneous Surfaces: Weak Heterogeneities." *Journal of Colloid and Interface Science* 104 (2): 477–88.

Prakash, V, N Priyadarshni, A Kumar Das, and S Chattopadhyay. 2022. "Fabrication of Hydrophobic Surface on Ti6Al4V by WEDM Process for Surgical Instruments and Bioimplants." *The International Journal of Advanced Manufacturing Technology* 118 (3):1–13.

Qi, D, N Lu, H Xu, B Yang, C Huang, M Xu, L Gao, Z Wang, and L Chi. 2009. "Simple Approach to Wafer-Scale Self-Cleaning Antireflective Silicon Surfaces." *Langmuir* 25 (14): 7769–72.

Qin, S, X Xu, Y Lu, L Li, T Huang, and J Lin. 2022. "Doping Gd3+ Ion in PDA-PHBV Coating on Ti6Al4V Alloy for Enhancing Corrosion Resistance and Proliferation of Human Gingival Fibroblasts and Human Umbilical Vein Endothelial Cells." *Acta Metallurgica Sinica (English Letters)* 35 (5): 1–13.

Qu, M, J He, and J Zhang. 2010. "Superhydrophobicity, Learn from the Lotus Leaf." In *Biomimetics Learning from Nature*, edited by Amitava Mukherjee, 1–325. IntechOpen.

Quéré, D. 2005. "Non-Sticking Drops." *Reports on Progress in Physics* 68 (11): 2495. https://doi.org/10.1088/0034-4885/68/11/R01.

Quéré, D. 2008. "Wetting and Roughness." *Annual Review of Materials Research* 38 (1): 71–99. https://doi.org/10.1146/annurev.matsci.38.060407.132434.

Quéré, D, M-J Azzopardi, and L Delattre. 1998. "Drops at Rest on a Tilted Plane." *Langmuir* 14 (8): 2213–16.

Raeymaekers, B, I Etsion, and FE Talke. 2007. "A model for the magnetic tape/guide interface with laser surface texturing." *International Joint Tribology Conference* 48108: 669–71.

Rajesh, PSM, C Delaroa, M Gagné, JE Klemberg-Sapieha, F Sirois, and D Therriault. 2019. "Continuous and Selective-Area Coating of Silver on Fiber-Reinforced Polymer Composites for Aerospace Applications." *Materials Today Communications* 18:206–12.

Ramachandran, R, and M Nosonovsky. 2016. "Vibrations and Spatial Patterns Change Effective Wetting Properties of Superhydrophobic and Regular Membranes." *Biomimetics* 1 (1): 4.

Raphaël, E. 1989. "Capillary Rise of a Wetting Fluid in a Semi-Circular Groove." *Journal de Physique* 50 (4): 485–91.

Rathgen, H, K Sugiyama, C-D Ohl, D Lohse, and F Mugele. 2007. "Nanometer-Resolved Collective Micromeniscus Oscillations through Optical Diffraction." *Physical Review Letters* 99 (21): 214501.

Rayleigh, L. 1914. "XXII. The Equilibrium of Revolving Liquid under Capillary Force." *The London, Edinburgh, and Dublin Philosophical Magazine and Journal of Science* 28 (164): 161–70. https://doi.org/10.1080/14786440808635198.

Reyssat, M, A Pépin, F Marty, Y Chen, and D Quéré. 2006. "Bouncing Transitions on Microtextured Materials." *Europhysics Letters* 74 (2): 306.

Reyssat, M, JM Yeomans, and D Quéré. 2007. "Impalement of Fakir Drops." *Europhysics Letters* 81 (2): 26006.

Richard, D, C Clanet, and D Quéré. 2002. "Contact Time of a Bouncing Drop." *Nature* 417 (6891): 811. https://doi.org/10.1038/417811a.

Richard, D, and D Quéré. 2000. "Bouncing Water Drops." *Europhysics Letters* 50 (6): 769.

Richardson, S. 1973. "On the No-Slip Boundary Condition." *Journal of Fluid Mechanics* 59 (4): 707–19.

Roach, P, G McHale, CR Evans, NJ Shirtcliffe, and MI Newton. 2007. "Decoupling of the Liquid Response of a Superhydrophobic Quartz Crystal Microbalance." *Langmuir* 23 (19): 9823–30.

Romero-Gavilán, F, J Carlos-Almeida, A Cerqueira, M Gurruchaga, I Goni, IM Miranda-Salvado, MH Vaz Fernandes, and J Suay. 2020. "Sol-Gel Coatings Made Using Methyl-Modified Alkoxysilanes: The Balance between Protection and Bioactivation." *Progress in Organic Coatings* 147:105770.

Rowlinson, JS, and B Widom. 2013. *Molecular Theory of Capillarity*. Courier Corporation.

Roy, P, S Berger, and P Schmuki. 2011. "TiO2 Nanotubes: Synthesis and Applications." *Angewandte Chemie International Edition* 50 (13): 2904–39.

Rukosuyev, MV, J Lee, S Jin Cho, G Lim, and MBG Jun. 2014. "One-Step Fabrication of Superhydrophobic Hierarchical Structures by Femtosecond Laser Ablation." *Applied Surface Science* 313:411–17.

Sadeghi, M, M Kharaziha, HR Salimijazi, and E Tabesh. 2019. "Role of Micro-Dimple Array Geometry on the Biological and Tribological Performance of Ti6Al4V for Biomedical Applications." *Surface and Coatings Technology* 362:282–92.

Samanta, A, W Huang, H Chaudhry, Q Wang, SK Shaw, and H Ding. 2020. "Design of Chemical Surface Treatment for Laser-Textured Metal Alloys to Achieve Extreme Wetting Behavior." *ACS Applied Materials & Interfaces* 12 (15): 18032–45.

Santos, L, JP Neto, A Crespo, P Baião, P Barquinha, L Pereira, R Martins, and E Fortunato. 2015. "Electrodeposition of WO3 Nanoparticles for Sensing Applications." *Electroplating of Nanostructures*, 1–22.

Sbragaglia, M, AM Peters, C Pirat, BM Borkent, RGH Lammertink, M Wessling, and D Lohse. 2007. "Spontaneous Breakdown of Superhydrophobicity." *Physical Review Letters* 99 (15): 156001.

Secundus, CP. 2004. *Natural History: A Selection*. Penguin Books.

Shibe, V, and V Chawla. 2014. "A Review of Surface Modification Techniques in Enhancing the Erosion Resistance of Engineering Components." *International Journal of Research in Mechanical Engineering & Technology* 4 (2): 92–95.

Shibuichi, S, T Onda, N Satoh, and K Tsujii. 1996. "Super Water-Repellent Surfaces Resulting from Fractal Structure." *The Journal of Physical Chemistry* 100 (50): 19512–17.

Shin, KR, YG Ko, and DH Shin. 2011. "Effect of Electrolyte on Surface Properties of Pure Titanium Coated by Plasma Electrolytic Oxidation." *Journal of Alloys and Compounds* 509: S478–81.

Shin, Y, Y Qiao, N Canfield, Z Yu, HM Meyer III, DR Merkel, EK Nickerson, NS Kanbargi, A Ortiz, and AK Naskar. 2022. "Significant Slowdown of Plasma-Optimized Surface Energy Deactivation by Vacuum Sealing for Efficient Adhesive Bonding." *Composites Part B: Engineering* 240:110001.

Shirtcliffe, NJ, G McHale, MI Newton, CC Perry, and FB Pyatt. 2006. "Plastron Properties of a Superhydrophobic Surface." *Applied Physics Letters* 89 (10): 104106.

Song, H-J, X-Q Shen, H-Y Ji, and X-J Jing. 2010. "Superhydrophobic Iron Material Surface with Flower-like Structures Obtained by a Facile Self-Assembled Monolayer." *Applied Physics A* 99:685–89.

Song, W, VS Gaware, ÖV Rúnarsson, M Másson, and JF Mano. 2010. "Functionalized Superhydrophobic Biomimetic Chitosan-Based Films." *Carbohydrate Polymers* 81 (1): 140–44.

Song, Y, RP Nair, M Zou, and Y Wang. 2009. "Superhydrophobic Surfaces Produced by Applying a Self-Assembled Monolayer to Silicon Micro/Nano-Textured Surfaces." *Nano Research* 2 (2): 143–50. https://doi.org/10.1007/s12274-009-9012-0.

Spori, DM, T Drobek, S Zürcher, M Ochsner, C Sprecher, A Mühlebach, and ND Spencer. 2008. "Beyond the Lotus Effect: Roughness Influences on Wetting over a Wide Surface-Energy Range." *Langmuir* 24 (10): 5411–17.

Steinberger, A, C Cottin-Bizonne, P Kleimann, and E Charlaix. 2007. "High Friction on a Bubble Mattress." *Nature Materials* 6 (9): 665–68.

Sun, L, J Guo, H Chen, D Zhang, L Shang, B Zhang, and Y Zhao. 2021. "Tailoring Materials with Specific Wettability in Biomedical Engineering." *Advanced Science* 8 (19): 2100126.

Sun, T, G Wang, L Feng, B Liu, Y Ma, L Jiang, and D Zhu. 2004. "Reversible Switching between Superhydrophilicity and Superhydrophobicity." *Angewandte Chemie International Edition* 43 (3): 357–60.

Sun, Y, H Li, R Huang, X Song, H Zhao, H Xia, D Zhu, B Chen, and C Tan. 2023. "Improved Wettability of AlSi5 on DP980 Steel during Laser-Induced Heating by Surface Texture Preparation." *Journal of Manufacturing Processes* 89:111–23. https://doi.org/10.1016/j.jmapro.2023.01.066.

Swain, PS, and R Lipowsky. 1998. "Contact Angles on Heterogeneous Surfaces: A New Look at Cassie's and Wenzel's Laws." *Langmuir* 14 (23): 6772–80.

Taggart, AF, TC Taylor, and CR Ince. 1929. *Experiments with Flotation Reagents*. American Institute of Mining and Metallurgical Engineers, Incorporated.

Tam, HL, KW Cheah, DTP Goh, and JKL Goh. 2013. "Iridescence and Nano-Structure Differences in Papilio Butterflies." *Optical Materials Express* 3 (8): 1087–92. https://doi.org/10.1364/OME.3.001087.

Tan, C, J Su, Y Liu, Z Feng, X Song, X Wang, B Chen, and H Xia. 2022. "Enhanced Interfacial Bonding Strength of Laser Bonded Titanium Alloy/CFRTP Joint via Hydrogen Bonds Interaction." *Composites Part B: Engineering* 239:109966.

Thorpe, WH. 1950. "Plastron Respiration in Aquatic Insects." *Biological Reviews* 25 (3): 344–90.

Tong, W, and D Xiong. 2022. "Direct Laser Texturing Technique for Metal Surfaces to Achieve Superhydrophobicity." *Materials Today Physics* 23 (March):100651. https://doi.org/10.1016/j.mtphys.2022.100651.

Tuteja, A, W Choi, M Ma, JM Mabry, SA Mazzella, GC Rutledge, GH McKinley, and RE Cohen. 2007. "Designing Superoleophobic Surfaces." *Science* 318 (5856): 1618–22.

Udduttula, A, J Li, Z Ma, B Teng, JV Zhang, AM Ferreira, P Gentile, G Wang, X Zhao, and P-G Ren. 2022. "A Novel Apatite-Inspired Sr5(PO4)2SiO4 Plasma-Sprayed Coating on Ti Alloy Promoting Biomineralization, Osteogenesis and Angiogenesis." *Ceramics International* 48 (8): 10979–89. https://doi.org/10.1016/j.ceramint.2021.12.317.

Vargaftik, NB, BN Volkov, and LD Voljak. 1983. "International Tables of the Surface Tension of Water." *Journal of Physical and Chemical Reference Data* 12 (3): 817–20.

Vialetto, J, N Nussbaum, J Bergfreund, P Fischer, and L Isa. 2022. "Influence of the Interfacial Tension on the Microstructural and Mechanical Properties of Microgels at Fluid Interfaces." *Journal of Colloid and Interface Science* 608:2584–92.

Vishwakarma, V, G Saravanan Kaliaraj, and KKA Mosas. 2022. "Multifunctional Coatings on Implant Materials—A Systematic Review of the Current Scenario." *Coatings* 13 (1): 69.

Vogler, EA. 1998. "Structure and Reactivity of Water at Biomaterial Surfaces." *Advances in Colloid and Interface Science* 74 (1–3): 69–117.

Volpe, CD, D Maniglio, M Brugnara, S Siboni, and M Morra. 2004. "The Solid Surface Free Energy Calculation: I. In Defense of the Multicomponent Approach." *Journal of Colloid and Interface Science* 271 (2): 434–53.

Vrij, A. 1976. "Polymers at Interfaces and the Interactions in Colloidal Dispersions." *Pure and Applied Chemistry* 48 (4): 471–83.

Wagner, T, C Neinhuis, and W Barthlott. 1996. "Wettability and Contaminability of Insect Wings as a Function of Their Surface Sculptures." *Acta Zoologica* 77 (3): 213–25.

Walsh, FC, CTJ Low, RJK Wood, KT Stevens, J Archer, A Rl Poeton, and A Ryder. 2009. "Plasma Electrolytic Oxidation (PEO) for Production of Anodised Coatings on Lightweight Metal (Al, Mg, Ti) Alloys." *Transactions of the IMF* 87 (3): 122–35.

Wang, F, K Zhao, J Cheng, and J Zhang. 2011. "Conciliating Surface Superhydrophobicities and Mechanical Strength of Porous Silicon Films." *Applied Surface Science* 257 (7): 2752–55.

Wang, H, Y Li, and H Zhu. 2016. "Experimental Study on the Tribological Performance of Fractal-like Textured Surface under Mixed Lubrication Conditions." *Surface and Coatings Technology* 307:220–26. https://doi.org/10.1016/j.surfcoat.2016.08.071.

Wang, L, S Guo, and S Dong. 2008. "Facile Electrochemical Route to Directly Fabricate Hierarchical Spherical Cupreous Microstructures: Toward Superhydrophobic Surface." *Electrochemistry Communications* 10 (4): 655–58.

Wang, Q, W Hou, and Y Zhang. 2009. "Superhydrophobic and Luminescent Methylsilicone Resin Film." *Applied Surface Science* 256 (3): 664–67.

Wang, R, K Hashimoto, A Fujishima, M Chikuni, E Kojima, A Kitamura, M Shimohigoshi, and T Watanabe. 1998. "Photogeneration of Highly Amphiphilic TiO_2 Surfaces." *Advanced Materials* 10 (2): 135–38.

Wang, X, W Liu, F Zhou, and D Zhu. 2009. "Preliminary Investigation of the Effect of Dimple Size on Friction in Line Contacts." *Tribology International* 42 (7): 1118–23. https://doi.org/10.1016/j.triboint.2009.03.012.

Wang, Y, W Wang, L Zhong, J Wang, Q Jiang, and X Guo. 2010. "Super-Hydrophobic Surface on Pure Magnesium Substrate by Wet Chemical Method." *Applied Surface Science* 256 (12): 3837–40.

Wang, Y, H Yu, C Chen, and Z Zhao. 2015. "Review of the Biocompatibility of Micro-Arc Oxidation Coated Titanium Alloys." *Materials & Design* 85:640–52.

Wark, IW, and AB Cook. 1934. "An Experimental Study of the Effect of Xanthates on Contact Angles at Mineral Surfaces." *Transactions of the American Institute of Mining and Metallurgical Engineers* 112:189–244.

Weast, RC. 1988. *Handbook of Chemistry and Physics*. Springer.

Wei, ZJ, WL Liu, D Tian, CL Xiao, and XQ Wang. 2010. "Preparation of Lotus-like Superhydrophobic Fluoropolymer Films." *Applied Surface Science* 256 (12): 3972–76.

Wenzel, RN. 1936. "Resistance of Solid Surfaces to Wetting by Water." *Industrial & Engineering Chemistry* 28 (8): 988–94.

Wolansky, G, and A Marmur. 1998. "The Actual Contact Angle on a Heterogeneous Rough Surface in Three Dimensions." *Langmuir* 14 (18): 5292–97.

Wolansky, G, and A Marmur. 1999. "Apparent Contact Angles on Rough Surfaces: The Wenzel Equation Revisited." *Colloids and Surfaces A: Physicochemical and Engineering Aspects* 156 (1–3): 381–88.

Woo, SH, J Park, and BR Min. 2015. "Relationship between Permeate Flux and Surface Roughness of Membranes with Similar Water Contact Angle Values." *Separation and Purification Technology* 146:187–91.

Woodward, I, WCE Schofield, V Roucoules, and JPS Badyal. 2003. "Super-Hydrophobic Surfaces Produced by Plasma Fluorination of Polybutadiene Films." *Langmuir* 19 (8): 3432–38.

Wu, J, J Xia, W Lei, and B-p Wang. 2010. "Fabrication of Superhydrophobic Surfaces with Double-Scale Roughness." *Materials Letters* 64 (11): 1251–53.

Wu, S. 1971. "Calculation of Interfacial Tension in Polymer Systems." *Journal of Polymer Science Part C: Polymer Symposia* 34:19–30. Wiley Online Library.

Wu, S. 1982. "Modifications of Polymer Surfaces: Mechanism of Wettability and Bondability Improvements." *Polymer Interface and Adhesion*, 279–336.

Wu, W, RF Jr Giese, and CJ Van Oss. 1995. "Evaluation of the Lifshitz-van Der Waals/Acid-Base Approach to Determine Surface Tension Components." *Langmuir* 11 (1): 379–82.

Wu, Y, J Du, G Liu, D Ma, F Jia, JJ Klemeš, and Jin Wang. 2022. "A Review of Self-Cleaning Technology to Reduce Dust and Ice Accumulation in Photovoltaic Power Generation Using Superhydrophobic Coating." *Renewable Energy* 185:1034–61.

Wu, Z, D Wu, W Yang, and R Jin. 2006. "Preparation of Highly Reflective and Conductive Metallized Polyimide Films through Surface Modification: Processing, Morphology and Properties." *Journal of Materials Chemistry* 16 (3): 310–16.

Xiong, Y, F Zhang, T Dai, C Shang, and Q Wan. 2022. "Crystal Growth Mechanism and Mechanical Properties of Ti-6Al-4V Alloy during Selective Laser Melting." *Materials Characterization* 194:112455. https://doi.org/10.1016/j.matchar.2022.112455.

Xu, B, and Z Cai. 2008. "Fabrication of a Superhydrophobic ZnO Nanorod Array Film on Cotton Fabrics via a Wet Chemical Route and Hydrophobic Modification." *Applied Surface Science* 254 (18): 5899–5904.

Xu, Q-F, and J-N Wang. 2010. "Superhydrophobic and Transparent Coatings Prepared by Self-Assembly of Dual-Sized Silica Particles." *Frontiers of Materials Science in China* 4:180–88.

Xu, Y, T Pan, F Liu, P Zhao, X Jiang, and C Xiong. 2022. "Surface Coating and Doping of Single-Crystal LiNi0.5Co0.2Mn0.3O2 for Enhanced High-Voltage Electrochemical Performances via Chemical Vapor Deposition." *Journal of Electroanalytical Chemistry* 914:116286. https://doi.org/10.1016/j.jelechem.2022.116286.

Yan, D, N Qu, H Li, and X Wang. 2010. "Significance of Dimple Parameters on the Friction of Sliding Surfaces Investigated by Orthogonal Experiments." *Tribology Transactions* 53 (5): 703–12. https://doi.org/10.1080/10402001003728889.

Yan, YY, N Gao, and W Barthlott. 2011. "Mimicking Natural Superhydrophobic Surfaces and Grasping the Wetting Process: A Review on Recent Progress in Preparing Superhydrophobic Surfaces." *Advances in Colloid and Interface Science* 169 (2): 80–105.

Yang, J, P Pi, X Wen, D Zheng, M Xu, J Cheng, and Z Yang. 2009. "A Novel Method to Fabricate Superhydrophobic Surfaces Based on Well-Defined Mulberry-like Particles and Self-Assembly of Polydimethylsiloxane." *Applied Surface Science* 255 (6): 3507–12. https://doi.org/10.1016/j.apsusc.2008.09.092.

Yang, S, D Wu, S Qi, G Cui, R Jin, and Z Wu. 2009. "Fabrication of Highly Reflective and Conductive Double-Surface-Silvered Layers Embedded on Polymeric Films through All-Wet Process at Room Temperature." *The Journal of Physical Chemistry B* 113 (29): 9694–701.

Yao, Z, D Wu, J Tang, W Wu, JYY Heng, and H Zhao. 2016. "A Novel Colored Talc Filler: Preparation and Surface Property Determination Using Two Distinct Methods." *Chemometrics and Intelligent Laboratory Systems* 155:54–61.

Ybert, C, C Barentin, C Cottin-Bizonne, P Joseph, and L Bocquet. 2007. "Achieving Large Slip with Superhydrophobic Surfaces: Scaling Laws for Generic Geometries." *Physics of Fluids* 19 (12): 123601.

Yerokhin, AL, X Nie, A Leyland, A Matthews, and SJ Dowey. 1999. "Plasma Electrolysis for Surface Engineering." *Surface and Coatings Technology* 122 (2–3): 73–93.

Yoon, R-H, DH Flinn, and YI Rabinovich. 1997. "Hydrophobic Interactions between Dissimilar Surfaces." *Journal of Colloid and Interface Science* 185 (2): 363–70.

Young, TR. 1805. "An Assay on the Cohesion of Fluids." *Philosophical Transactions of the Royal Society* 95:65–87.

Yusof, NN, SM Aziz, F Mohd Noor, SN Syed Yaacob, and S Hashim. 2022. "A Novel Borate-Based 45S5 Bioglass®: In Vitro Assessment in Phosphate-Buffered Saline Solution." *Journal of Non-Crystalline Solids* 596:121843.

Zhan, N, Y Li, C Zhang, Y Song, H Wang, L Sun, Q Yang, and X Hong. 2010. "A Novel Multinozzle Electrospinning Process for Preparing Superhydrophobic PS Films with Controllable Bead-on-String/Microfiber Morphology." *Journal of Colloid and Interface Science* 345 (2): 491–95.

Zhang, K, H Li, C Zhang, Y Han, X Guo, and T Liu. 2023. "Effect of Ion Beam Etching on the Tribological Performance of Laser Textured Co-Cr-Mo Alloy." *Optics & Laser Technology* 160:109097. https://doi.org/10.1016/j.optlastec.2022.109097.

Zhang, XT, O Sato, and A Fujishima. 2004. "Water Ultrarepellency Induced by Nanocolumnar ZnO Surface." *Langmuir* 20 (14): 6065–67.

Zhao, T, and L Jiang. 2018. "Contact Angle Measurement of Natural Materials." *Colloids and Surfaces B: Biointerfaces* 161:324–30.

Zhao, Y, Y Tang, X Wang, and T Lin. 2010. "Superhydrophobic Cotton Fabric Fabricated by Electrostatic Assembly of Silica Nanoparticles and Its Remarkable Buoyancy." *Applied Surface Science* 256 (22): 6736–42. https://doi.org/10.1016/j.apsusc.2010.04.082.

Zhitomirsky, I. 2002. "Cathodic Electrodeposition of Ceramic and Organoceramic Materials. Fundamental Aspects." *Advances in Colloid and Interface Science* 97 (1–3): 279–317.

Zhu, X, Z Zhang, X Men, J Yang, and X Xu. 2010. "Rapid Formation of Superhydrophobic Surfaces with Fast Response Wettability Transition." *ACS Applied Materials & Interfaces* 2 (12): 3636–41.

Zisman, WA. 1963. "Influence of Constitution on Adhesion." *Industrial & Engineering Chemistry* 55 (10): 18–38.

7 Modelling and Simulation of Tribological Responses Related to Specific Manufacturing Processes

Adesh Kumar Tomar and Krishnkant Sahu

7.1 INTRODUCTION

Global technical advancements over the past few decades made it necessary to design machine parts for precision, accuracy, improved performance, longer life, etc. As a result, in recent years, advanced manufacturing processes have gained a lot of momentum to produce textured surfaces to harness the better behaviour of tribo-pairs. The application of textured surfaces on various tribological systems to improve their performance is basically inspired by biomimetics (Dean and Bhushan, 2010). The nature of surface roughness is random, whereas the provision of textured surfaces by the advanced manufacturing processes is well organized or deterministic in nature (Brizmer et al., 2003; Kligerman et al., 2005). The fabrication of micro-textured patterns on the surfaces of various tribological systems is in different forms such as discrete patterns (micro-dimples) and continuous forms (micro-grooved surfaces) (Sahu et al., 2023; Tomar et al., 2023). Additional hydrodynamic action is provided by the numerous micro-dimples that serve as micro-fluid film bearings (Ganai et al., 2021; Sharma and Tomar, 2021). Thus, micro-texturing improves the performance of tribological systems, such as larger load-carrying capacity. These intentionally produced surfaces by using various manufacturing processes provide various advantages, for example, these surfaces act as lubricant reservoirs and offer oil for spreading at the tribo-pair contact. These lubricant reservoirs also entrap wear debris, protecting the surface from damage (Lu and Khonsari 2007; Papadopoulos et al., 2011; Tomar and Sharma, 2020a). The behaviour of micro-dimples on the hydrodynamic thrust bearings and mechanical seal was studied (Anno et al., 1968; Anno et al., 1969). They reported that the application of textured patterns on hydrodynamic thrust bearings and mechanical seal is an effective way to improve the load-carrying capacity. Etsion group in 1996 (Etsion and Burstein, 1996; Etsion et al., 1999; Etsion, 2005) performed important studies dealing with textured surfaces. They fabricated the hemispherical dimples on the mechanical seals and theoretically investigated their performance. It was stated that proper values of texture density and size are important factors to harness better performance. Tønder (2001) and Cupillard et al. (2008) also found that the fabrication of textured surfaces on tribo-pairs working under

 DOI: 10.1201/9781003359333-7

sliding action produces additional local pressures and, consequently, enhances the load-carrying capacity. Using the FDM, Tala-Ighil et al. (2007) and Tala-Ighil et al. (2011) considered the effect of the textured surface locations and different geometric shapes of textured arrangements on the behaviour of hydrodynamic bearings. They found a positive effect of the partial texturing configuration with respect to the full texturing configuration. It was also revealed that the parallelepipedal-textured shape has a comparatively more positive influence on the bearing performance than spherical and cylindrical textured shapes. The influence of trapezoidal and saw-shaped texture profiles using the neural network technique on the hydrodynamic journal bearing was studied (Sinanoğlu et al., 2005). They revealed that a saw-shaped texture profile offers a lower loading capacity as compared to a trapezoidal-shaped texture pattern. In an investigation (Qiu et al. 2012), the behaviour of a parallel sliding bearing configuration was examined while taking into account various geometrically textured surface shapes, including triangular, spherical, circular, elliptical, chevron, and ellipsoidal-shaped textured patterns. In the comparison analysis, it was observed that an ellipsoidal-shaped texture profile provides an improved load-carrying capacity compared to other geometrically textured surface shapes. The narrow groove theory was incorporated to examine the behaviour of partially textured hydrodynamic journal bearings (Rao et al., 2014). It was noticed that the partially textured bearings reduce the value of the frictional losses and significantly advance the loading capacity. The influence of sinusoidal and full-wave patterns of rough surfaces on a hydrodynamic circular bearing system was analysed (Kango and Sharma, 2010). It was observed that the sinusoidal roughness profile provides a reduced loading capacity as compared to the full-wave profiles.

Recent years have witnessed a rise in demand for an effective, reliable, and robust design of tribological systems. Grooved bearings offer attractive advantages such as lower noise operations, better fluid film stiffness, improved stability characteristics, and an extended life span in comparison to conventional circular journal bearings (Hirs, 1965; Vohr and Chow, 1965; Tomar and Sharma, 2020b). Miniaturisation and high-speed operations in precision machining processes have led to an increase in the need for grooved bearings in recent times. This type of bearing has important applications in magnetic disks, computer drives, and motors in DVD players (Kawabata et al., 1989; Yoshimoto et al., 1999; Ji et al., 2014). In modern-day machinery, the reduction in friction and control of wear of mechanical elements is essential to extend the life of mechanical systems with efficient operation. Using the FDM, Kang et al. (1996) studied the grooved circular journal bearings and reported that these bearings provide better stability than a plain journal-bearing system. Theoretical and experimental works were performed on micro-groove hydrodynamic circular bearings (Adatepe et al., 2011; Adatepe et al., 2013). They used circumferential and transversal micro-grooves on the bearing surfaces and observed a larger value of frictional losses for transversal micro-grooves than circumferential micro-groove bearings. Gong et al. (2019) developed a model of micro-groove journal bearing operated with water. It was found that micro-groove bearings offer enhanced bearing performance. A thorough review of the published literature reveals that textured bearing surfaces greatly affect the behaviour of tribo-pairs. The lubricating behaviour of tribological systems owing to the application of textured surfaces may decrease the value of frictional losses, enhance loading capacity, reduce energy consumption, and enhance reliability. The provision of textured surfaces in the form of discrete patterns, i.e., micro-dimples, and continuous form, i.e., grooved surfaces, is shown in Figures 7.1 and 7.2, respectively.

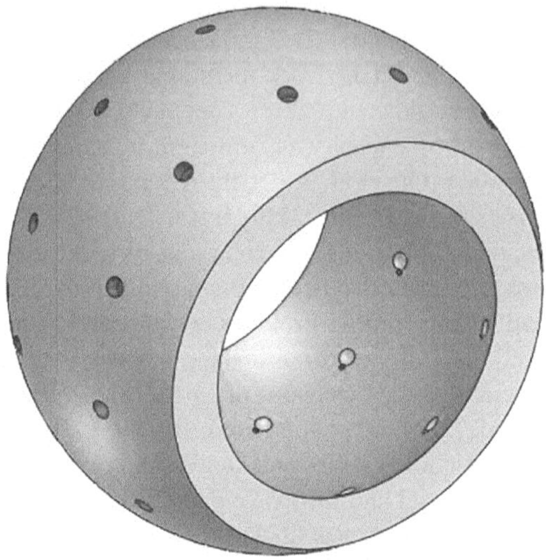

FIGURE 7.1 Spherical bearing with textured surfaces. (Reproduced with permission from Copyright Tribology International, Elsevier (Tomar and Sharma 2020a).)

FIGURE 7.2 Spherical bearing with groove surfaces. (Reproduced with permission from Copyright *International Journal of Mechanical Sciences*, Elsevier (Sharma and Tomar, 2021).)

7.2 ROLE OF SURFACE TOPOGRAPHY

Unprecedented technical advancements over the past few years have made it necessary for machines to offer improved performance, longer service life, and lower friction as well as wear (Isaksson et al., 2011; Gautam et al., 2016; Prajapati and Tiwari, 2019). As a result, the new generation of the tribological system must be developed using more exact and realistic design data in order to effectively meet these requirements. Perfectly smooth surfaces are actually quite challenging to create in reality. The smooth surface is ideal, without any peaks and valleys (Elrod, 1973, 1979; Tønder, 1996; Sharma and Kushare, 2015). Even though sophisticated precision manufacturing techniques are available, there will always be some degree of roughness in tribological systems. Additionally, it is also challenging to sustain the necessary thickness of a lubricating oil film on smooth tribo-surfaces, which could result in a seizure. Surface roughness heights are about equal in the order of magnitude to lubricant film thickness in various tribological systems. It is therefore desirable to conduct a comprehensive numerical investigation of the tribological system to evaluate the accurate performance parameters or realistic design data.

Various researchers employed the stochastic model to study the influence of rough surfaces on the tribo-pairs produced by various manufacturing processes. The notable work in the field of the influence of rough surfaces on the behaviour of tribo-pairs was performed by Christensen (1971). They presented a stochastic model to deal with the directional roughness of the tribological system. However, the stochastic model has a limitation as it represents only longitudinal and transverse roughness profiles. The influence of roughness on the behaviour of finite-width hydrodynamic bearings was further studied by Tonder using stochastic models (Tønder, 1986). According to their findings, a number of parameters, including lubricant film thickness, fluid film pressure, pressure gradients, lubricant flow, and others, have an important role in the working of striated rough surfaces having high roughness densities. The stochastic modelling for the elasto-hydrodynamic lubrication issues in the rollers was studied (Chow and Cheng, 1976). It was difficult to consider the influence of 3D surface roughness using stochastic theory as the stochastic modelling was restricted to a certain type of roughness profiles. Later, Patir and Cheng (1978) and Patir and Cheng (1979) proposed an average flow modelling to scrutinize the tribological systems considering the influence of roughness. The model presented by Patir and Cheng, which can assess longitudinal, isotropic, and transverse surface patterns, can also analyse the influence of 3D roughness that is both isotropic and non-isotropic. Several researchers have investigated the effects of surface roughness on various tribological systems using the average flow model (Tomar and Sharma, 2022b). An average flow model was used to study the behaviour of a rough hydrodynamic bearing considering mixed lubrication regimes (Ai et al., 1998). The stability analysis of the hydrodynamic bearing was conducted (Ramesh et al., 1997) using an average flow model. According to their investigation, the bearing performance is significantly affected by the various roughness patterns. The steady-state performance of a hydrodynamic circular journal bearing was investigated while taking into account the isotropic surface roughness pattern and the presence of misalignment (Guha, 2000). They reported that roughness considerably alters the bearing performance.

The effects of roughness on hydrostatic thrust bearing, considering mixed lubrication conditions, were also examined (Kazama and Yamaguchi, 1993). The influence of fluid inertia and rough surfaces on the performance of hybrid recessed journal bearings was studied (San Andrés, 1990). They reported that the effects of roughness improve the dynamic performance of tribological systems. A thorough scan of the published literature reported that surface roughness significantly influences the performance of the tribological system.

7.3 VARIOUS MANUFACTURING TECHNIQUES

The array of straight lines, a sinusoidal pattern, a positive full wave pattern, an array of curved lines, and a crosshatch pattern falls in the category of continuous texture, as shown in Table 7.1. The differences in the design of surface texture patterns can alter the behaviour of tribo-pairs in terms of pressure regeneration and loading capacity. Here, the major task is to determine the critical (optimum) dimensional relationships among the different geometric parameters of texture surfaces. These geometric parameters are the length, depth, and width of the textured surface; the distance between textured surfaces, i.e., array, pitch, and density, as well as the angle of texture surfaces, play a significant role in enhancing the performance of various tribological systems.

Recently, the advent of sophisticated manufacturing technologies, including chemical etching (Lu and Khonsari, 2007), laser surface texturing (LST) (Etsion

TABLE 7.1
Various Surface Textures and Shapes of Feature

Surface Texture	Example of Surface Texture	Shapes of Feature
Discrete texture	Spherical	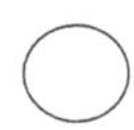
	Rectangle	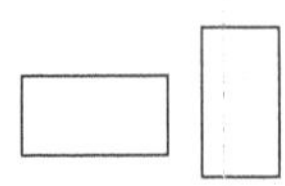
	Square	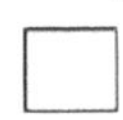
	Triangle	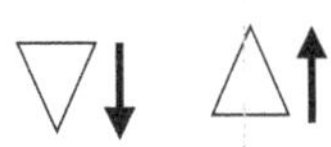
	Ellipse	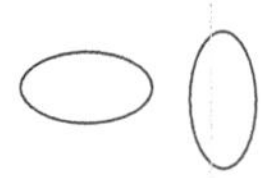
Continuous texture	An array of straight lines	
	Crosshatch pattern	
	An array of curved lines	
	Sinusoidal, positive full wave	

et al., 2004; Brizmer and Kligerman, 2012), vibro-machining texturing (Coblas et al., 2015), novel dressing technique (Oliveira et al., 2010), and micro-stereolithography (Yang et al., 2009), facilitates the manufacturing of textured surfaces on the various tribo-pairs with the precision of the higher-order level up to micro-scale. The salient features of several surface texturing techniques are discussed below:

- **LST technique:** This technique is the most advanced and useful technique to produce textured surfaces. This method is very fast and produces the textured surfaces on the various tribo-contacts with an accuracy of a higher-order level up to the micro-scale. The intensity and pulsation of laser radiation can be adjusted very accurately to produce micro-dimples of the desired dimensions. LST may produce a larger number of textured surfaces on the tribological systems, and these textured surfaces act as a micro-hydrodynamic bearing in cases of mixed or full lubrication. In this method, various laser systems are used for surface texturing, for example, CO_2 and Nd:YAG lasers (Etsion, 2004a,b; Sudeep et al., 2015).
- **Vibro-mechanical texturing (VMT) technique:** This texturing technique is based on standard turning operations. This technique performs the process on a computer numerically controlled lathe machine and is governed by a piezoelectric actuated tool positioning stage. The VMT technique uses an oscillated single-point cutting tool that engages with a rotating workpiece. The localized removal of surface material takes place and forms the micro-dimples on the workpiece due to the oscillation of the cutting tool. The base of the tunable oscillation tool path controls the size of the textured surfaces, and textured surfaces of various dimensions can be produced by controlled vibratory motion. This surface texturing approach is easy to use and has a number of advantages, such as cost-effective manufacturing and high-precision texturing (Greco et al., 2009; Costa and Hutchings, 2015).
- **Novel dressing technique:** This method is also known as the pattern grinding technique and is an efficient method to produce regular textured surfaces. This method is simple, productive, and inexpensive. Grinding with the wheel-shaped in a specific way is used to produce regular patterns with different shapes on flat and cylindrical surfaces (Stępień, 2009; Costa and Hutchings, 2015). Stepien and Szafarczyk were the first to demonstrate this method in 1989 (Stepien and Szafarczyk, 1989).
- **Chemical etching texturing technique:** This method is popularly known as the chemical milling process in industries. Acids or other chemicals are used for dissolving unwanted materials, i.e., metals, glasses, or semiconductor materials. Such a technique is used for various metals having depths of metal removal as large as 12 mm. Such a technique is successfully used in various applications, such as in fabrication and aerospace industries, to remove the shallow layers of material from large aircraft components and missile skin panels (Yang and Kim, 2002; Ibatan et al., 2015).
- **Mechanical indentation technique:** Such a method is a simple and convenient way to produce the textured surface on the tribo-contacts. The indenter is pressed into the surface of the component, which causes plastic

deformation and produces the required textured surfaces. Hydraulic devices or screw jack arrangements are used to press indenters, which produce the textured surface (Coblas et al., 2015; Costa and Hutchings, 2015).

- **Abrasive jet machining technique:** In this technique, initially, a mask is placed over the surface of the part to be textured. Then, using a jet nozzle, high-velocity air with fine abrasive particles is projected at the sample. This technique has the main advantage that no deburring process is required around the textured surfaces (Coblas et al., 2015; Costa and Hutchings, 2015). Wakuda et al. (2003) used this technique to produce a spherical shape textured pattern with a dimple diameter of 5 μm and a width of 40–120 μm on a silicon nitride ceramic plate.
- **Photolithography texturing technique:** This technique removes parts of a thin film or the bulk of a substrate and is used for microfabrication. This method has specific advantages, such as exact control over the dimension of the textured surface, and can accurately and precisely create patterns across a whole surface. Ultraviolet light is used to transfer the geometric patterns from a photomask to a substrate. Several steps, such as cleaning, coating, exposure, developing, banking, and etching, are performed in this technique (Venkatesan and Stephens, 2005; Coblas et al., 2015).
- **LIthographie Galvanoformung Adformung (LIGA) texturing technique:** This technique may be used to produce either positive or negative asperities on the sample. Polymethyl methacrylate layer is bonded on the substrate at a certain asperity height, which is exposed to the radiation of X-rays by applying a mask. In this process, large textured surfaces may be produced with very high accuracy, contrary to the LST technique. The textured surfaces produced may range from 2 to 1000 μm in diameter and 800 nm to 2 mm in depth with standard shapes (circular, square, triangular, hexagonal, etc.) (Venkatesan and Stephens, 2005; Coblas et al., 2015; Costa and Hutchings, 2015). Stephens et al. (2002) used this method to produce a hexagonal array of positive asperities on thrust bearings.

7.4 ANALYSIS AND MATHEMATICAL MODELLING

The analysis and mathematical modelling are used to carry out the numerical simulation of a tribological system. Detailed mathematical formulations concerning the specific tribo-pair are performed considering the various geometric and operating conditions. The influence of different textured surfaces, i.e., micro-dimples and micro-grooves and surface roughness patterns, is analysed. Therefore, the governing equations are required to consider various factors such as geometric and operational parameters. The governing equations are derived for the different applications. In general, the governing equations for complex engineering problems are partial differential equations, which are non-linear in nature. Therefore, mathematical modelling is performed to deal with the non-linearity that arises in the governing equations. In recent times, the development of a new class of bearings has had a significant effect on the performance of machines. Nowadays, textured spherical bearings are fabricated by advanced manufacturing processes. These bearing configurations provide various advantages such as the ability to sustain both radial and axial loadings and sustain the misalignment. However, the developed model for this class of bearings involves high mathematical complexity owing to the spherical geometry of the bearing. Therefore, finite element techniques can be applied to get the numerical solution to these complex engineering problems. FE analysis is carried out to handle non-linearity, which arises in the governing equations. The source code in MATLAB software is therefore developed to obtain the performance indicators of tribological systems using different mathematical expressions.

7.4.1 Finite Element Modelling

The governing equations for complex engineering problems are non-linear in nature. Thus, governing equations for various tribological systems are solved by numerical techniques. The FEM approach is a numerical method that can be used for obtaining the appropriate solution of boundary value-based partial differential equations. In this technique, the whole problem domain may be divided into a number of small parts, known as finite elements. Various numerical methods are proposed by many researchers to solve engineering problems, but the FEM is the most flexible one in terms of dealing with complicated boundary value-based partial differential equations and gives approximate solutions by minimizing the error function. The different types of elements can be used to discretize the computational domain of the tribological system. The field variable within an element is interpolated by Lagrange interpolation. For example, in the case of a circular journal bearing, assuming a linear distribution within a four-noded quadrilateral element, the unknown lubricant pressure is given as

$$\bar{p} = \sum_{i=1}^{4} \bar{p}_i N_i \tag{7.1}$$

Here, N_i is the shape function relating the nodal pressure $(\bar{p}_i)$ to pressure in (e^{th}) element.

For a four-noded flow domain element of lubricant, Equation 7.1 may be written as

$$\bar{p}^e = \begin{bmatrix} N_1 & N_2 & N_3 & N_4 \end{bmatrix} \begin{bmatrix} \bar{p}_1 \\ \bar{p}_2 \\ \bar{p}_3 \\ \bar{p}_4 \end{bmatrix} \tag{7.2}$$

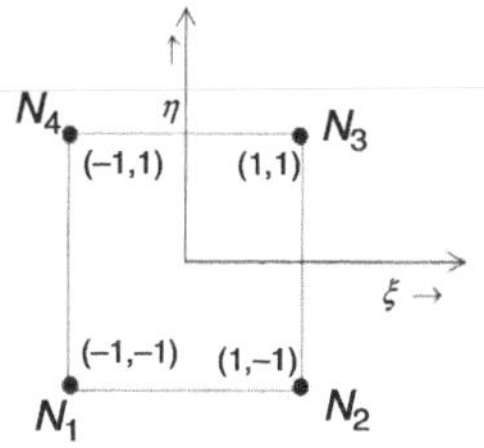

The shape function of the isoparametric flow domain element in terms of local coordinates (ξ,η) is given as

$$N_1 = \frac{1}{4}(1-\xi)(1-\eta) \tag{7.3}$$

$$N_2 = \frac{1}{4}(1+\xi)(1-\eta) \tag{7.4}$$

$$N_3 = \frac{1}{4}(1+\xi)(1+\eta) \tag{7.5}$$

$$N_4 = \frac{1}{4}(1-\xi)(1+\eta) \tag{7.6}$$

The geometry of the fluid film domain is approximated as follows:

$$\alpha = \sum_{i=1}^{4}\left[N_i(\xi,\eta),\, \alpha_i^e\right] \tag{7.7}$$

$$\text{and } \beta = \sum_{i=1}^{4}\left[N_i(\xi,\eta),\, \beta_i^e\right] \tag{7.8}$$

Further, it will have to replace the local co-ordinates (ξ and η) with global co-ordinates (α and β) with the help of the chain rule of partial differentiation as follows:

$$\frac{\partial N_i}{\partial \alpha} = \frac{\partial N_i}{\partial \xi}\frac{\partial \xi}{\partial \alpha} + \frac{\partial N_i}{\partial \eta}\frac{\partial \eta}{\partial \alpha} \tag{7.9}$$

$$\text{and } \frac{\partial N_i}{\partial \beta} = \frac{\partial N_i}{\partial \xi}\frac{\partial \xi}{\partial \beta} + \frac{\partial N_i}{\partial \eta}\frac{\partial \eta}{\partial \beta} \tag{7.10}$$

But in this case, it is not easy to calculate $\frac{\partial \xi}{\partial \alpha}, \frac{\partial \eta}{\partial \alpha}, \frac{\partial \xi}{\partial \beta}$, and $\frac{\partial \eta}{\partial \beta}$ directly from the transformation expressions (Equations 7.9 and 7.10). So it will have to apply the chain rule in the following manner:

$$\frac{\partial N_i}{\partial \xi} = \frac{\partial N_i}{\partial \alpha}\frac{\partial \alpha}{\partial \xi} + \frac{\partial N_i}{\partial \beta}\frac{\partial \beta}{\partial \xi} \tag{7.11}$$

$$\text{and } \frac{\partial N_i}{\partial \eta} = \frac{\partial N_i}{\partial \alpha}\frac{\partial \alpha}{\partial \eta} + \frac{\partial N_i}{\partial \beta}\frac{\partial \beta}{\partial \eta} \tag{7.12}$$

Equations 7.11 and 7.12 can be given in matrix form as follows:

$$\begin{Bmatrix} \dfrac{\partial N_i}{\partial \xi} \\ \dfrac{\partial N_i}{\partial \eta} \end{Bmatrix} = \begin{bmatrix} \dfrac{\partial \alpha}{\partial \xi} & \dfrac{\partial \beta}{\partial \xi} \\ \dfrac{\partial \alpha}{\partial \eta} & \dfrac{\partial \beta}{\partial \eta} \end{bmatrix} \begin{Bmatrix} \dfrac{\partial N_i}{\partial \alpha} \\ \dfrac{\partial N_i}{\partial \beta} \end{Bmatrix} = [\mathbf{J}] \begin{Bmatrix} \dfrac{\partial N_i}{\partial \alpha} \\ \dfrac{\partial N_i}{\partial \beta} \end{Bmatrix} \tag{7.13}$$

The matrix $[\mathbf{J}]$ is known as the Jacobian Matrix. This matrix plays a very important role in Gauss Quadrature. Now one can easily evaluate the above expressions:

$$\begin{Bmatrix} \dfrac{\partial N_i}{\partial \alpha} \\ \dfrac{\partial N_i}{\partial \beta} \end{Bmatrix} = \mathbf{J}^{-1} \begin{Bmatrix} \dfrac{\partial N_i}{\partial \xi} \\ \dfrac{\partial N_i}{\partial \eta} \end{Bmatrix} = \begin{bmatrix} J_{11}^* & J_{12}^* \\ J_{21}^* & J_{22}^* \end{bmatrix} \begin{Bmatrix} \dfrac{\partial N_i}{\partial \xi} \\ \dfrac{\partial N_i}{\partial \eta} \end{Bmatrix} \tag{7.14}$$

Now for evaluation $[J]^{-1}$, the matrix $[J]$ should be a non-singular matrix, i.e., the determinants $[J]$ should be greater than zero.

$$[\mathbf{J}] = \begin{bmatrix} \dfrac{\partial \alpha}{\partial \xi} & \dfrac{\partial \beta}{\partial \xi} \\ \dfrac{\partial \alpha}{\partial \eta} & \dfrac{\partial \beta}{\partial \eta} \end{bmatrix} \tag{7.15}$$

$$\det[J] = \frac{\partial \alpha}{\partial \xi}\frac{\partial \beta}{\partial \eta} - \frac{\partial \alpha}{\partial \eta}\frac{\partial \beta}{\partial \xi} > 0 \tag{7.16}$$

This is the prerequisite condition for evaluating an integral using Gauss Quadrature. This determinant of the matrix $[J]$ is defined by the ratio of the area element in a real element to the area element in the master element. Thus, we can use $d\alpha d\beta = J d\xi d\eta$.

Furthermore, differentiating Equation 7.16 with respect to α and β, respectively, the following expressions are obtained:

$$\frac{\partial \bar{p}^e}{\partial \alpha} = \begin{bmatrix} \dfrac{\partial N_1}{\partial \alpha} & \dfrac{\partial N_2}{\partial \alpha} & \dfrac{\partial N_3}{\partial \alpha} & \dfrac{\partial N_4}{\partial \alpha} \end{bmatrix} \begin{Bmatrix} \bar{p}_1 \\ \bar{p}_2 \\ \bar{p}_3 \\ \bar{p}_4 \end{Bmatrix} \tag{7.17}$$

$$\frac{\partial \bar{p}^e}{\partial \beta} = \begin{bmatrix} \dfrac{\partial N_1}{\partial \beta} & \dfrac{\partial N_2}{\partial \beta} & \dfrac{\partial N_3}{\partial \beta} & \dfrac{\partial N_4}{\partial \beta} \end{bmatrix} \begin{Bmatrix} \bar{p}_1 \\ \bar{p}_2 \\ \bar{p}_3 \\ \bar{p}_4 \end{Bmatrix} \tag{7.18}$$

7.5 NUMERICAL SOLUTION SCHEME

Due to the non-linearity in the governing equation of the tribological system, the iterative scheme is needed to obtain the solution. The finite element method (FEM) can be applied to solve partial differential equations to establish the field variable in the computational domain so as to compute the performance parameters (Tomar and Sharma, 2019, 2022a). Suitable boundary conditions are used to solve the governing equation of a tribological system. The FEM is a versatile numerical technique to handle non-linear differential equations and solve complex engineering problems. FEM helps gain a general understanding of the physical phenomena occurring in the tribological system. The computer program is developed in MATLAB software, which is quite generalized and is capable of studying the behaviour of tribological systems under different geometrical and operational conditions. The developed program numerically simulates the performance characteristics of the tribological system, considering the different issues such as the effects of textured, non-textured, and roughened surfaces in the tribological system. The appropriate convergence criteria are needed to terminate the iterative loops in order to examine the behaviour of the tribological system. When the convergence criterion is achieved, the performance parameters of tribological systems are obtained using mathematical expressions. Figure 7.3 depicts the flow chart of the procedure used in numerical simulation. The numerical simulation used to determine the performance parameters of the tribological system comprises the following steps:

i. The computational domain is discretized by using finite elements, and the values of operational and geometrical parameters of the tribological system are initialized.
ii. Computate the field variable by solving the governing equations.
iii. Generate the elemental matrices for the tribological system.
iv. Using the usual assembly procedure, the equation of the global system is obtained.
v. Suitable boundary conditions are incorporated to solve the global system equation.
vi. The iterative numerical solution scheme such as the Gauss–Siedel method is applied to solve the equation of system.
    i. The convergence criterion is defined to compute the converged solution of the tribological system.
    ii. The performance indicators of the tribological systems are computed by using the mathematical expressions after satisfying the defined convergence criterion.

7.6 MESH SENSITIVITY ANALYSIS

The numerical simulation of tribological responses related to specific manufacturing processes is a further challenging task as it needs high computational power and time. A fine FEM grid of the fluid film domain is required to obtain the accurate influence of surface topography. Before proceeding to compute numerical results

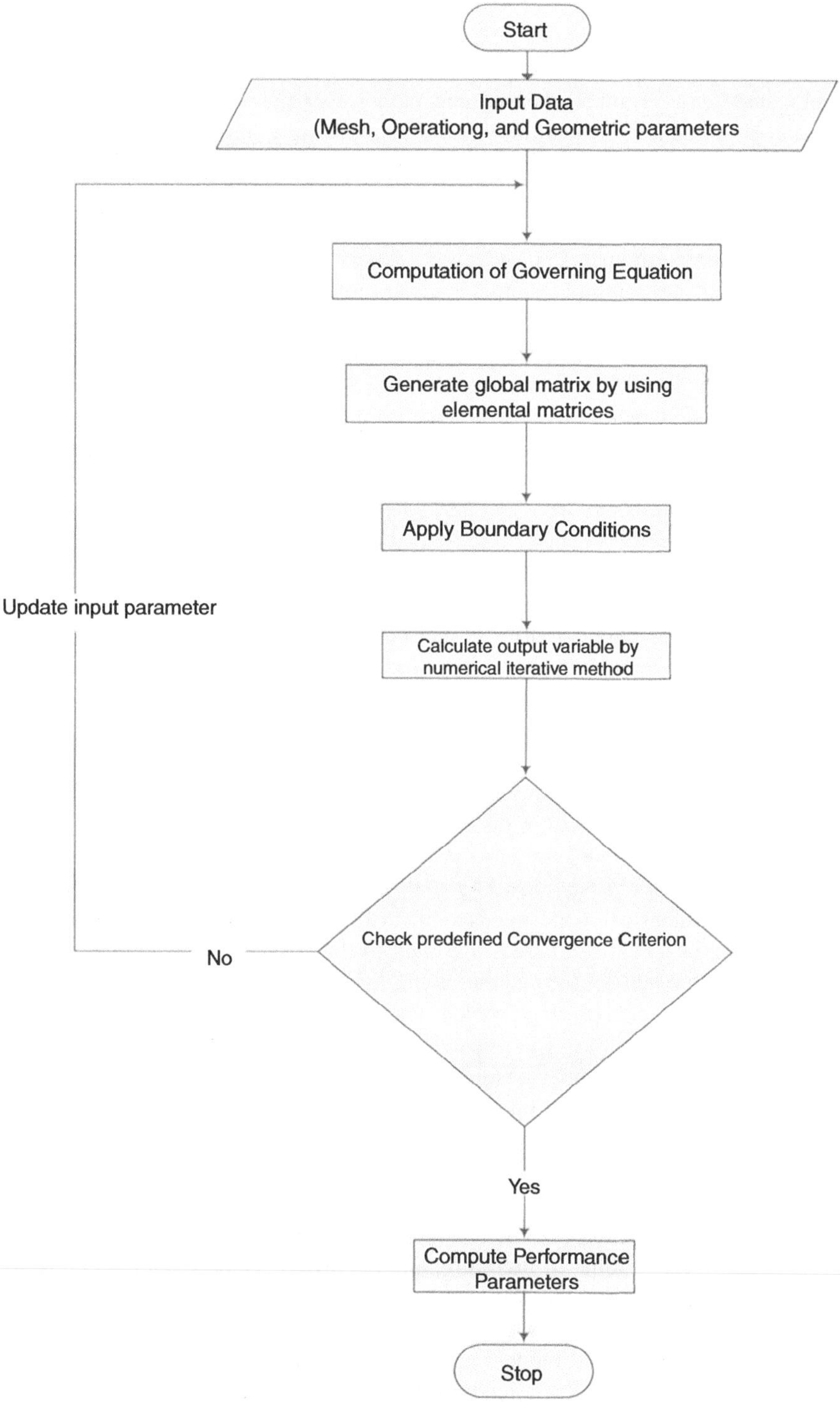

FIGURE 7.3 Flow chart for the numerical solution scheme.

from the developed FEM model, it becomes imperative to ascertain the accuracy of the developed model to generate reliable and accurate design data for the tribological system. The improvement in the accuracy of the numerically simulated results is not significant after a certain level of refinement; however, the time required for the computation of the numerical solution increases significantly. Therefore, for the numerical simulation of the tribological system, the mesh convergence test is performed to obtain a proper mesh size by using different mesh sizes.

In the mesh sensitivity test, the performance parameters are computed as a function of the mesh size. The successive refinement in mesh size, along with the convergence of the solution scheme, leads to more accurate results. The mesh sensitivity test is said to be completed once the percentage difference between the primary variable between the two successive mesh sizes drops below a pre-defined tolerance (%) value. This predefined tolerance value significantly affects the computation time and memory requirements of the CPU. Hence, a pre-defined tolerance (%) value must be suitably/judiciously selected in order to attain a good compromise between the accuracy and computation time of numerical results. The influence of the mesh size in the numerical simulation of a textured spherical journal bearing was studied (Tomar and Sharma 2020a). The value of the radial load remains unchanged after 11,000 elements. The proper mesh size for numerical simulation has been found to be 213 by 55 nodes. In such a way, the mesh sensitivity test needs to be conducted for numerical simulation of tribological responses related to specific manufacturing processes.

7.7 VALIDATION OF THE DEVELOPED NUMERICAL MODEL

In the numerical simulation, the development of an appropriate model and its validation are essential to ascertain the correctness of the developed model. The mesh sensitivity analysis concerning the appropriate mesh sizes is carried out for the tribological system. The mesh sensitivity analysis ensures grid independence; however, the developed numerical model needs to be further validated with available published theoretical and experimental studies related to the particular tribological problem. This is required to ensure the authenticity of the developed model to obtain realistic design data for the engineering problem undertaken in the analysis. In some of the engineering problems, the experimental results are rather very limited. Therefore, the results obtained from the developed numerical model can be validated with both theoretical and available experimental results wherever possible. The theoretical and experimental validations of the developed numerical model may be performed with similar studies available in the literature. The developed numerical model is modified as per the case reported by the reference studies by selecting similar parameters used in the reference studies. The results from the developed model are compared with the previously published literature, and the difference between the two results of the reference study and the developed model is observed. If a small difference is obtained between the two results, then the developed numerical model is considered correct, and the developed model can be used for the analysis of the design engineering problem. A small deviation between the two results may be observed due to variations in the numerical scheme, assumptions, convergence criterion, and mesh grids. The numerical results obtained from the developed model can be validated with their own experimentation performed by the same researcher. However, due to financial constraints, it is not possible to develop the setup for experimentation for every tribological system. Therefore, under such circumstances, the validation can be performed with the earlier published theoretical and experimental studies. Once the validation of the developed numerical model is performed, the performance parameters of the tribological system are obtained.

7.8 NUMERICAL SIMULATIONS

Numerical simulations are required to investigate the behaviour of tribological systems having mathematical models that are too complex for analytical solutions, such as the majority of non-linear systems. Numerous numerical techniques are employed for the simulation of engineering issues. The finite element technique and the finite difference method are the two methods that are mostly employed. The FEM model approximates the problems with a collection of elements dividing the geometry along boundaries while FDM approximates using a grid of points dividing the geometry along the coordinate axis. The FDM can function relatively quickly because the governing equations are solved by direct differentiation along each coordinate axis. Typically, the FEM requires more time to complete than the FDM since it solves the governing equations by discretizing the domain using parts of a chosen shape and assembling them into the entire system. The numerical simulation is performed to compute the various performance parameters of the tribological system. The performance parameters are obtained by using the mathematical expressions derived for the particular tribological components. The parametric studies are also performed to obtain the optimum design parameters for the tribological system. The operating and geometric parameters can be modified for the particular tribological system in the developed source code. The performance parameters of the tribological system are obtained by numerical simulation. During simulations, the processing time is split into discrete time steps to calculate the dynamic behaviour of the system. FEM calculations will be performed gradually. The accuracy of the findings for the dynamically changing variables can be improved with a lower time step. However, it results in more calculations and longer computation times. The transient analysis of the tribological system is very crucial for the smooth functioning of the tribological system. The manufacturing process of the tribo-pairs produces the surface of different features, which ultimately affects the steady and transient behaviour of the tribological system. The thermal effects are the main cause of rupture of fluid film thickness in tribo-pairs, and metal-to-metal contact occurs. When tribo-components operate under severe operating conditions of heavy loading and high speed, it results in the temperature rise of the lubricant fluid film and bearing surface. As a result, lubricant viscosity gets reduced and the tribo-pairs are susceptible to operate at lower values of minimum fluid film thickness. Therefore, the chances of metal-to-metal contact rise, and hence the failure of the tribological system occurred. Under such a situation, the analysis based on isothermal conditions may not give accurate results.

Therefore, the thermal issues need to be considered, and the analysis requires the simultaneous solution of energy, conduction, and Reynolds equations. Therefore, a comprehensive numerical study considering thermal effects may be carried out for the tribological system. The thermal analysis can be studied along with considering the different issues, such as the non-Newtonian behaviour of lubricants and surface topography features, to generate comprehensive design data for the tribological system. Therefore, numerical simulation is very important to determine the performance of a tribological system.

7.9 SUMMARY

This chapter provides extensive updates on mathematical modelling and numerical simulation of tribological responses related to specific manufacturing processes. The advanced manufacturing processes produce tribo-components of high accuracy and precision. However, the surface of the mechanical components is bound to have some degree of roughness or surface irregularities. The surface topography of a tribological system significantly affects the behaviour of the tribological system. Therefore, numerical simulation is very important to accurately predict the performance of tribological systems. Further, the operating and geometric parameters of the tribological system can be optimized by performing numerical simulations. The numerical simulation of the tribological system is a further challenging task as it needs a high computational time and a fine FEM grid of the computational domain to compute the accurate numerically simulated results. Overall, the modelling and simulation of tribo-pairs related to particular manufacturing processes are important to obtain the realistic design data of the tribological system.

REFERENCES

Adatepe, H., Bıyıklıoglu, A. & Sofuoglu, H., 2011. An experimental investigation on frictional behavior of statically loaded micro-grooved journal bearing. *Tribology International,* 44 (12), 1942–1948.

Adatepe, H., Bıyıklıoglu, A. & Sofuoglu, H., 2013. An investigation of tribological behaviors of dynamically loaded non-grooved and micro-grooved journal bearings. *Tribology International,* 58, 12–19.

Ai, X., Cheng, H.S., Hua, D., Moteki, K. & Aoyama, S., 1998. A finite element analysis of dynamically loaded journal bearings in mixed lubrication. *Tribology Transactions,* 41 (2), 273–281.

Anno, J., Walowit, J. & Allen, C., 1969. Load support and leakage from microasperity-lubricated face seals. *Journal of Lubrication Technology,* 91, 726–31.

Anno, J.N., Walowit, J. & Allen, C., 1968. Microasperity lubrication. *Journal of Lubrication Technology,* 90, 351–5.

Brizmer, V. & Kligerman, Y., 2012. A laser surface textured journal bearing. *Journal of Tribology,* 134 (3), 031702.

Brizmer, V., Kligerman, Y. & Etsion, I., 2003. A laser surface textured parallel thrust bearing. *Tribology Transactions,* 46 (3), 397–403.

Chow, L.S.H. & Cheng, H.S., 1976. The effect of surface roughness on the average film thickness between lubricated rollers. *Journal of Lubrication Technology,* 98 (1), 117–124.

Christensen, H., 1971. Some aspects of the functional influence of surface roughness in lubrication. *Wear,* 17 (2), 149–162.

Coblas, D.G., Fatu, A., Maoui, A. & Hajjam, M., 2015. Manufacturing textured surfaces: State of art and recent developments. *Proceedings of the Institution of Mechanical Engineers, Part J: Journal of Engineering Tribology,* 229 (1), 3–29.

Costa, H. & Hutchings, I., 2015. Some innovative surface texturing techniques for tribological purposes. *Proceedings of the Institution of Mechanical Engineers, Part J: Journal of Engineering Tribology,* 229 (4), 429–448.

Cupillard, S., Glavatskih, S. & Cervantes, M.J., 2008. Computational fluid dynamics analysis of a journal bearing with surface texturing. *Proceedings of the Institution of Mechanical Engineers, Part J: Journal of Engineering Tribology,* 222 (2), 97–107.

Dean, B. & Bhushan, B., 2010. Shark-skin surfaces for fluid-drag reduction in turbulent flow: A review. *Philosophical Transactions of the Royal Society A: Mathematical, Physical and Engineering Sciences,* 368 (1929), 4775–4806.

Elrod, H. G., 1973. Thin-film lubrication theory for Newtonian fluids with surfaces possessing striated roughness or grooving. *Journal of Lubrication Technology,* 95 (4), 484–489.

Elrod, H. G., 1979. A general theory for laminar lubrication with Reynolds roughness. *Journal of Lubrication Technology,* 101 (1), 8–14.

Etsion, I., 2004a. Improving tribological performance of mechanical components by laser surface texturing. *Tribology Letters,* 17 (4), 733–737.

Etsion, I., 2004b. State of the art in laser surface texturing. *ASME 7th Biennial Conference on Engineering Systems Design and Analysis.* Manchester: American Society of Mechanical Engineers, 585–593.

Etsion, I., 2005. State of the art in laser surface texturing. *Journal of Tribology,* 127 (1), 248–253.

Etsion, I. & Burstein, L., 1996. A model for mechanical seals with regular microsurface structure. *Tribology Transactions,* 39 (3), 677–683.

Etsion, I., Halperin, G., Brizmer, V. & Kligerman, Y., 2004. Experimental investigation of laser surface textured parallel thrust bearings. *Tribology Letters,* 17 (2), 295–300.

Etsion, I., Kligerman, Y. & Halperin, G., 1999. Analytical and experimental investigation of laser-textured mechanical seal faces. *Tribology Transactions,* 42 (3), 511–516.

Ganai, P., Pandey, R.K. & Dutt, J.K., 2021. Performance improvement of foil air journal bearing employing micro-pocket and textures on top compliant surface. *Surface Topography: Metrology and Properties,* 9 (2), 025045.

Gautam, G., Kumar, N., Mohan, A., Gautam, R.K. & Mohan, S., 2016. Tribology and surface topography of tri-aluminide reinforced composites. *Tribology International,* 97, 49–58.

Gong, J., Jin, Y., Liu, Z., Jiang, H. & Xiao, M., 2019. Study on influencing factors of lubrication performance of water-lubricated micro-groove bearing. *Tribology International,* 129, 390–397.

Greco, A., Raphaelson, S., Ehmann, K., Wang, Q.J. & Lin, C., 2009. Surface texturing of tribological interfaces using the vibromechanical texturing method. *Journal of Manufacturing Science and Engineering,* 131 (6), 061005.

Guha, S.K., 2000. Analysis of steady-state characteristics of misaligned hydrodynamic journal bearings with isotropic roughness effect. *Tribology International,* 33 (1), 1–12.

Hirs, G., 1965. The load capacity and stability characteristics of hydrodynamic grooved journal bearings. *ASLE Transactions,* 8 (3), 296–305.

Ibatan, T., Uddin, M. & Chowdhury, M., 2015. Recent development on surface texturing in enhancing tribological performance of bearing sliders. *Surface and Coatings Technology,* 272, 102–120.

Isaksson, P., Nilsson, D., Larsson, R. & Almqvist, A., 2011. The influence of surface roughness on friction in a flexible hybrid bearing. *Proceedings of the Institution of Mechanical Engineers, Part J: Journal of Engineering Tribology,* 225 (10), 975–985.

Ji, J., Fu, Y. & Bi, Q., 2014. Influence of geometric shapes on the hydrodynamic lubrication of a partially textured slider with micro-grooves. *Journal of Tribology,* 136 (4), 041702.

Kang, K., Rhim, Y. & Sung, K., 1996. A study of the oil-lubricated herringbone-grooved journal bearing—part 1: Numerical analysis. *Journal of Tribology,* 118 (4), 906–911.

Kango, S. & Sharma, R.K., 2010. Studies on the influence of surface texture on the performance of hydrodynamic journal bearing using power law model. *International Journal of Surface Science and Engineering,* 4 (4–6), 505–524.

Kawabata, N., Ozawa, Y., Kamaya, S. & Miyake, Y., 1989. Static characteristics of the regular and reversible rotation type herringbone grooved journal bearing. *Journal of Tribology,* 111 (3), 484–490.

Kazama, T. & Yamaguchi, A., 1993. Application of a mixed lubrication model for hydrostatic thrust bearings of hydraulic equipment. *Journal of Tribology,* 115 (4), 686–691.

Kligerman, Y., Etsion, I. & Shinkarenko, A., 2005. Improving tribological performance of piston rings by partial surface texturing. *Journal of Tribology,* 127 (3), 632–638.

Lu, X. & Khonsari, M., 2007. An experimental investigation of dimple effect on the Stribeck curve of journal bearings. *Tribology Letters,* 27 (2), 169–176.

Oliveira, J., Bottene, A. & Franca, T., 2010. A novel dressing technique for texturing of ground surfaces. *CIRP Annals,* 59 (1), 361–364.

Papadopoulos, C., Efstathiou, E., Nikolakopoulos, P. & Kaiktsis, L., 2011. Geometry optimization of textured three-dimensional micro-thrust bearings. *Journal of Tribology,* 133 (4), 041702.

Patir, N. & Cheng, H.S., 1978. An average flow model for determining effects of three-dimensional roughness on partial hydrodynamic lubrication. *Journal of Lubrication Technology,* 100 (1), 12–17.

Patir, N. & Cheng, H.S., 1979. Application of average flow model to lubrication between rough sliding surfaces. *Journal of Lubrication Technology,* 101 (2), 220–229.

Prajapati, D.K. & Tiwari, M., 2019. Effect of topography parameter, load, and surface roughness on friction coefficient in mixed lubrication regime. *Lubrication Science,* 31 (5), 218–228.

Qiu, M., Delic, A. & Raeymaekers, B., 2012. The effect of texture shape on the load-carrying capacity of gas-lubricated parallel slider bearings. *Tribology Letters,* 48 (3), 315–327.

Ramesh, J., Majumdar, B.C. & Rao, N.S., 1997. Thermohydrodynamic analysis of submerged oil journal bearings considering surface roughness effects. *Journal of Tribology,* 119 (1), 100–106.

Rao, T.V.V.L.N., Rani, A.M.A., Nagarajan, T. & Hashim, F.M., 2014. Analysis of couple stress fluid lubricated partially textured slip slider and journal bearing using narrow groove theory. *Tribology International,* 69, 1–9.

Sahu, K., Sharma, S.C. & Tomar, A.K., 2023. Analysis of MR fluid lubricated slot entry hybrid conical journal bearing with texturing arrangements. *Tribology International,* 188, 108788.

San Andrés, L., 1990. Turbulent hybrid bearings with fluid inertia effects. *Journal of Tribology,* 112 (4), 699–707.

Sharma, S.C. & Kushare, P.B., 2015. Two lobe non-recessed roughened hybrid journal bearing – a comparative study. *Tribology International,* 83, 51–68.

Sharma, S.C. & Tomar, A.K., 2021. Study on MR fluid hybrid hole-entry spherical journal bearing with micro-grooves. *International Journal of Mechanical Sciences,* 202–203, 106504.

Sinanoğlu, C., Nair, F. & Karamış, M.B., 2005. Effects of shaft surface texture on journal bearing pressure distribution. *Journal of Materials Processing Technology,* 168 (2), 344–353.

Stephens, L.S., Siripuram, R., Hayden, M. & Mccartt, B., 2002. Deterministic micro asperities on bearings and seals using a modified LIGA processed. *ASME Turbo Expo 2002: Power for Land, Sea, and Air*. Amsterdam: American Society of Mechanical Engineers, 573–580.

Stępień, P., 2009. Regular surface texture generated by special grinding process. *Journal of Manufacturing Science and Engineering,* 131 (1), 011015.

Stepien, P. & Szafarczyk, M., 1989. Generation of regular patterns on ground surfaces. *CIRP Annals-Manufacturing Technology,* 38 (1), 561–566.

Sudeep, U., Tandon, N. & Pandey, R., 2015. Tribological studies of lubricated laser-textured point contacts in rolling/sliding reciprocating motion with investigations of wettability and nanohardness. *Tribology Transactions,* 58 (4), 625–634.

Tala-Ighil, N., Fillon, M. & Maspeyrot, P., 2011. Effect of textured area on the performances of a hydrodynamic journal bearing. *Tribology International,* 44 (3), 211–219.

Tala-Ighil, N., Maspeyrot, P., Fillon, M. & Bounif, A., 2007. Effects of surface texture on journal-bearing characteristics under steady-state operating conditions. *Proceedings of the Institution of Mechanical Engineers, Part J: Journal of Engineering Tribology,* 221 (6), 623–633.

Tomar, A.K. & Sharma, S.C., 2019. Finite element analysis of multirecess hybrid spherical journal bearing system. *Proceedings of the Institution of Mechanical Engineers, Part J: Journal of Engineering Tribology,* 234 (11), 1798–1821.

Tomar, A.K. & Sharma, S.C., 2020a. An investigation into surface texture effect on hole-entry hybrid spherical journal bearing performance. *Tribology International,* 151, 106417.

Tomar, A.K. & Sharma, S.C., 2020b. A study of hole-entry grooved surface hybrid spherical journal bearing operating with electrorheological lubricant. *Journal of Tribology,* 142 (11), 111802.

Tomar, A.K. & Sharma, S.C., 2022a. Non-Newtonian lubrication of hybrid multi-recess spherical journal bearings with different geometric shapes of recess. *Tribology International,* 171, 107579.

Tomar, A.K. & Sharma, S.C., 2022b. Study on surface roughness and piezo-viscous shear thinning lubricant effects on the performance of hole-entry hybrid spherical journal bearing. *Tribology International,* 168, 107349.

Tomar, A.K., Sharma, S.C. & Sahu, K., 2023. Performance analysis of textured spherical hybrid journal bearings operated with magnetorheological fluid. *Journal of Tribology,* 146 (1), 014602.

Tønder, K., 1986. The lubrication of unidirectional striated roughness: Consequences for some general roughness theories. *Journal of Tribology,* 108 (2), 167–170.

Tønder, K., 1996. Dynamics of rough slider bearings: Effects of one-sided roughness/waviness. *Tribology International,* 29 (2), 117–122.

Tønder, K., 2001. Inlet roughness tribodevices: Dynamic coefficients and leakage. *Tribology International,* 34 (12), 847–852.

Venkatesan, S. & Stephens, L.S., 2005. Surface textures for enhanced lubrication: Fabrication and characterization techniques. World Tribology Congress III. American Society of Mechanical Engineers, 521–522.

Vohr, J.H. & Chow, C.Y., 1965. Characteristics of herringbone-grooved, gas-lubricated journal bearings. *Journal of Basic Engineering,* 87 (3), 568–576.

Wakuda, M., Yamauchi, Y., Kanzaki, S. & Yasuda, Y., 2003. Effect of surface texturing on friction reduction between ceramic and steel materials under lubricated sliding contact. *Wear,* 254 (3), 356–363.

Yang, H., Ratchev, S., Turitto, M. & Segal, J., 2009. Rapid manufacturing of non-assembly complex micro-devices by microstereolithography. *Tsinghua Science & Technology,* 14, 164–167.

Yang, J.-C. & Kim, D.-E., 2002. A study on the characteristics of stiction and friction of texture surface. *Journal of the Korean Society for Precision Engineering,* 19 (7), 51–58.

Yoshimoto, S., Ito, Y. & Takahashi, A., 1999. Pumping characteristics of a herringbone-grooved journal bearing functioning as a viscous vacuum pump. *Journal of Tribology,* 122 (1), 131–136.

8 Manufacturing and Surface Treatment Strategies for Corrosive and Tribocorrosive Environments

A Case Study on Titanium Alloys for Biomedical Applications

Jibin T. Philip, M. Gururaj Acharya, and K. S. Abhimanyu

8.1 INTRODUCTION

Human beings who long for quick healing have made researchers improve in the field of biomaterials, which has immense importance for bone replacement and stents. Significant research has been done to improve the mechanical and electrochemical properties of different fabricated titanium (Ti)-based alloys for biomedical applications. The world demand for bioimplants has risen continuously at a profound rate of 66% according to a conservative projection model in Australia alone between the years 2013 and 2046 (Zhu and Wang, 2021). People have benefited from implants such as a tooth, hip replacement, and spinal fixation devices (Talha et al., 2019).

Biosafety and biofunctionality are important concepts of biocompatibility. It has been further grouped as (1) mechanical biocompatibility (stability of implant at the surgery sites), (2) biological compatibility (bone tissue–material interaction), and (3) histocompatibility (body response depending upon the human body). This is the interaction of material attached to tissue which increases the healing of the affected area. Therefore, we concentrate on metal implants that are nontoxic, inert to moving environments, and efficient in their implant applications. Various studies are focusing on the improvement of the Ti-6Al-4V alloy to improve corrosion resistance as there is a revelation that 41% of Ti-6Al-4V fails in biomedical implants because of the corrosion effect (Manivasagam et al., 2010; Mahlobo et al., 2022). Implants that

DOI: 10.1201/9781003359333-8

are used in the human body are always in contact with blood that contains inorganic ions and organic molecules, which initiates corrosion on the alloy used (Contu et al., 2005; Tang et al., 2006; Burstein and Liu, 2007). Ti-6Al-4V is a bioneutral material that is connected to the bone through mechanical means rather than directly to the bone tissues. Moreover, it releases aluminum (Al) and vanadium (V) over a period of time, which can cause certain health problems, such as an inflammatory reaction that leads to osteolysis and loss of the prosthesis.

Various manufacturing techniques have been utilized to produce Ti alloys including casting or solidification, powder metallurgy (P/M), additive manufacturing (AM) techniques, viz., electron beam melting (EBM) and direct metal laser sintering, hot pressing, fast localized melting, and fast cooling, etc. (Bolzoni et al., 2012; Vrancken et al., 2012; Attar et al., 2015; Dai et al., 2016; Toh et al., 2016; Zhang and Attar, 2016). With the development of these technologies, it has become easier to manufacture Ti-6Al-4V parts with reduced manufacturing time, greater material accessibility, and improved ductility compared to current commercial implants (Ghods et al., 2020; Tareq et al., 2021).

Surface treatment methods are rising that consist of surface texturing and surface coating methods (Pfleging et al., 2015; Kaliaraj et al., 2021; Liu et al., 2021). To address the issue of the stress shielding effect in Ti alloys, Aufa et al. have proposed a surface treatment method that was found to decrease corrosion, enhance cell adherence, and promote apatite formation on the bone remodeling surface (Aufa et al., 2022). Some of the surface texturing methods were particularly used because of their eco-friendly, easy, and controlled reactions. One of the methods was laser texturing, which decreases corrosion and improves tribological behavior, electrochemical corrosion performance, and biological activity (Chen et al., 2019; Ding et al., 2020; Ju et al., 2021).

This chapter critically examines the advancements in the production and processing of Ti6Al4V alloy implants. Key areas of focus include manufacturing strategies, surface engineering, corrosion behavior, and characterization techniques. Additionally, the review identifies current challenges in the field and explores existing research gaps. The primary goal is to provide a comprehensive evaluation of the potential of Ti6Al4V alloy implants and the impact of various production and processing methods.

8.2 Ti6Al4V AS A BIOMATERIAL

In recent decades, there has been a high demand for cellular-structured biomedical Ti alloys. Figure 8.1 shows some of the applications of biomaterials in the human body. The exceptional biocompatibility of Ti and its alloys makes them a preferred choice of material for dental and hip implants. It is widely reported that it promotes osseointegration with host bone (Cosyn and De Rouck, 2009; Gulati and Ivanovski, 2017). Figure 8.2 represents some of the alloys used in medical implant devices. Based on market evaluation, titanium-based alloys hold the largest market share compared to metal alloys like stainless steel and cobalt-chromium. This is due to their superior biocompatibility, high strength-to-weight ratio, and excellent resistance to fatigue and corrosion (Bandyopadhyay et al., 2022).

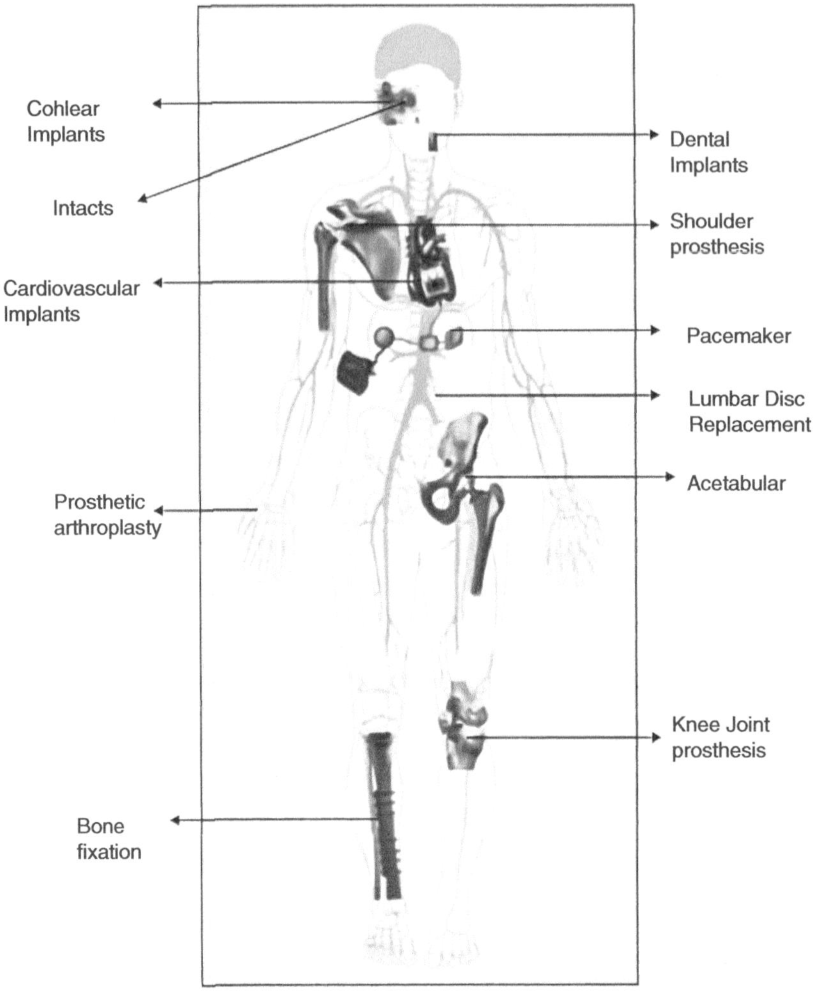

FIGURE 8.1 Biomaterial for human application. (Reproduced with permission from Ref. (Moghadasi et al., 2022). Copyright Elsevier, 2022.)

Ti alloys are generally differentiated or divided based on the microstructures (governed by the phases), i.e., alpha (α), beta (β), and alpha + beta ($\alpha+\beta$). α phase can be stabilized using alloying elements such as carbon (C), nitrogen (N), aluminum (Al), and gallium (Ga), whereas the β-phase is stabilized using molybdenum (Mo), vanadium (V), tantalum (Ta), and niobium (Nb) (Chen and Thouas, 2015). Ti6Al4V alloy is one of the most widely used materials for medical implants. The alloy is stiffer than the actual bone as it exhibits Young's modulus of 100–500 GPa, while the human bone is around 10–30 GPa. This causes a major problem when it comes to hip replacement. The alloy makes the bone difficult to carry a massive load, which ends in stress, and bone responses such as resorption and loss of density (Chopski et al., 2016).

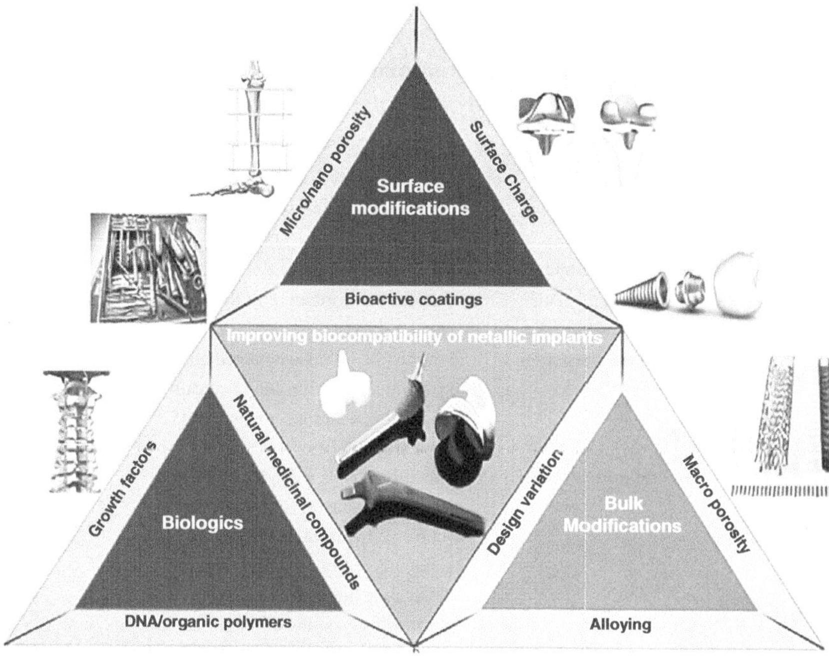

FIGURE 8.2 Some current-generation metallic implants used in the biomedical devices industry are represented. (Reproduced with permission from Ref. (Bandyopadhyay et al., 2022). Copyright Elsevier, 2023.)

But the specified value is less than that of stainless steel and chromium alloys, which range from 200 to 550 GPa.

Alloys such as Ti6Al4V, TiMo12Zr6Fe2, or TiNb13Zr13 are some of the less allergic responders, and some of the reported allergies are due to the byproducts such as Ni (nickel) (Khodaei et al., 2023). To increase the strength and durability of these materials, alloying elements, viz., Ni, cobalt (Co), and chromium (Cr), are added. Also, Mo and manganese (Mn) are some of the other alloying elements that are used. These, along with their salts, cause inflammatory responses. Based on the information discussed, both acute and chronic inflammatory reactions can be caused by implants. However, the data indicate that titanium alloy elicits a superior immune response compared to other alloys. Table 8.1 shows different Ti alloys used in medical implants.

8.3 CORROSION BEHAVIOR OF Ti ALLOYS

Figure 8.3 shows the corrosion mechanism of Ti6Al4V in a corrosive NaCl medium at room temperature. Ti-6Al-4V is stated to be one of the stable metals and has higher corrosion resistance in atmospheric conditions because of the TiO_2 oxide layer formation. Yongqiang Fu et al. (2021) have shown corrosion studies of Ti6Al4V coated with CrMoSiCN nanocomposites evaluated through aerated artificial seawater at

TABLE 8.1
Different Ti Alloys Considered for Bioimplants

Test Material	Atmospheric Condition	Outcomes	References
TiN/ZrN-coated nanotubular on Ti-35Ta-xHf alloy	HF, HCl, and HNO_3	The alloy showed a uniform and stable α+β phase, with ZrN and TiN films.	Moon et al. (2012)
TNZT alloy (β-Ti-type Ti-35Nb-7Zr-5Ta)	Room temperature high-pressure torsion	No change in the Bcc β-Ti phase. Uniform deformation of the material.	Maity et al. (2020)
Ti-Zr-Hf-Nb-Ta	Hank's solution	Low corrosion rates obtained at 10^{-4} mm/year, no pitting occurring in Hank's solution showing good biocompatibility.	Yang et al. (2022)
TiZrHfNbFe	Phosphate-buffered saline solution	The alloy shows a dendritic structure with dual phases. Initially, the corrosion current density of the alloy decreases as the Fe content increases, but then it rises. Promising biocompatible implantation materials.	Wang et al. (2022)
Ti-xCu alloy	Phosphate-buffered saline solution	Passivity on Ti-xCu alloys, an increase in Cu wt% results in a rise in charge transfer resistance and double-layer capacitance, accompanied by a decrease in the inner barrier and outer hydroxide layer.	Siddiqui et al. (2021)
Ti-Zr-Nb-Ta-Mo	Phosphate-buffered saline solution	The corrosion occurs by an increase in porosity and corrosion product with deterioration of bonds by grain boundary corrosion.	Pellier et al. (2011), Hua et al. (2021)
Hierarchical micro/nano Ti6Al4V	Phosphate-buffered saline solution	Showed cell adhesion and expansion, promoting cell proliferation. In vivo experiments showed new bone formation around the titanium rods.	Ren et al. (2021)
Ti-Cu-Zr-Fe-Sn-Si-Ag system	Phosphate-buffered saline solution	Good passive film formation gives corrosion resistance to the alloy. Cell culture indicates biocompatible material.	Yang et al. (2021)
Ti-24Nb-4Zr-8Sn alloy	Kokubo's m-simulated body fluid (m-SBF)	The alloy's surface chemistry and increased hydrophilicity contributed to the observed cell adhesion, proliferation, and protein synthesis.	Nune et al. (2017)
Ti-35Nb-10Zr	Physiological saline solution	Layer composition with TiO_2, Nb_2O_5, and ZrO_2 was obtained on the alloy.	Jeong et al. (2013)

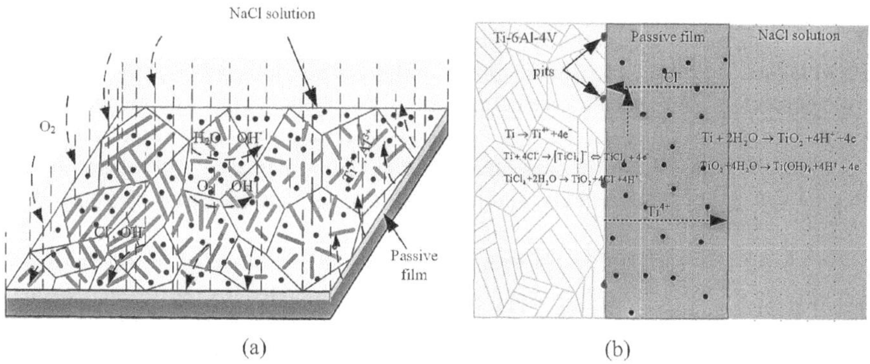

FIGURE 8.3 (a) Schematic diagram of the coating layer in 3.5 wt% NaCl solution during the pitting process. (b) Reaction mechanism in the interface region. (Reproduced with permission from Ref. (Huang et al., 2022). Copyright Elsevier, 2022.)

room temperature. According to ASTM D 11498, the seawater was prepared, and the results demonstrated that the coating functioned as a physical barrier that prevented seawater from penetrating during the initial immersion period. Also, antibacterial tests were carried out that suggested that the coating greatly enhanced the surface and showed antibacterial rates of 98.4% and 96.1% against *E. coli* and *S. aureus*, respectively. Qin et al. (2021) examined the corrosion behavior and mechanism of Ti35Nb produced through SLM when exposed to Hank's solution at 37°C ± 0.5°C. Figures 8.4 and 8.5 show a graphical representation of the Ti35Nb alloy in the presence of simulated body fluids (SBFs). The results of this research work revealed that the corrosion mechanism of SLM-produced Ti35Nb samples shows oxide layers of Nb and TiNb oxide films.

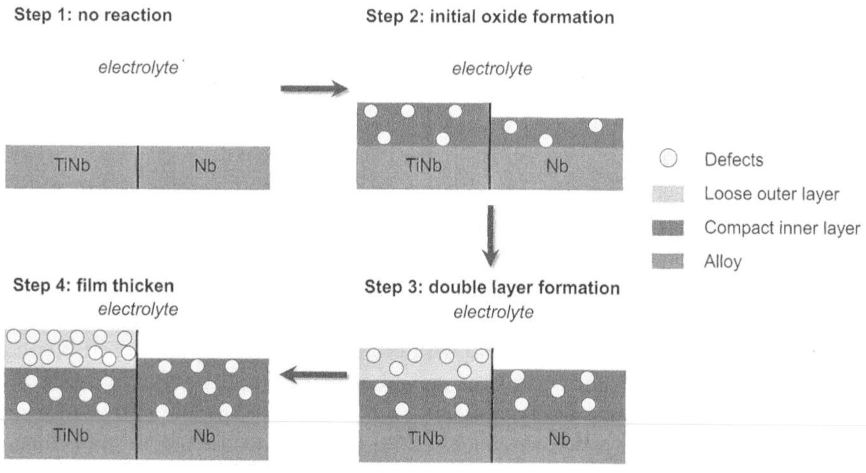

FIGURE 8.4 The process of oxide film formation and distribution of defects in the oxide film of SLM-manufactured Ti35Nb-M alloy is shown in schematic illustrations. (Reproduced with permission from Ref. (Qin et al., 2021). Copyright Elsevier, 2021.)

Ti6Al4V alloys coated with alternating current-gas tungsten arc welding (AC-GTAW) method have been investigated for their corrosion mechanism using 3.5 wt% NaCl solution, and it was emphasized that the oxides and hydroxides formed after corrosion, and the microstructure and alloy elements have contributed a significant anticorrosion property. The AC-GTAW can improve corrosion resistance by changing the transformation between the alpha and beta phases. The massive alpha phase consisted of more defects, and its corrosion resistance was poor (Huang et al., 2022).

Xi et al. (Li et al. 2022) showed that corrosion resistance of Ti-Al-Si coating was applied on TiBw/Ti6Al4V composites and evaluated for hot corrosion resistance in the presence of NaCl and Na_2SO_4 solutions. The study reported that the outermost layer of the coating exhibited dense and continuous Al_2O_3 scales. The effect of Na_2SO_4 salt on the material was found to be less severe compared to NaCl salt. Furthermore, small amounts of silicates were observed on the Ti-Al-3Si coating during the hot corrosion process. The interfacial reaction can be shown as below.

$$4Ti_3AlSi_5 + 13Ti \rightarrow 5Ti_5Si_4 + 4Al \tag{8.1}$$

It can be seen in Equation 8.1 that the presence of the Ti_5Si_4 layer can prevent the surface reaction between the Ti(Al, Si)$_3$ layer and Ti6Al4V, leading to an improvement in the hot corrosion resistance of the Ti-Al-3Si coating.

Xiaonqing et al. (Ni et al., 2023) conducted electrochemical experiments (Figure 8.5) on Ti alloys in the presence of 0.1 M NaF and 3.5 wt% NaCl mixed solution (pH = 4). The titanium alloy Ti6Al4V has a higher alpha martensite phase that is supposed to show low corrosion resistance to stress corrosion rate, and if there is an increase in the concentration of oxygen, there is a decrease in hardness. Heat treatment can increase the stability of β phase and alloy homogeneity. It is seen that removing the α martensite phase can increase the corrosion resistance of the material. The corrosion behavior of Ti alloys rises as there is an increase in NaF concentration, which leads to the formation of weak repassive layers.

Much research has been conducted to improve the resistance toward corrosion and mechanical strength of Ti6Al4V alloy with different alloying elements during fabrication. Some of the alloying elements, like Nb (niobium), show a layer that inhibits material corrosion in medical applications. Siu et al. conducted potentiodynamic polarization research showing that 5 wt% of Nb in the fabrication of Ti6Al4V provides an outstanding corrosion resistance compared to wrought Ti-6Al-4V alloy. But, there was a decrease in the strength and microhardness of the alloy sample because of the presence of a higher beta phase than the alpha martensite phase. Also, the study showed that it is compatible with biomedical implants. Under SBF conditions, the Ti alloy undergoes corrosion and wear that decreases with an increase in the scratching rate and applied potential.

On the whole, the corrosion resistance of Ti6Al4V was associated with the formation of alpha martensite. It is important to concentrate on developing or fabricating a novel alloy in order to optimize implant corrosion behavior as other methods show minimal improvements.

8.3.1 Mechanisms of Corrosion

Ti6Al4V alloy undergoes formidable oxidation layer depletion during different environmental condition tests, and an initial layer formed of TiO_2, Al_2O_3, and V_2O_5 on its surface. Different studies (Fan et al., 2020; Dai et al., 2021; Wu et al., 2021; Qu et al., 2022) mentioned that the tendency of significant corrosion is not only caused by types of electrolyte or medium temperature but also initiates an eutectic reaction between salts present in the electrolyte and diffuse inside the pores of the oxide layers and further deteriorates. Hulin et al. (Wu et al., 2021) studied the corrosion of Ti6Al4V alloy in NaCl and Na_2SO_4 media where the mechanism is as shown below in the equation.

$$Na_2SO_4 \rightarrow Na_2O + SO_3 \rightarrow Na_2O + SO_2 + \frac{1}{2}O_2 \tag{8.2}$$

$$2NaCl + O_2 + SO_2 \rightarrow Na_2SO_4 + \frac{1}{2}Cl_2 \tag{8.3}$$

Further, it passes to the surface of the alloy and reacts with Al and Ti to form the following compounds:

$$2Al + 3Cl_2 \rightarrow 2AlCl_3 \tag{8.4}$$

$$Ti + 2Cl\_2 \rightarrow TiCl\_4 \tag{8.5}$$

$$Ti + 2Cl_2 \rightarrow TiCl_2 \tag{8.6}$$

The chlorine in this research plays an important role by carrying out oxide growth that forms a long strip texture. Also, Al_2O_3 forms in the oxide layer. TiO_2 forms a feeble interlaced structure on the microstructure (Xiong et al., 2008; Dai et al., 2021). The cracking induced by this oxide formation also leads to lamellar cleavages. The alloy undergoes corrosion degradation and material loss because of the presence of a corrosive medium, which leads to oxidation. The degradation of the oxide layer on the surface and alpha matrix leads to severe corrosion. Therefore, oxidation medium affects the degradation of oxide layer and alloy surface, making the formation of strips (Lee et al., 2007; Ciszak et al., 2016; Wu et al., 2021).

Therefore, the discussed research article above showed that a mixture of salts and oxide layer causes more corrosion of Ti alloys compared to single oxide layer formation. This provides a better understanding of the mechanism of corrosion-induced degradation in titanium alloys under corrosive environmental conditions.

Zhang et al. (2021) discussed the corrosion mechanism of the Ti alloy in artificial saliva with varied concentrations of F^- ions and pH values. The researchers present that F^- ions break the formed passive film by oxygen ions.

$$TiO_2 + 4H^+ + 6F^- \rightarrow TiF_{6sol}^{2-} + 2H_2O \tag{8.7}$$

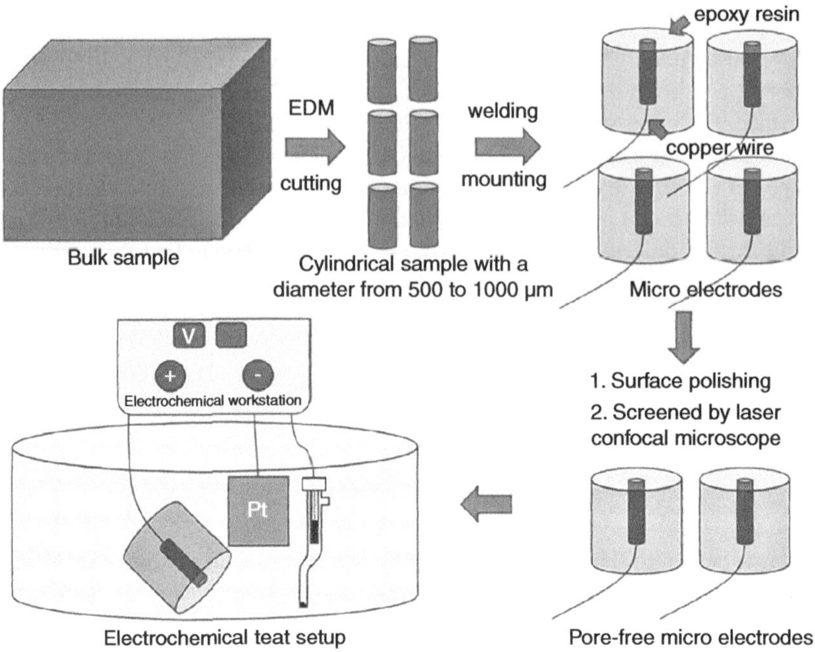

FIGURE 8.5 Procedure for preparing microelectrodes without pores, to be used in electrochemical testing. (Reproduced with permission from Ref. (Ni et al., 2023). Copyright Elsevier, 2023.)

The dissolution of Ti6Al4V takes place more rapidly as the destruction of the passive film continues via fluoride ions reacting

$$Ti + 6F^- \leftrightarrow TiF_{6sol}^{3-} + 3e^- \quad (8.8)$$

$$TiF_{6ads}^{3-} \rightarrow TiF_{6sol}^{2-} + e^- \quad (8.9)$$

$$Al + 3F^- \rightarrow AlF_3 + 3e^- \quad (8.10)$$

Where "sol" represents intermediate species dissolved in the solution and "ads" represents the adsorbed species. The influence of pH is discussed with three cases of values pH=2, 4, and 6. The alloy has a similar NaF concentration in all the pH values mentioned. The passive film stability decreases as the pH value increases, primarily relying on the adsorption of TiF_6^- and OH^-. The electrochemical measurements show that there is minimal variation in the measured values due to the increase in NaF concentration. A higher pH value results in higher OH^- ions that could participate in the reaction process and provide passivating species to be adsorbed on Ti6Al4V alloy. Also, there is a decrease in the current density rate with the increase in NaF concentration at a high pH value.

8.4 TRIBOCORROSION BEHAVIOR

Sliding, fretting, rolling, and impingement are some of the conditions in a corrosive environment that lead to wear and corrosion, which is generally defined as tribocorrosion.

The surface treatments are the important areas where the tribocorrosion focuses on improving the wear and corrosion performance of the given sample. This is done through the analysis of changes in the environment (pH and temperature). These studies are conducted in different mechanical conditions because of the flexibility present in the tribometers. The tribometer maintains an average speed ranging from 0.1 to 1000 mm/s, and an average pressure ranging from 23 MPa to 3.15 GPa. Different types of corrosive media and body fluid electrolytes are utilized, such as NaCl, NaF, and Hank's solution. Though mechanical scratch creates tribocorrosion damage in the presence of SBFs. The addition of corrosive ions such as chloride and fluoride increases the influence of corrosion in the surroundings. Therefore, pH and temperature cannot be important aspects of determining the tribocorrosion. Figure 8.6 shows a schematic presentation of the tribocorrosion mechanism.

Referring to several papers (Komotori et al., 2007; More et al., 2011; Diomidis et al., 2012; Murphy et al., 2012; Dimah et al., 2012; Licausi et al., 2013; Doni et al., 2013; Acharya and Shetty, 2021a,b) in bioimplant alloys carried out in electrochemical conditions by viewing their open-circuit potential (OCP). Vieira et al. (2012) described that there is a breaking of the oxide layer formed on the Ti alloy when sliding between the worn area and the unworn area.

Jun Komotori et al. (2007) discussed tribo corrosion of Ti6Al4V alloy with different scratching frequencies. The chemical electrolyte used for immersion was a combination of salts such as NaCl, KCl, $CaCl_2$, and $NaHCO_3$. Ti alloy has a good corrosion resistance due to the formation of TiO_2 oxide film. But, when this film deteriorates in the presence of high scratching rates, the corrosion increases and causes

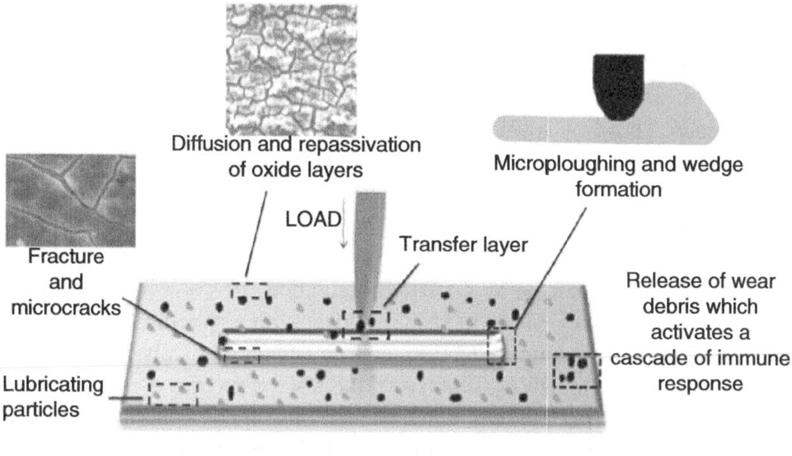

FIGURE 8.6 Schematic representation of the tribocorrosion mechanism. (Reproduced with permission from Ref. (Revathi et al., 2017). Copyright Elsevier, 2017.)

pits with abrasive wear and pits. Also, as the scratching rate increases, the surface morphology transitions from wear to corrosion wear and then to adhesive wear.

A stable passive film is formed when the Ti alloy implant is kept inside the human body. It was seen that, as we increase the scratching rate, the surface wear mechanism of the alloy shifts from abrasive wear to corrosion wear with increasing scratching. Moreover, scratching-induced damage hinders the self-regeneration ability of the passive film on the titanium alloy surface.

Therefore, the range of depletion of the oxide layer that protects the surface depends on the surface modification of the sample, electrolyte, and thc load and sliding/scratching frequencies.

8.5 MANUFACTURING ROUTES FOR THE PRODUCTION OF Ti6Al4V

The process of manufacturing involves transforming raw materials into final products. Among the various manufacturing techniques, AM stands out. Its application in important fields of engineering and medicine such as aerospace, automotive and biomedical engineering, construction, and dentistry has resulted in significant advancements.

8.5.1 Liquid State Fabrication

8.5.1.1 Casting

Titanium is traditionally used as a dental implant because of its greater melting point. Also, at high temperatures, titanium oxidizes and remains inert during the casting process. It reacts with oxides like SiO_2 at high temperatures. The casting is done in five important stages such as (1) pattern making, (2) mold making, (3) casting, (4) divesting, and (5) surface finishing. The drawback of casting technology is the low flexibility in terms of geometry and design of intricate parts, which often need thermal and mechanical processing after manufacturing. Ti6Al4V in this process has drawbacks such as a proper crucible and mold to cast along with high reactivity. The wear of Ti6Al4V is continuously decreased due to the presence of a feeble oxide layer. Also, the material undergoes gross plastic deformation with deep grooves even under low loads. The alloy is worn out in the surface area and causes severe adhesion (Alam and Haseeb, 2002; Borgioli et al., 2005; Ming et al., 2006; Zivic et al., 2011; Song et al., 2012; Doni et al., 2013). The method involves liquid metal converted to solid with a slow cooling rate (~0.5 K/s), thus leading to a high concentration of β phase (Miranda et al., 2013; Sampaio et al., 2016).

Another study by Lazić et al. (2024) investigated the corrosion resistance and biocompatibility of nickel–titanium (NiTi) alloys, with a specific focus on samples manufactured through vacuum induction melting and an innovative continuous casting technique known as NiTicc. Analysis of the microstructure reveals a finer grain size in NiTicc, suggesting possible enhancements in superelastic properties. Acidic corrosion tests demonstrate a notable increase in the release of Ni and Ti ions from the as-cast NiTi alloy, surpassing established safety thresholds. Electrochemical assessments unveil the superior corrosion resistance of NiTicc, attributed to its more uniform oxide layer. Evaluation of cytotoxicity reveals a transient increase in metabolic activity of fibroblasts exposed to alloy extracts, followed by a decline but still maintaining levels above the cytotoxic threshold. These findings underscore the clinical potential of continuous casting in augmenting the properties of NiTi alloys, particularly in bolstering corrosion resistance, a pivotal consideration for medical and dental implementations.

8.5.1.2 Electrodeposition

It is very important to improve the engineering components and economically superior quality to develop novel materials. One such technique is electrodeposition, which can produce a straightforward, inexpensive procedure suitable for commercial processing. Electrodeposition is the most widely applied technique for multilayer coatings. It is particularly employed to obtain materials with ultrafine structures in the industrial sector (Haseeb et al., 1993; Detor and Schuh, 2007; Rupert and Schuh, 2010; Allahyarzadeh et al., 2017; Zsurzsa et al., 2017). Physical vapor deposition, chemical vapor deposition, physical coating spraying, and roll-to-roll coating methods are the most commonly used deposition techniques.

The parameters of an electrodeposition bath include pH, anode and cathode, temperature, bath composition, and plating time. Research has been conducted on multilayer deposits that exhibit increased tensile strength and ductility, as well as improved tribological and corrosion resistance. The whole setup is facile with the simplicity of a single electroplating bath.

Another study by Boraei et al. (2024) delves into exploring the corrosion characteristics of β-titanium alloy, titanium, and Ti-6Al-4V alloy when exposed to Hank's physiological solution (HPS) using diverse electrochemical techniques. The findings reveal that the β-titanium alloy displays superior resistance against corrosion in comparison to the other alloys due to the formation of a protective passive film enriched with ZrO_2, TiO_2, Ta_2O_5, and Nb_2O_5. Various electrochemical analyses, including potentiodynamic polarization curves, electrochemical impedance spectroscopy, and OCP measurements, consistently underscore the β-titanium alloy's enhanced corrosion resistance, which stems from its distinctive microstructural features and alloying elements like niobium, zirconium, and tantalum. Additionally, the alloy's resistance to corrosion is explored under the influence of hydrogen peroxide (H_2O_2), elucidating the effects of immersion durations and H_2O_2 concentrations on the stability and efficacy of the passive film. These insights into the stability and protective properties of the β-titanium alloy's passive film in HPS hold significant implications for potential biomedical applications, particularly in contexts where corrosion resistance is pivotal for the long-term performance and biocompatibility of implants. Scanning electron microscopy (SEM) observations further confirm the formation of a durable passive film on the alloy's surface, offering valuable insights into the underlying mechanisms governing its corrosion resistance and its suitability for biomedical implant materials (Figure 8.7).

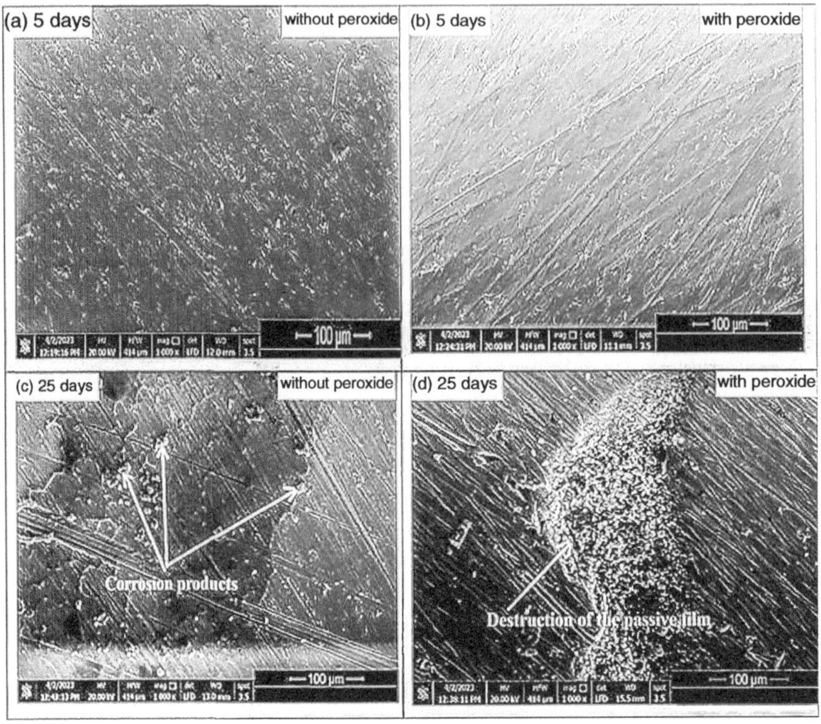

FIGURE 8.7 SEM images of surface of β-Ti alloy postimmersion in HPS: (a) 5 days sans peroxide. (b) 5 days with peroxide. (c) 25 days sans peroxide. (d) 25 days with peroxide at 310 K. (Reproduced with permission from Ref. (Boraei et al., 2024). Copyright Springer Nature, 2024.)

8.5.2 Solid State Fabrication

8.5.2.1 Heat Treatment

Under heat treatments, Ti and its alloys undergo structural modification in several stages. The oxide film on the surface partially breaks into the environment in the presence of a corrosive medium and leads to the hydration of Ti. The hydrogel layer is dried and densified during the thermal process to create a durable amorphous or crystalline alkali titanate layer. The in vitro tests demonstrate that after chemical and thermal treatment, apatite with a rounded design can form on the surface. Shi et al. (2009) showed studies on heat-treated Ti6Al4V implants to develop a bioactive surface on the alloy and understand in vitro bioactivity. The heat treatments cause a change in surface chemistry, surface roughness, and topography of the material that eventually affects osteogenic cells. The results also explain that due to the treatment on the surface, there is a generation of moderate bioactive surface and porous titania gel layer, which may help explain the significantly increased cell proliferation.

Studies have shown that increasing the β-phase content and reducing the high-angle grain-boundary length per unit area significantly enhance the corrosion resistance of Ti-6Al-4V alloy. The β-phase, being more stable and less reactive in corrosive environments compared to the α-phase, contributes to the formation of a more uniform and protective oxide layer on the alloy surface. This phase stability mitigates localized corrosion and helps maintain the integrity of the material in harsh conditions. Furthermore, high-angle grain boundaries are typically more susceptible to corrosion due to their higher energy states and the presence of defects that act as initiation sites for corrosion processes. By reducing the length of these high-angle grain boundaries per unit area, the alloy experiences fewer sites for corrosion initiation, thereby decreasing the overall corrosion rate (Huang et al., 2024).

8.5.2.2 Powder Metallurgy

It is a process for manufacturing metal components by heating the metal alloy powders below up to their melting point. This is discussed by taking a research article by Solorio et al. (2022) who studied the effect of the Ag metal addition on Ti6Al4V by pressing and sintering. An atomized metal alloy powder with a size of 45 μm was mixed with an Ag powder of size 5 μm. The results showed that 15 vol% of Ag and above during sintering is due to interconnected pores by liquid Ag. The viscosity of the liquid is reduced as the temperature increases. The sintered material composed of alpha Ti, Ag, and Ti_2Ag, and beta Ti and TiAg was found at low temperatures. The improvement of antibacterial activity and microhardness is obtained by the formation of Ti_2Ag and TiAg.

A study conducted by Çaha et al. (2020) investigated the corrosion and tribocorrosion behavior of Ti-40Nb and Ti-25Nb-5Fe alloys processed by powder metallurgy and revealed that the Ti-25Nb-5Fe alloy exhibited superior corrosion resistance in a saline solution compared to the Ti-40Nb alloy, with both alloys demonstrating enhanced resistance due to improved surface homogeneity. During tribocorrosion tests, the Ti-25Nb-5Fe alloy showed consistent wear volume loss across different electrochemical conditions (OCP, anodic, and cathodic), indicating that mechanical wear was the dominant degradation mechanism. In contrast, the Ti-40Nb alloy showed significantly lower wear volume loss under anodic potentiostatic conditions, attributed to the formation of a denser tribolayer on the worn surfaces. This suggests an antagonistic effect between wear and corrosion that enhances the alloy's resistance to further wear.

8.5.2.3 Friction Stir Welding

Friction stir welding is one type of solid-state welding method, used in the field of lightweight alloy weldings such as Al, Mg, or their alloys (Wang et al., 2018). It has high expectations for welding dissimilar metals. Similarly, this method has given positive results for Ti6Al4V and NiTi alloys due to their high melting point and high strength (Gangwar and Ramulu, 2018). The advantage of this process to manufacture Ti6Al4V has been clearly explained by Hauibo Deng et al. (2021). The study was conducted between two dissimilar alloys, NiTi and Ti6Al4V, which were welded at conventional friction stir welding and back-heated at 200°C. Layered Ti2Ni existed between the NiTi and Ti6Al4V interface. With a rotation speed of 475 rpm and 235 mm/min travel speed, defect-free joints were obtained. The tensile stress of the obtained joints was 269 MPa, and its elongation was 3.8%.

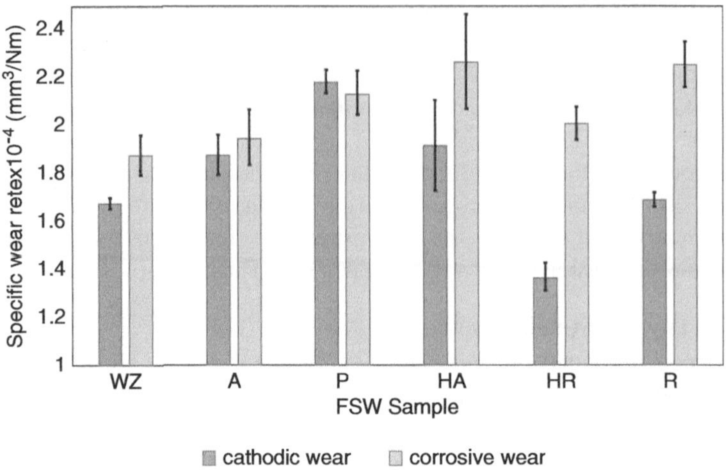

FIGURE 8.8 Specific wear rates for friction-stir-welded samples under cathodic protection and open-circuit voltage (OCV) conditions. (Reproduced with permission from Ref. (Davoren et al., 2017). Copyright Elsevier, 2017.)

Another study conducted by Davoren et al. (2017) investigated the tribocorrosion behavior of friction-stir-welded titanium alloy, which shows significant variations due to the welding process. During wear experiments under open-circuit voltage (OCV) conditions, the FSW process had little impact on the average potential. However, after the wear, the weld zone (WZ) re-passivated quickly, adhering to a second-order kinetic model. In a wear-free environment, the corrosion rates for weld zones ranged from 0.75 to 2.51 μm/yr, slightly higher than the parent material. Cathodic wear analysis revealed that the heat-affected zone on the retreating side (HR and R) had lower corrosion rates than the advancing side (HA and A) (Figure 8.8).

Under OCV conditions, the WZ exhibited lower corrosion rates than the parent material, attributed to faster passivation. The WZ demonstrated the least annual material loss, indicating unique tribocorrosion properties. The study also highlighted the abrasive nature of passivation products under OCV, stressing the importance of understanding specific wear environments to identify potential corrosion vulnerabilities in welded areas.

8.5.3 Additive Manufacturing (AM)

AM or rapid prototyping is the best in class to produce customized implants with high-level accuracy in complex mechanical geometries and short production times. Figure 8.9 shows different AM processes with powder, solid, and liquid-based processes. AM of Ti6Al4V is considered adopted by many industries, particularly in aerospace where the requirement is low weight, high strength, and safety. It reduces the weight of the alloy components and the overall shape with fewer joints and difficult geometry. In this way, the technology in manufacturing Ti alloys can significantly

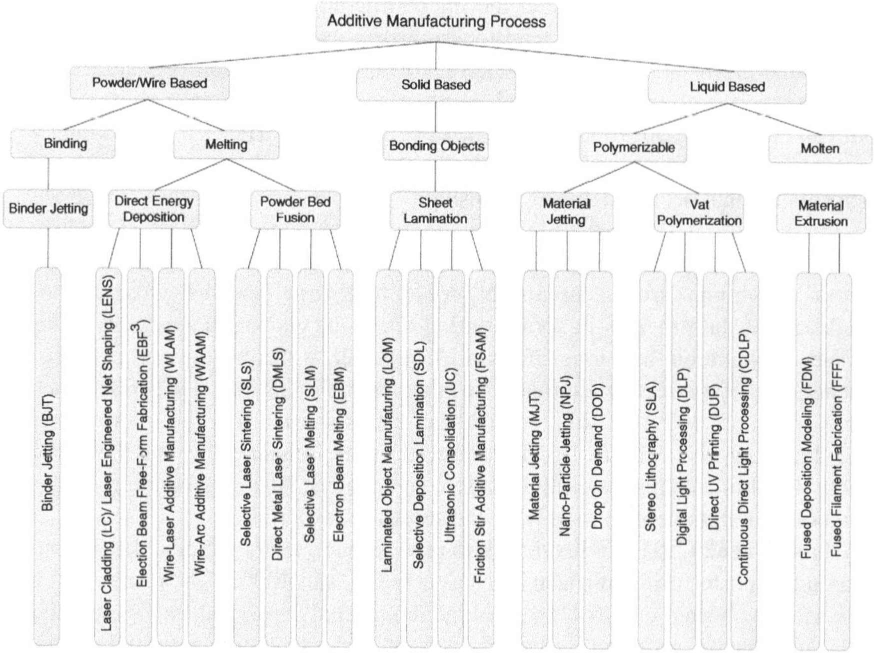

FIGURE 8.9 Classification of additive manufacturing processes. (Reproduced with permission from Ref. (Kanishka and Acherjee, 2023). Copyright Elsevier, 2023.)

change and improve the produced goods. Thus, the future developments of AM will be an important concern for the future Fourth Industrial Revolution.

8.5.3.1 Cold Spray

Lorena et al. (Perez-Andrade et al., 2022), in their work, analyzed the microstructure evolution of Ti6Al4V using the Armstrong process, which was deposited via a cold spray of N and He. The cold gas dynamic spray process, also known as cold spray, is an AM technique (Pattison et al., 2007; Villafuerte, 2015; Raoelison et al., 2017; Yin et al., 2018). Oxygen-sensitive materials such as Ti and Ti-based alloys, which lack material melting during the process, tend to prefer the cold spray method (Rafaja et al., 2009; Cinca et al., 2010; Goldbaum et al., 2011; Li et al., 2020). In this process, accelerated heated powder particles with pressurized carrier gas are passed through a de-Laval nozzle. Subsequently, these particles impact a substrate at high velocities, causing plastic deformation of the substrate as well as the sprayed particles. Due to continuous collisions at high velocities, the sprayed powder particles undergo compaction, resulting in a deposited material with metallurgical bonding at the interfaces due to the severe plastic deformation experienced during deposition. The key observations in this study were the production of highly dense Ti6Al4V deposits through cold spraying of coral-like morphology powders, which exhibited good grain refinement and comparable microhardness.

Kim et al. (2008) investigated the microstructure development of cold-sprayed Cp-Ti and Ti6Al4V spherical feedstock powders and discovered the formation of nano-sized grains at the substrate due to adiabatic shear instabilities and rotational dynamic recrystallization. Lek et al. (2018) also reported the observation of grains with sizes less than 50 nm at the interfaces between the particles and substrate, with martensitic lathes breaking into finer grains at these points due to the high strain rate deformation that occurred during deposition.

Vidaller et al. (2015) worked on the solid-state spraying technique of cold spraying by accelerating a metal powder, preferably with a size distribution of 20–40 μm, within a high-pressure gas stream (N_2 or He) to deposit Ti6Al4V. This method has made significant progress in spray coating due to its economic feasibility, efficient surface preparation, and long lifespan. As a result, it has become an ecologically friendly and cost-effective alternative to other methods (Gärtner et al., 2006; Ghelichi and Guagliano, 2009). In this method, the results showed that the microstructures and flattening ratios of the Ti6Al4V substrate were influenced by the thermal conductivity, ultimate tensile strength (UTS), and density. Also, the bonding, adhesion, and bond strength depend mainly on the substrate properties.

Hussain et al. (2011) studied how two process parameters affected the porosity of titanium deposits, using nitrogen (N_2) gas at 600°C and 800°C. The results indicated that using N_2 heated to 800°C reduced interconnected porosity due to increased particle deformation, resulting in a denser deposit. Porosity significantly influenced the corrosion behavior, assessed through OCPs and rest potentials (E_corr). Even after vacuum heat treatment, OCP values for free-standing titanium deposits were lower than those for bulk titanium, indicating remaining porosity (Figure 8.10).

Electrochemical tests on titanium coatings on carbon steel showed that the substrate significantly affected corrosion behavior, with porosity from both deposition temperatures leading to OCPs and passive current densities that differed from bulk titanium. Salt spray tests confirmed the negative impact of porosity on corrosion resistance.

8.5.3.2 Electron Beam Melting (EBM)

Figure 8.11 shows a schematic representation of the EBM process. Electron beam powder bed fusion is a metal AM process that builds unique metal parts that are hard to achieve through usual manufacturing routes (Körner, 2016; Bäreis et al., 2022). The advantages of using the EBM process are lower residual stress, reduced support structures, tidy processing environment, high production rate, and easy production of processable brittle metals that are hard to process by other manufacturing methods (Murr et al., 2012; Lodes et al., 2015; Sochalski-Kolbus et al., 2015; Yang et al., 2019; Tarafder et al., 2021; Bäreis et al., 2022; Ren et al., 2022). But, the improvisation of this process is much lagging compared to laser-based beam powder bed fusion. Luis et al. (Escano et al., 2022) conducted research on an EBM system that utilized a synchrotron X-ray monitoring method. In this process, a high-energy X-ray beam is directed through a sample containing a substrate and a powder layer that is connected to the electrical ground. An imaging system captures images of the X-ray as it passes through the sample, while diffraction patterns are obtained by analyzing the diffracted X-ray using detectors. Additionally, the system is equipped with a

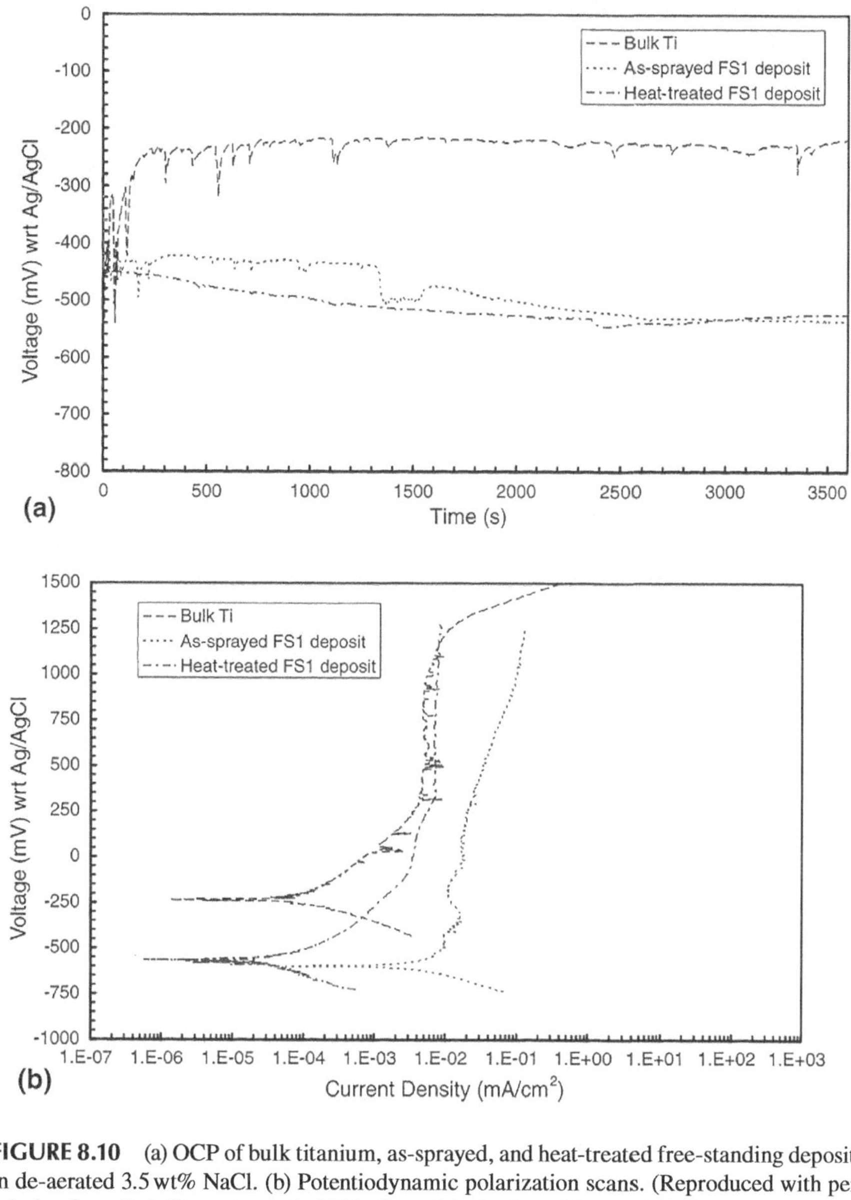

FIGURE 8.10 (a) OCP of bulk titanium, as-sprayed, and heat-treated free-standing deposits in de-aerated 3.5 wt% NaCl. (b) Potentiodynamic polarization scans. (Reproduced with permission from Ref. (Hussain et al., 2011). Copyright Springer Nature, 2010.)

visible light imaging system and an IR imaging system, which are used to monitor the sample's morphology and the temperature of the sample surface, respectively. The studies gave several understanding, such as the instrument is portable with high operability for beamline experiments. It can be used for both high vacuum and a good amount of X-ray transparency. Especially, Ti6Al4V samples were successfully manufactured using this method.

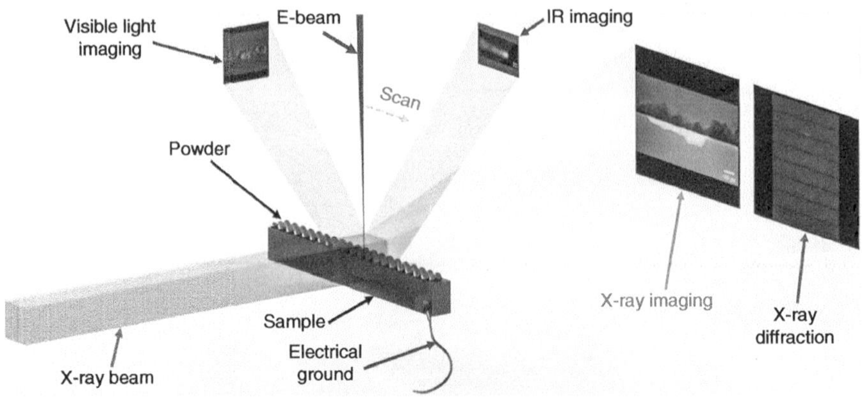

FIGURE 8.11 The in situ X-ray imaging and diffraction system, which also incorporates visible light imaging and infrared (IR) imaging, is designed to capture the dynamics of the EBM AM process. The system's principle is illustrated in the schematic. (Reproduced with permission from Ref. (Escano et al., 2022). Reusable under Creative Commons CC-BY license.)

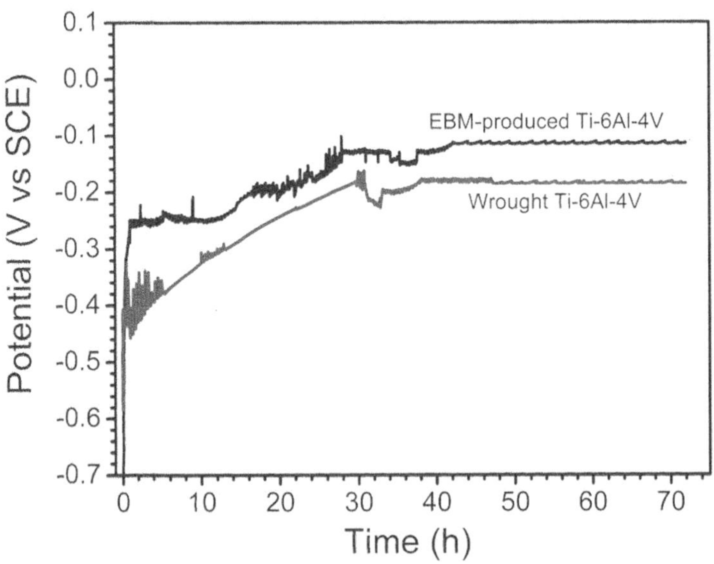

FIGURE 8.12 EBM and wrought Ti-6Al-4V's open-circuit potential over time in buffered saline. (Reproduced with permission from Ref. (Bai et al., 2017). Copyright Elsevier, 2017.)

Another study by Bai et al. (2017) delved into how EBM affects the corrosion behavior of Ti-6Al-4V, comparing it with the conventional wrought form in a saline solution. Findings suggest that EBM-made Ti-6Al-4V displays slightly better resistance against corrosion than its traditional counterpart (Figure 8.12).

This is credited to its higher proportion of β phase and finely structured α/β phases. These structural aspects impede charge transfer and slow down metal-dissolving rates. Moreover, the finely structured α and β phases in EBM-produced Ti-6Al-4V help balance the distribution of alloying elements, reducing the galvanic effect. The study highlights the potential of EBM-manufactured Ti-6Al-4V to not only perform better mechanically but also resist corrosion more effectively than the standard wrought version, a contrast to the reported corrosion issues of selective laser melting (SLM) Ti-6Al-4V alloys.

8.5.3.3 Selective Laser Melting (SLM)

The process of SLM is complex and influenced by numerous parameters that can result in significant variations in product quality. This method employs high-powered thermal energy to melt the metal powder into its desired shape (Lopez et al., 2016). It uses high-powered thermal energy to melt the metal powder into its final form. The technique consists of using high energy to melt the powder into a desirable shape, especially metals. This technique offers more freedom and eco-friendly benefits for manufacturing metal products with unique shapes. Metals such as Cu, Al, Ti, and Ta are processed through this technology. But, the uncertainty that occurs can potentially affect the obtained product development in AM. The first SLM-based Ti6Al4V was manufactured in 2011 by Vilaro et al. (2011), where the group observed the various microstructures produced by postfabrication heat treatments of Ti alloy below and above the β-trans. The results indicated that the material exhibited good yield and strength, although ductility was lower than industry standards. However, optimized heat treatments were found to increase the ductility of the material. Increasing the laser power resulted in a higher solidification of the melting pool, which drastically reduced the defects (Vilaro et al., 2011). Figure 8.13 shows SLM-based Ti6Al4V material used for biomedical applications.

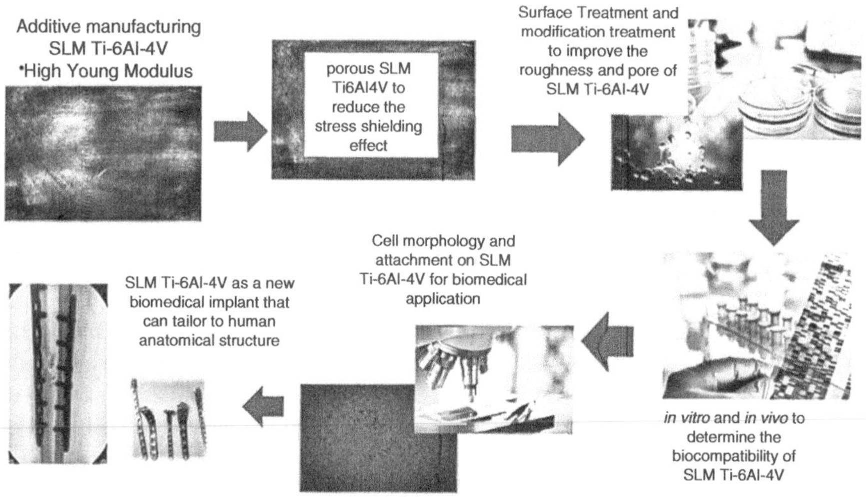

FIGURE 8.13 Additive manufacturing of Ti6Al4V for biomedical applications. (Reproduced with permission from Ref. (Aufa et al. 2022). Copyright Elsevier, 2022.)

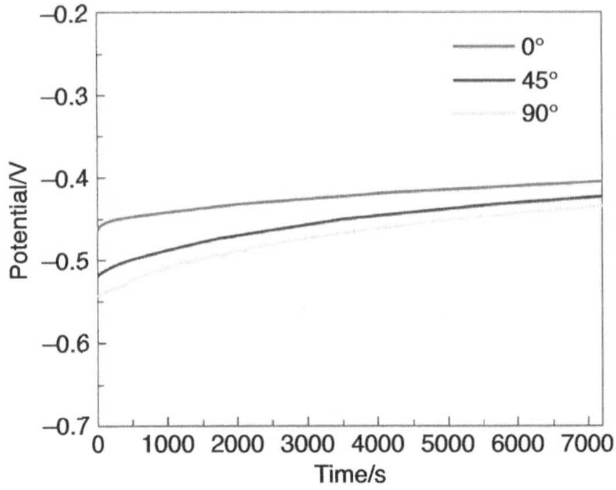

FIGURE 8.14 OCP vs. time for 0°, 45°, and 90° planes in Hank's solution at 37°C. (Reproduced with permission from Ref. (Chen et al., 2022). Elsevier, 2022.)

A study by Chen et al. (2022) delved into assessing how SLM-fabricated Ti-15Mo alloy behaves in terms of corrosion and tribocorrosion, particularly considering its different building directions. It is observed that there are noticeable differences in the microstructure among samples positioned at 0°, 45°, and 90° angles, with the 0° sample showing primarily equiaxed grains oriented along the (001) texture and having smaller average grain sizes compared to others. Corrosion tests conducted in Hank's solution suggest that the 0° plane offers better corrosion resistance than the 45° and 90° planes, a trend that persists under combined wear-corrosion conditions (Figure 8.14).

The study proposes several mechanisms to explain this anisotropic behavior, including the presence and impact of ω phase, grain characteristics, and texture orientation. While the ω phase has a limited influence on corrosion resistance due to its tiny size, it significantly affects tribocorrosion resistance due to its hardness. Additionally, refined grain size and increased grain boundary density facilitate faster passive film formation but may reduce corrosion resistance due to higher molten boundary density. Ultimately, finer grain size improves tribocorrosion resistance, with (001) texture orientation showing superior corrosion resistance due to higher surface energies.

8.5.3.4 Directed Energy Deposition

The Ti material, either in the form of powder or wire, is introduced through a nozzle in the Directed Energy Deposition (DED) process. This nozzle can perform multi-axial displacement on the component design, and the material is melted using a heat source. Figures 8.15 and 8.16 provide pictorial representations of the DED process and the texturing applied to the surface material. During the DED process, a heat source is applied to a specific volume, and element activation is controlled through the use of the birth-death element technique. Two assumptions were made in this case: first, any heat generated by plastic deformation is disregarded compared to the

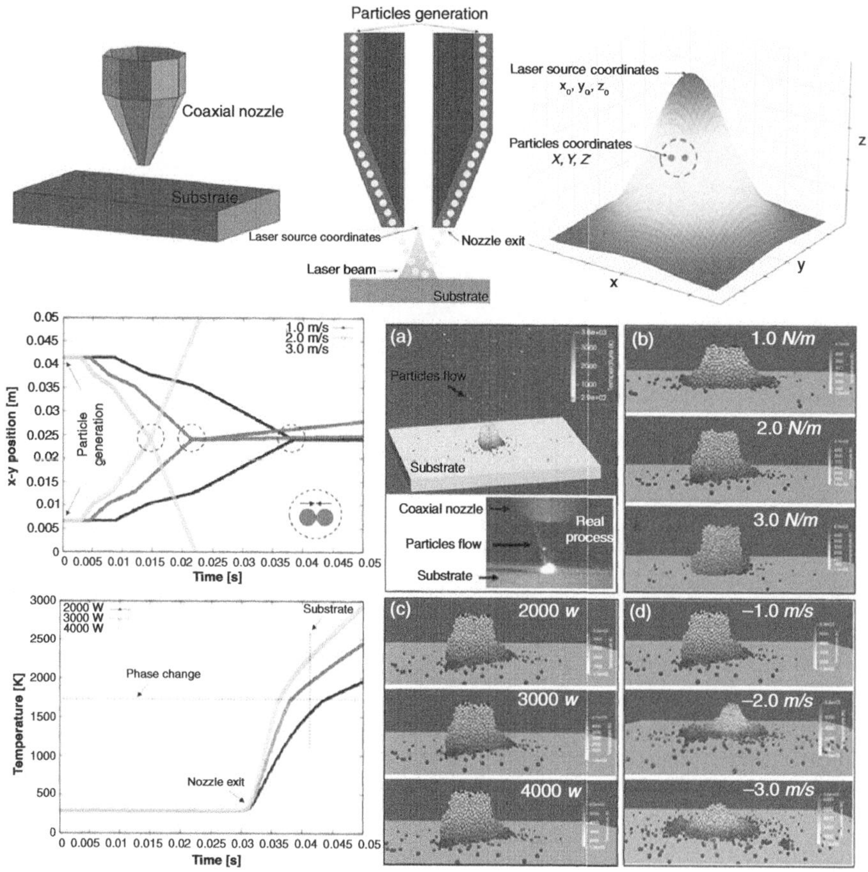

FIGURE 8.15 DED process. (Reproduced with permission from Ref. (Pacheco and Cardoso, 2022). Copyright Elsevier, 2022.)

laser input, and second, latent heat is not taken into account since the absorbed and released heat is assumed to be equal.

Equiaxed prior-β grains, rather than long columnar grains, are anticipated to be necessary for the AM of Ti alloys to achieve isotropic mechanical properties. Therefore, increasing the powder feed rate in the DED process will yield partially or fully equiaxed microstructures (Wang et al., 2015; Zhang et al., 2016). But, this method is difficult to use in some alloys where a steep depth or solidification is needed.

Zhang et al. showcased the research on the DED process, which shows that the usage of this process results in the full equiaxed grains in Ti6Al4V, and the isotropy of mechanical properties is seen in its ductility and Young's modulus. The main advantage of this process is the flexibility of AM to control the grain structures and strengths of other metallic materials. Also, by continuously altering the process parameters, anisotropy in additive-manufactured Ti alloy is resolved, without changing the alloy composition or by following an extra process (Zhang et al., 2023).

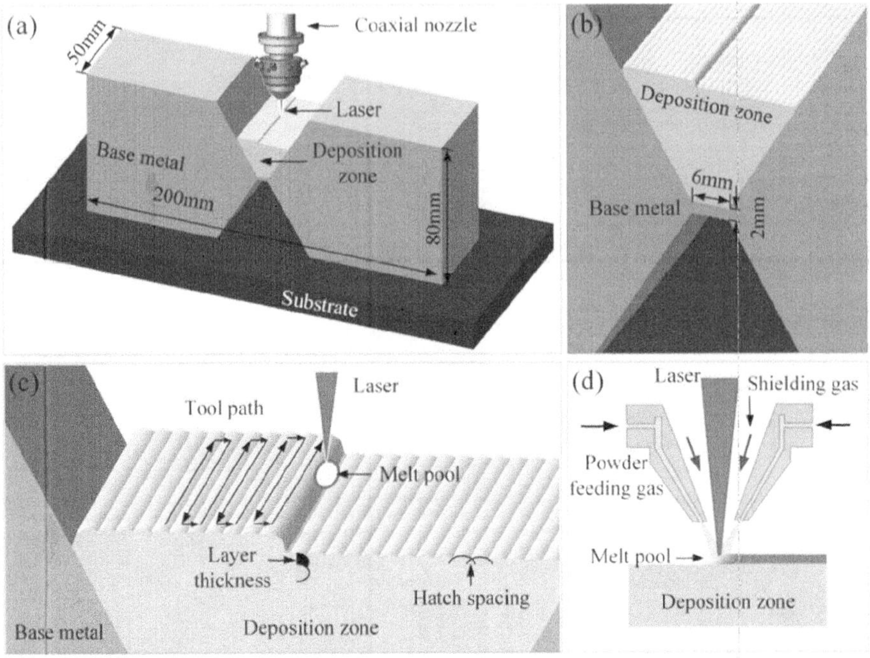

FIGURE 8.16 A schematic diagram is presented for the laser melting deposition (LMD) of thick titanium alloys, which includes the following illustrations: (a) the principle of laser additive manufacturing. (b) The welding groove diagram. (c) The details of laser additive manufacturing. (d) The LMD diagram. (Reproduced with permission from Ref. (Ling et al., 2023). Copyright Elsevier, 2023.)

The visual representation depicts the DED AM process, featuring a coaxial nozzle depositing material onto a substrate. It includes details of particle generation paths, temperature distribution on the substrate, efficiency graphs concerning powder catchment, and comparisons between simulated and actual outcomes, showcasing temperature distributions at various distances from the nozzle.

In a study, Li et al. (2021) delved into the corrosion characteristics of direct energy deposited (DED) titanium alloy vis-à-vis forged titanium alloy when subjected to a 15-wt% NaCl solution. The results indicated that the DED Ti64 alloy exhibited heightened corrosion resistance but diminished electrochemical machinability compared to its forged counterpart (Figure 8.17)

These dissimilarities stem from variations in microstructure, notably a higher β phase content, decreased solute composition disparity, and finer α and β phase lamellae within the DED Ti64 alloy. These structural disparities contribute to the enhanced corrosion resistance and escalated critical electrochemical machining voltage observed in the DED Ti64 alloy. Furthermore, surface topography and dissolution mechanisms emerged as pivotal factors, with the DED Ti64 alloy presenting a more irregular surface, albeit lower roughness, in contrast to the forged Ti64 alloy. This study underscores the pivotal role of microstructural features and surface characteristics in dictating the corrosion dynamics of DED titanium alloys.

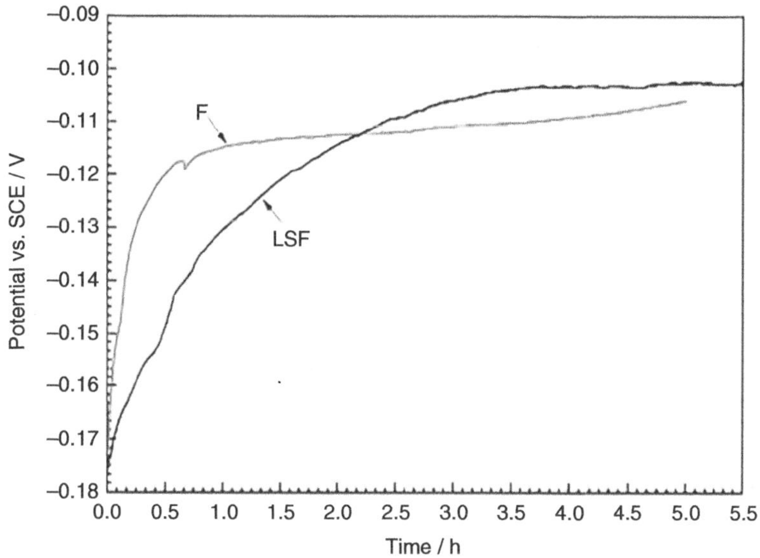

FIGURE 8.17 OCPs of Ti64 alloys from laser solid forming and forging techniques in a 15-wt.% NaCl solution were compared. (Reproduced with permission from Ref. (Li et al., 2021). Copyright Elsevier, 2021.)

8.6 SURFACE ENGINEERING FOR CORROSIVE ENVIRONMENTS

8.6.1 Coatings

Bai et al. (2016) investigated the effects of carbon coating on the corrosion and tribocorrosion behavior of Ti6A14V alloy in Hank's solution. The results indicated that the use of carbon multilayer films led to improved resistance to both corrosion and wear in comparison to single-layer films. Aufa et al. conducted a comprehensive review about utilizing the SLM technique for the fabrication of prosthetic implants and for implementation in various medical fields (Aufa et al., 2022). The method showed an increase in the structural porosity ranging from 65% to 75%, which shows a similar osteogenic activity to that of human bone. The SLM-processed alloy is considered excellent for tissue growth when it is chemically treated with HNO_3, HCl, and HF.

There are different types of spray coating techniques, namely (1) plasma spraying, (2) high-velocity oxygen-fuel spraying, and (3) cold spraying. These coating techniques can be applied using thermal spray processes. These are the best methods to obtain excellent mechanical and surface properties. According to the available data, it should also be noted that biomedical implants have not been developed using this process.

The best method available in the coating is a high entropy alloy that has high mixing entropy to form stable single-phase microstructures. Sometimes, it forms base-centered cubic (BCC) or hexagonal closely-packed crystal structures. They have superior wear resistance, high compressive strength, stiffness, hardness values, and better fracture toughness. Also, the change in the element arrangement in the structure results in a distorted chemical environment, which leads to high corrosion resistance. Comparing other alloys, alloys manufactured through this method have excellent surface chemistry, tunable biocompatibility, and scope for other structural parameter designs. Different types of coatings created for different environments are discussed below.

8.6.1.1 Single and Double-Layer Nitride Coatings via Cathodic Plasma Electrolytic Deposition

Cathodic plasma electrolytic deposition (CPED) is an innovative technique utilized to apply single and double-layer nitride coatings on medical implants. In an experiment involving Ringer's solution—a SBF—the coated samples demonstrated superior corrosion resistance (Figure 8.18).

This high corrosion resistance is pivotal for the durability and biocompatibility of implants, reducing the likelihood of degradation and subsequent release of harmful ions into the body. The CPED process forms a robust protective layer that can withstand the harsh physiological environment, thereby extending the functional lifespan of the implants (Kamkar et al., 2022).

8.6.1.2 CuNP-Coated TiO_2 Nanotube Surfaces through Anodic Oxidation

Anodic oxidation has been employed to create copper nanoparticle (CuNP)-coated TiO_2 (titanium dioxide) nanotube surfaces, which exhibit lower wetting angle values. This characteristic indicates enhanced hydrophilicity, beneficial for cell adhesion and proliferation. In SBF conditions, these coatings also displayed excellent corrosion resistance. CuNP-coated nanotube coatings on Ti6Al4V significantly improve

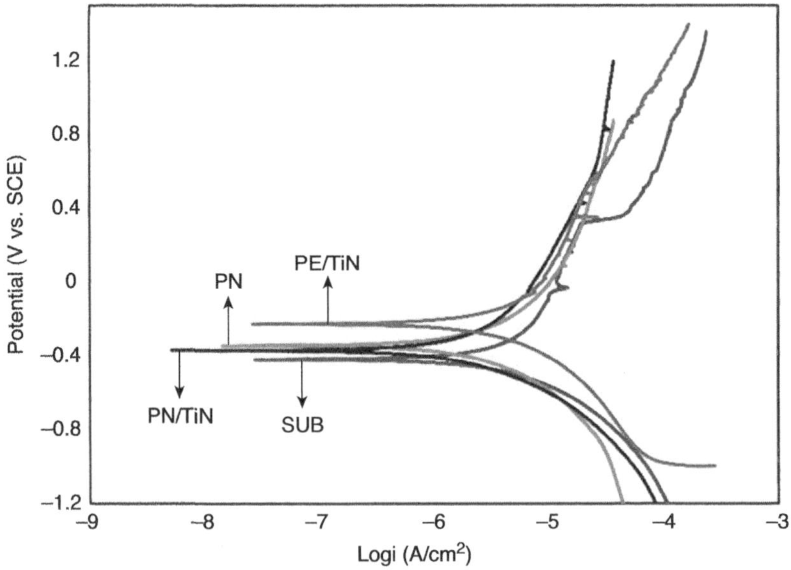

FIGURE 8.18 Tafel plot of the uncoated and coated samples in Ringer solution. (Reproduced with permission from Ref. (Kamkar et al., 2022). Copyright Elsevier, 2022.)

corrosion resistance by creating a protective barrier that minimizes ion release, leading to lower corrosion current density than uncoated Ti6Al4V.

Among various coating durations, the 60-second coating demonstrated the most effective reduction in corrosion current density and enhancement in corrosion resistance. The phase structure of the oxide film, coating thickness, surface morphology, and structural defects also play crucial roles in determining the corrosion resistance. The integration of CuNPs not only improves the mechanical properties but also endows the surface with antimicrobial properties, making it highly suitable for orthopedic and dental implants where infection prevention is paramount (Durdu et al., 2022).

8.6.1.3 Phosphate-Based Electrolytes with EDTA in Plasma Electrolytic Oxidation

The plasma electrolytic oxidation (PEO) method, when combined with phosphate-based electrolytes containing different concentrations of ethylenediaminetetraacetic acid (EDTA), significantly influences the coating properties. The electrolyte composition can alter the growth rate and pore size of the coating, impacting its overall performance. The potential-time response for coatings on Ti6Al4V alloy, shown in Figure 8.19, highlights three distinct regions. In Region I, there is a linear potential increase during conventional anodizing up to around 180 V, mainly due to gas evolution. Region II displays a decreased slope as micro-discharges begin above the breakdown potential, aiding in coating growth. In Region III, the potential reaches approximately 460 V, with ongoing micro-discharges on the surface. The energy required for coating rupture increases with thickness, resulting in larger micro-discharges. All coatings, regardless of varying EDTA concentrations, show similar potential-time responses, with consistent initial linear behavior and final potentials during the PEO process.

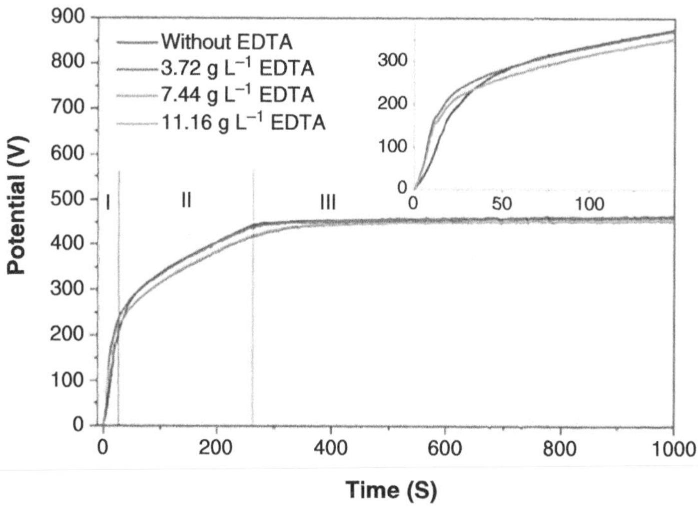

FIGURE 8.19 Potential-time response of samples without EDTA and with different concentrations of EDTA. (Reproduced with permission from Ref. (Ríos et al., 2022). Copyright Elsevier, 2022.)

In terms of corrosion resistance, the PEO process produces thick, durable coatings that enhance the Ti6Al4V alloy's ability to resist corrosion by forming a protective barrier. The formation of micro-discharges contributes to a uniform oxide layer, ensuring improved corrosion resistance, unaffected by variations in electrolyte composition. This versatility allows for tailored coatings to meet specific clinical requirements (Ríos et al., 2022).

8.6.1.4 Bioactive Glass-Zirconia Core–Shell Composite Coatings via Electrophoretic Deposition

Electrophoretic deposition (EPD) has been utilized to create bioactive glass (BG)-zirconia core–shell composite coatings, which are particularly effective for load-bearing implants. The zirconia core provides the necessary mechanical strength and durability needed to withstand the stresses experienced in orthopedic and dental applications. The BG shell promotes bone bonding and osseointegration, demonstrated by the formation of calcium–phosphate (Ca–P) deposits in SBF. This combination ensures that implants offer both mechanical support and biological compatibility. Additionally, the zirconia core's corrosion resistance is enhanced by the BG shell, which protects it from direct exposure to bodily fluids, thus improving the implant's overall stability. The composite coatings also show excellent tribocorrosion resistance; the zirconia core reduces mechanical wear, while the BG layer prevents chemical degradation at wear sites. This dual protection against mechanical and chemical stresses ensures the long-term performance and integrity of the implants. Consequently, these EPD-fabricated coatings are a promising solution for biomedical implants, offering improved bioactivity, mechanical robustness, and resistance to degradation in physiological environments (Hoomehr et al., 2022).

8.6.1.5 Hyaluronic Acid-Based Hydrogel Coatings for Long-Term Implantation

Surface modification techniques have been utilized to develop hyaluronic acid (HA)-based hydrogel coatings, which have undergone detailed physicochemical and nanomechanical assessments to determine their appropriateness for long-term implantation. HA-based hydrogels are highly regarded for their superior biocompatibility and their capacity to replicate the natural extracellular matrix, facilitating cell growth and tissue integration. Additionally, the inclusion of multifunctional antibacterial strategies within these coatings is crucial in preventing infections related to implants, thereby improving the overall success rates of surgical procedures. Importantly, these HA-based hydrogel coatings also address the challenges of corrosion and tribocorrosion in implant materials. Corrosion involves the chemical degradation of materials due to interactions with the bodily environment, potentially compromising implant integrity and lifespan. Tribocorrosion, the synergistic effect of mechanical wear and corrosion, further accelerates material degradation under physiological conditions. By acting as a protective layer, HA-based hydrogel coatings help minimize metal ion release and wear particle generation, thus reducing both corrosion and tribocorrosion, which enhances the durability and reliability of the implants (Andrade del Olmo et al., 2023a).

8.6.1.6 Multilayer HA-Ag/TiO_2/TiN/Ti Coatings via RF Magnetron Sputtering

RF magnetron sputtering is an adaptable method for depositing multilayer coatings such as hydroxyapatite–silver (HA–Ag) on TiO_2/TiN/Ti substrates, which are essential for orthopedic applications. This technique impacts the structural and morphological characteristics of the coatings, including grain size and surface roughness. The silver component imparts antimicrobial properties, while HA boosts osteoconductivity, aiding bone integration. Furthermore, these multilayer coatings exhibit improved resistance to corrosion and tribocorrosion. Corrosion resistance is crucial for preventing implant degradation in the body, thereby increasing its longevity. Tribocorrosion resistance, on the other hand, is essential for ensuring the implant's durability under mechanical stress and friction, minimizing wear and the release of harmful metallic ions. Combined, these properties make HA-Ag-coated TiO_2/TiN/Ti substrates highly effective at preventing infections, promoting bone integration, and maintaining structural integrity, thus enhancing the overall success and longevity of orthopedic implants (Lenis et al., 2022).

8.6.1.7 Multilayered Porous Hydroxyapatite via Plasma Spray Coating

Plasma spray coating has been utilized to create multilayered porous HA coatings. Although there is a slight decrease in adhesion strength, hardness, and modulus, these properties still exceed those of human cortical bone. The coating's porosity improves bioactivity and antibacterial properties, promoting bone ingrowth and decreasing the risk of infection. This makes plasma-sprayed HA coatings particularly suitable for orthopedic implants, where mechanical support and biological interaction are crucial. Additionally, the porous structure of these coatings significantly reduces the effects of corrosion and tribocorrosion. The HA layer serves as a protective barrier, shielding the underlying metal substrate from corrosive bodily fluids and mechanical wear. This dual functionality not only extends the implant's lifespan but also preserves its structural integrity under the cyclic loading and corrosive conditions typical in the human body. Therefore, the combination of mechanical strength, bioactivity, and resistance to corrosion makes plasma-sprayed HA coatings ideal for orthopedic applications (Rajesh et al., 2022).

8.6.1.8 Zinc Oxide Coatings through Electrodeposition

The utilization of electrodeposition techniques has played a pivotal role in advancing zinc oxide (ZnO) coatings, particularly in conjunction with BG, resulting in a notable improvement in adhesion strength. This combination not only guarantees the robustness of the coating but also fosters substantial bioactivity, as evidenced by prolonged bioactivity experiments conducted in SBF, showcasing the formation of calcium–phosphate (Ca–P) deposits. Such bioactivity holds paramount importance in fostering bone regeneration and integration, thereby making ZnO coatings highly suitable for a myriad of implant applications. However, it is imperative to acknowledge the potential challenges associated with corrosion and tribocorrosion in practical scenarios. Corrosion, which stems from chemical reactions between the coating and its environment, has the propensity to compromise the coating's integrity over time. Additionally, in tribocorrosive environments characterized by simultaneous mechanical wear and corrosion, the degradation of the coating may be hastened. Thus, while the augmented adhesion strength and bioactivity of ZnO-BG composite coatings offer promising benefits, addressing concerns related to corrosion and tribocorrosion is crucial to ensure their sustained performance in implant applications (Moradi et al., 2022).

8.6.1.9 MgO-Ag_2O-HA Coatings via Laser and Plasma Spray Techniques

Laser and plasma spray techniques have been utilized to create coatings comprising MgO (magnesium oxide), Ag_2O (silver oxide), and HA. Although the presence of MgO and Ag_2O did not significantly impact adhesive bond strength or osteoblast proliferation, they facilitated sustained silver release over time. This sustained release remains below toxic levels, providing long-term antimicrobial effects that can prevent nonaseptic loosening of implants. This makes MgO–Ag_2O–HA coatings particularly valuable for applications where long-term infection prevention is critical.

In addition to their antimicrobial properties, coatings comprising MgO, Ag_2O, and HA have demonstrated promising resistance against corrosion and tribocorrosion, making them advantageous for implant applications subjected to physiological environments. The presence of MgO and Ag_2O in these coatings contributes to the formation of a protective barrier, which helps mitigate the corrosive effects of bodily fluids and environmental factors. Furthermore, the incorporation of HA enhances the bioactivity and osseointegration of the implants, thus promoting their long-term stability. This synergistic combination not only facilitates sustained silver release for antimicrobial action but also ensures the durability and longevity of the implants by providing corrosion resistance, ultimately advancing the efficacy and reliability of biomedical devices in clinical settings (Ke et al., 2019).

8.6.1.10 Multifunctional Antibacterial Chitosan-Based Hydrogel Coatings

Surface modification techniques have been employed to develop chitosan (CHI)-based hydrogel coatings with multifunctional antibacterial properties. These coatings have been tested against *Staphylococcus aureus* and *Escherichia coli*, demonstrating the ability to repel bacteria and kill them upon contact. The bifunctional antibacterial properties make CHI-based hydrogel coatings highly effective in preventing implant-associated infections during the postoperative period. In addition to their antibacterial properties, CHI-based hydrogel coatings serve as protective barriers against corrosion and tribocorrosion, crucial factors in maintaining the integrity and functionality of implanted medical devices. By mitigating the effects of corrosion, these coatings contribute to the long-term success of surgical procedures by reducing the risk of implant-associated infections and complications, thereby enhancing patient outcomes (Andrade del Olmo et al., 2023b).

8.6.2 Surface Treatments

Titanium alloys, especially Ti6Al4V, are widely used in biomedical implants due to their excellent biocompatibility, mechanical properties, and corrosion resistance. However, the surface properties of these alloys often need to be modified to improve their performance in clinical applications. This analysis delves into various surface treatment methods, conditions, and outcomes, highlighting their significance in enhancing the properties of titanium alloy implants. Table 8.2 discusses about specific surface treatment methods employed on Ti6Al4V alloy in the past and their impacts.

8.6.2.1 Thermal Treatments

Thermal treatments are instrumental in modifying the microstructure of titanium alloys, enhancing their mechanical properties for biomedical applications. This research involved subjecting samples to heat treatment at 704°C for 120 minutes and 740°C for 130 minutes, resulting in the transformation of the α' phase into a combination of α and β phases. Despite variations in temperature and cooling rates, consistent levels of the β-phase were achieved, indicating the reliability of the transformation process. The presence of the β-phase is critical as it improves the ductility and toughness of the alloy, making it suitable for use in load-bearing implants. However, alongside these mechanical enhancements, it is essential to consider the impact of thermal treatments on corrosion and tribocorrosion behavior. While titanium alloys generally exhibit corrosion resistance, the introduction of the β-phase through thermal treatments may alter this property. Corrosion studies suggest that corrosion initiation may occur at interfaces between the α and β phases, underscoring the need to evaluate the corrosion implications of microstructural changes induced by thermal treatments. Furthermore, tribocorrosion, a combination of mechanical wear and corrosion, presents challenges for biomedical implants, potentially affecting their longevity and safety. Therefore, a comprehensive assessment of the corrosion effects of thermal treatment-induced microstructural changes is crucial for developing durable implant materials that ensure patient well-being (Cerri et al., 2022).

8.6.2.2 Ultrashort Pulse Lasers

Ultrashort pulse lasers have been explored for their ability to modify the surface wettability, morphology, and chemical composition of titanium alloys. Experimental and simulation methods showed that laser treatment increased the contact angle, which is primarily influenced by changes in surface chemical composition. This alteration in wettability can significantly impact the biological interactions at the implant surface, potentially improving cell adhesion and reducing bacterial colonization. In conjunction with addressing surface wettability, ultrashort pulse lasers have been investigated for their capacity to mitigate corrosion and tribocorrosion challenges in titanium alloys, critical aspects for ensuring the longevity and effectiveness of implants. Research suggests that laser-induced modifications to the surface not only improve wettability but also enhance corrosion resistance by facilitating the formation of protective oxide layers or modifying surface chemistry to deter the onset and spread of corrosion. Additionally, the controlled adjustment of surface morphology via laser processing has shown promise in reducing the adverse impacts of tribocorrosion, thereby diminishing wear rates and bolstering the material's resilience to mechanical deterioration in corrosive settings. This comprehensive approach, integrating advancements in surface wettability, corrosion resistance, and tribocorrosion mitigation, underscores the potential of ultrashort pulse lasers in refining the durability and biocompatibility of titanium alloy implants, thereby enhancing their clinical performance and extending their service life (Zhang et al., 2022).

TABLE 8.2
Summary of Surface Processing and Treatments in Ti6Al4V Alloy

Surface Treatment Method	Conditions	Outcomes	References
Thermal treatments	Heat treated at 704°C for 120 minutes and 740°C for 130 minutes.	Partial transformation of α′ phase into α+β phase; similar β-phase amounts; enhanced ductility and toughness.	Cerri et al. (2022)
Ultrashort pulse lasers	Examined wettability, morphology, and chemical composition	Increased contact angle; improved cell adhesion; reduced bacterial colonization.	Zhang et al. (2022)
Electron powder bed fusion (EBPBF)	Investigated adhesive bonding on vertical, horizontal, and trabecular surfaces	High bond strength; fractures within adhesive at ~5800 N; ensures structural integrity.	Schmidt and Chen (2022)
Ultrasonic rolling processing	Spindle speed as a crucial factor	Refined grains; residual compressive stress; enhanced hardness and fatigue resistance.	Pang et al. (2022)
Ultraviolet (UV) photo functionalization	Electron beam melting (EBM) additive manufacturing technology	The present method improved hydrophilicity; reduced carbon contamination; better cell attachment and biocompatibility.	Yin et al. (2022)
Thermal oxidation on the surface	Heat treatment at high temperatures for 2 hours, then gradual cooling	Fourfold increase in hardness; reduced wear rate; excellent adhesion of oxide layer; maintained friction coefficient.	García-Rueda et al. (2022)
Selective laser melting (SLM) for Ti6Al4V TPMS scaffolds	Laser power 240 W, scanning speed 240 mm/s, layer thickness 30 μm, hatch distance 50 μm	Unmelted powders; few micro-cracks; increased tensile ductility due to β phase formation and residual stress release.	Ge et al. (2023)
Scanning electron beam rotary polishing	Annealing and air cooling at 800°C	Effective polishing of curved surfaces; uniform hardness distribution; consistent mechanical properties.	Wang et al. (2023b)
Optimization and acid etching with MAO	The as-printed scaffold was annealed at 485°C for 4 hours	Systematic procedure for fabricating bone-adaptive 3D-printed Ti6Al4V scaffolds; enhanced bioactivity and osseointegration.	Wang et al. (2023a)
Deep eutectic solvent ethaline modification	Electrochemical dissolution of α and β phases of Ti6Al4V alloy	Creation of irregularly shaped pits filled with amorphous tubular units; enhanced bioactivity and cell proliferation.	Kityk et al. (2023)
Deep cryogenic treatment and ultrasonic rolling	Vacuum annealing for 1 hour at 890°C, followed by air cooling	Decreased surface roughness; formation of plastic deformation layer; improved strength-plasticity and mechanical properties.	Luo et al. (2022)

8.6.2.3 Electron Powder Bed Fusion (EBPBF)

Electron powder bed fusion (EBPBF) is an AM technique that emphasizes understanding adhesive bonding in different surface orientations—vertical, horizontal, and trabecular meshed surfaces. The study revealed that the bond strength in the interface region exceeded the shear strength of the adhesive, resulting in fracture within the adhesive at an average maximum force (FMax) of approximately 5800 N. Corrosion and tribocorrosion are vital considerations in the durability of implants manufactured through EBPBF. Corrosive environments, particularly in physiological conditions, can compromise the integrity of implant materials over time. Moreover, the combination of mechanical wear and corrosion, known as tribocorrosion, exacerbates material degradation, potentially leading to premature implant failure. Understanding the interplay between adhesive bonding, corrosion resistance, and tribological performance is crucial for developing robust implants that withstand both mechanical stresses and corrosive environments, ensuring long-term functionality and patient safety (Schmidt and Chen, 2022).

8.6.2.4 Ultrasonic Rolling Processing

Ultrasonic rolling processing is a surface treatment method that uses high-frequency vibrations to refine grains and induce residual compressive stress on the surface of titanium alloys. The spindle speed during this process plays a significant role in determining the extent of grain refinement and the level of compressive stress induced. These changes enhance the mechanical properties of the alloy, such as hardness and fatigue resistance, making it more suitable for load-bearing applications. The refined microstructure and induced compressive stress also improve the wear resistance and overall durability of the implants (Pang et al., 2022).

8.6.2.5 Ultraviolet Photo Functionalization

Ultraviolet (UV) photo functionalization is employed to enhance the hydrophilicity of Ti6Al4V implants produced using EBM AM technology. This method significantly improves the hydrophilicity of porous implants while reducing carbon contamination on the surface. Enhanced hydrophilicity promotes better cell attachment and proliferation, which is crucial for the successful integration of implants with the surrounding tissue. Incorporating UV photo functionalization in the production of Ti6Al4V implants via EBM not only enhances hydrophilicity but also addresses corrosion and tribocorrosion concerns. By reducing surface carbon contamination, this method mitigates corrosion susceptibility, ensuring the long-term structural integrity of the implants. Moreover, enhanced hydrophilicity fosters better osseointegration, further bolstering the implants' resistance to tribocorrosion-induced degradation. This comprehensive approach not only improves biocompatibility but also ensures the implants' durability under physiological conditions, thus advancing the efficacy and longevity of biomedical implants (Yin et al., 2022).

8.6.2.6 Thermal Oxidation

Thermal oxidation (TO) involves exposing titanium alloys to high temperatures for two hours, followed by gradual cooling. This treatment increases the hardness of the alloy fourfold compared to the initial value, which is beneficial for decreasing the wear rate while maintaining the friction coefficient. The formation of a robust oxide layer during heat treatment enhances the surface's hardness and wear resistance. This oxide layer acts as a protective barrier, minimizing corrosion susceptibility and mitigating tribocorrosion-induced degradation over time. The improved adhesion of the oxide layer ensures the long-term durability and performance of implants, safeguarding against both chemical and mechanical challenges in biomedical applications. This holistic approach enhances biocompatibility and ensures the reliability of implants in clinical settings (García-Rueda et al., 2022).

8.6.2.7 Selective Laser Melting for Ti6Al4V TPMS Scaffolds

Ti6Al4V triply periodic minimal surface (TPMS) scaffolds were produced using commercial Selective Laser Melting (SLM) equipment with specific parameters: a laser power of 240 W, a scanning speed of 240 mm/s, a layer thickness of 30 μm, a hatch distance of 50 μm, and a laser spot size of 100 μm. The resulting truss surfaces had significant amounts of unmelted powders and a few micro-cracks. The increased tensile ductility was attributed to the combined effects of β phase formation and residual stress release. These characteristics are vital for implants that require high tensile strength and ductility to withstand physiological load (Ge et al., 2023).

8.6.2.8 Scanning Electron Beam Rotary Polishing

Scanning electron beam rotary polishing, combined with annealing and air cooling at 800°C, is effective for polishing curved surfaces of titanium alloys. This method results in a uniform distribution of hardness thickness in each layer, demonstrating the evenness of the polishing effect. Uniform hardness distribution is essential for maintaining consistent mechanical properties across the implant surface, which is crucial for load-bearing applications. This technique ensures that the implants have a smooth and uniform surface, reducing the risk of wear and corrosion (Wang et al., 2023b).

8.6.2.9 Optimization and Acid Etching Combined with Micro-Arc Oxidation

The as-printed Ti6Al4V scaffold was subjected to optimization and acid etching followed by Micro-Arc Oxidation (MAO) and annealing at 485°C for 4 hours. This systematic procedure enhanced the scaffold's bone-adaptive properties. Acid etching creates a rough surface that promotes better bone cell attachment, while MAO provides a bioactive surface that enhances osseointegration. The annealing process improves the mechanical properties of the scaffold, making it more suitable for load-bearing applications (Wang et al., 2023a).

8.6.2.10 Deep Eutectic Solvent Ethaline for Electrochemical Surface Modification

Electrochemical surface modification using deep eutectic solvent ethaline facilitates the dissolution of α and β phases of the Ti6Al4V alloy. This process creates irregularly shaped pits filled with amorphous tubular units that grow uniformly. These surface features enhance the bioactivity of the implants by promoting better cell attachment and proliferation. Employing electrochemical surface modification with deep eutectic solvent ethaline on Ti6Al4V alloy not only enhances bioactivity but also addresses corrosion and tribocorrosion concerns. The dissolution of α and β phases generates irregular pits filled with amorphous tubular units, augmenting the surface area and bioactivity. This increased surface complexity fosters superior cell attachment and proliferation, facilitating robust osseointegration. Furthermore, the formation of a stable, uniform tubular structure enhances corrosion resistance by reducing the likelihood of localized corrosion initiation. Consequently, the modified surface exhibits improved biocompatibility and resistance to tribocorrosion-induced degradation, ensuring the long-term efficacy of biomedical implants (Kityk et al., 2023).

8.6.2.11 Deep Cryogenic Treatment and Ultrasonic Rolling

Deep cryogenic treatment combined with ultrasonic rolling was applied to Ti6Al4V alloy bars after vacuum annealing at 890°C for 1 hour and air cooling. This treatment decreased surface roughness and formed a plastic deformation layer with better strength plasticity. The reduction in surface roughness minimizes the risk of bacterial colonization and improves the wear resistance of the implants. By forming a plastic deformation layer with improved strength-plasticity characteristics, this treatment enhances the implants' resistance to both corrosion and tribocorrosion-induced degradation. Additionally, the reduction in surface roughness mitigates bacterial colonization risks and enhances the implants' wear resistance, further ensuring their longevity and performance in physiological environments. This integrated approach underscores the importance of comprehensive treatments in optimizing the biocompatibility and durability of biomedical implants (Luo et al., 2022).

8.6.3 Surface Texturing (Macro, Micro, and Nano)

The main motive of surface texturing is to increase the bonding ability in coatings, increase the life span, improve wettability, change electrical and thermal conductivity, and vary the friction coefficient. There are several types of surface texturing available, including micro-dimple, micro-grooves, nano-dots, micro-pits, micro-pillars, and inverted pyramids. Table 8.3 discusses some of the texturing studies that were researched for different conditions.

8.6.3.1 Laser Surface Texturing (LST)

Figure 8.20 shows a schematic representation of LST on Ti6Al4V alloy. It is the most used technique due to its controlling and accurate efficiency along with its eco-friendly nature under standard temperature and pressure (Calabrese et al., 2019). It involves the absorption of laser energy followed by the rapid melting or vaporization of the material. Also, it provides good repeatability with no contamination. It can be used with various pulses per second like nanoseconds, femtoseconds, picoseconds, etc. The technique is highly recommended in manufacturing prosthetic components. Some of the other applications of this technique are aging resistance, surface energy, superhydrophobic coatings, and improving bone implants (Sadeghi et al., 2019).

TABLE 8.3
Summary of Different Surface Texturing on Ti6Al4V

Surface Texturing Method	Conditions	Outcomes	References
Duplex treatment of LST-TO	Dry sliding conditions, ball-on-disk tribometer	Gradient TO-coating with Al_2O_3 and TiO_2; high hardness and elastic modulus; enhanced wear performance.	Yuan et al. (2020a)
Plasma molybdenizing and laser surface texturing	Double glow plasma alloying (DGPSA) with Mo layer on polished Ti6Al4V	Threefold increase in microhardness due to solid solution strengthening; improved surface strengthening and wear resistance.	Yuan et al. (2020b)
Micro conical dimple texturing via micro-milling	Mechanical micro-milling	Uniform, defect-free textures; enhanced hydrophilicity; decreased contact angle; and increased surface free energy with roughness.	Jain et al. (2020)
Hierarchical micro/submicro-textured 3D-printed Ti6Al4V surface	Etching in boiling acid bath with sulfuric and hydrochloric acids	Significantly promoted morphology, proliferation, and osteogenic differentiation of BMSCs; improved in vivo osseointegration.	Zhang et al. (2020)
Crystallographic texture through recrystallization annealing	Recrystallization annealing (RA) was applied to the as-received sample	Development of equiaxed α grains interspersed with intergranular β phase; comparable microstructure on three perpendicular surfaces.	Amirnejad et al. (2021)
Pulsed laser remelting	Overlapping multiple laser pulses	Formation of a thin layer of metallic glass with good wear resistance when using picosecond and femtosecond laser pulses.	Grabowski et al. (2021)
Groove-type laser engraved chip breaker design	Laser texturization to generate patterns on the substrate	Texturization did not affect final surface roughness at tested cutting speed, feed, and depth of cut.	Fernández-Lucio et al. (2021)
Powder bed-based laser additive manufacturing	Fabrication by SLM followed by annealing	Anisotropy in tensile properties related to texture and microstructure formation.	Chang et al. (2020)
Heat treatments	Various heat treatment schedules to produce distinct microstructures and textures	Surface hardness significantly influenced by grain size; improved hardness with finer grains.	Amirnejad et al. (2020)
Electron beam melting (EBM) and postheat treatment	EBM with conventional annealing and two-step thermal heat processing (THP)	Coarsening of Widmanstätten and GBα phases; alteration of α phase texture with basal planes rotated 30° around building direction after annealing.	Doğu et al. (2020)

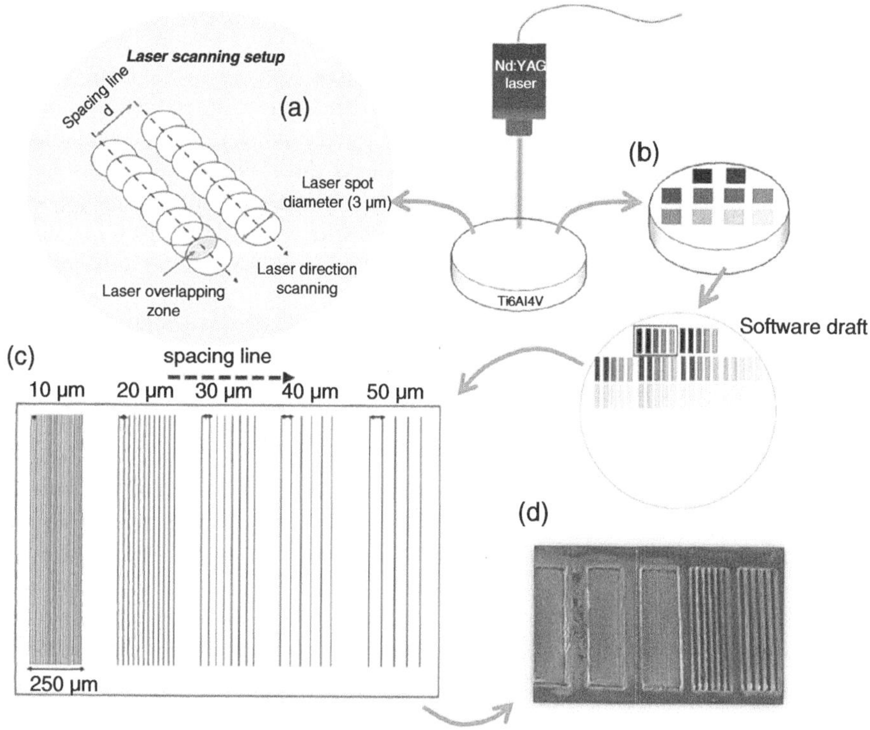

FIGURE 8.20 Scheme of laser surface-micro-grooved process. (Reproduced with permission from Ref. (Moura et al., 2019). Copyright Elsevier, 2019.)

8.6.3.2 Duplex Treatment of Laser Surface Texturing and Thermal Oxidation

Duplex treatment combining Laser Surface Texturing (LST) with Thermal Oxidation (TO) is employed to enhance the wear resistance of Ti6Al4V alloys. Under dry sliding conditions at room temperature, the ball-on-disk rotational motion tribometer (CFT-1) is used to evaluate wear performance. The TO-coating formed comprises an outer oxide layer containing Al_2O_3 and TiO_2 and an oxygen diffusion coating beneath the surface. The initial step involves creating a patterned surface using laser pulses. This increases the surface area and roughness, promoting better adhesion of subsequent coatings. The textured surface is then subjected to high temperatures, causing the formation of a hard, wear-resistant oxide layer. This layer significantly increases surface hardness and elastic modulus, enhancing wear performance. The combined effect of LST and TO results in a gradient coating that provides superior wear resistance due to the high hardness and modulus of the oxide layer. This makes the alloy suitable for applications where high durability is required under abrasive conditions (Yuan et al., 2020a).

8.6.3.3 Plasma Molybdenizing and Laser Surface Texturing

Combining plasma molybdenizing with LST significantly improves the surface strength and wear resistance of Ti6Al4V alloys. Using the double glow plasma alloying (DGPSA) technology, a molybdenum (Mo) layer is deposited onto the polished alloy surface. This technique involves the deposition of Mo onto the titanium alloy surface. Mo atoms diffuse into the alloy matrix, forming a solid solution with Ti, which significantly enhances microhardness. The surface is then textured using laser pulses, creating a fine pattern that enhances mechanical interlocking and wear resistance. The Mo layer increases microhardness threefold due to solid solution strengthening, while the laser texturing improves surface morphology and strength. This dual treatment makes the alloy highly resistant to wear and suitable for demanding applications (Yuan et al., 2020b).

8.6.3.4 Micro Conical Dimple Textured Surface via Micro-Milling

Micro-milling techniques are adeptly employed to engineer orderly conical dimple textures on Ti6Al4V surfaces, fostering uniformity and defect-free textures. Precise modulation of milling parameters like tool speed, feed rate, and cutting depth guarantees the generation of consistent and uniform patterns. While these textures elevate surface roughness, thereby augmenting hydrophilicity and reducing contact angle values, their impact on tribocorrosion is noteworthy. The increased surface free energy not only promotes cell adhesion and proliferation but also contributes to heightened resistance against tribocorrosion-induced degradation. This approach, tailored for biomedical implants, underscores the synergy between enhanced cell interaction, fluid dynamics, and tribocorrosion resistance, pivotal for their successful integration and sustained performance in physiological environments (Jain et al., 2020).

8.6.3.5 Hierarchical Micro/Submicro-Textured 3D-Printed Surfaces

Creating hierarchical textures on 3D-printed Ti6Al4V surfaces through acid etching in a boiling acid bath containing sulfuric and hydrochloric acids for two minutes not only enhances biological activity but also addresses tribocorrosion concerns. This selective material removal process generates a hierarchical structure with micro and submicro features, improving surface roughness and promoting better cell morphology, proliferation, and osteogenic differentiation. The resulting modified surface mimics the natural extracellular matrix, fostering in vivo osseointegration. Moreover, the enhanced surface texture contributes to improved tribocorrosion resistance, ensuring the long-term stability and performance of orthopedic implants in physiological environments. This multifaceted approach underscores the significance of surface modification techniques in optimizing both biological and mechanical properties of biomedical implants (Zhang et al., 2020).

8.6.3.6 Crystallographic Texture through Recrystallization Annealing

Recrystallization annealing (RA) is instrumental in fostering a homogeneous microstructure within the Ti6Al4V alloy, essential for addressing tribocorrosion concerns. By subjecting the alloy to controlled heating within the $\alpha+\beta$ phase field and subsequent controlled cooling, RA achieves a stable microstructure characterized by equiaxed α grains interspersed with β phase. This treatment not only enhances anisotropy and mechanical properties, ideal for multi-directional stress applications, but also fortifies the alloy against tribocorrosion-induced degradation. The uniform microstructure significantly improves the material's resistance to both mechanical wear and corrosive attack, ensuring its reliability and longevity in demanding environments subjected to varying stress and corrosive conditions (Amirnejad et al., 2021).

8.6.3.7 Pulsed Laser Remelting

Pulsed laser remelting involves overlapping multiple laser pulses to texture large surfaces of Ti6Al4V alloy elements. When picosecond and femtosecond laser pulses are used, the formation of a frozen thin layer of metallic glass is likely. This process dynamics entails rapid melting and quenching of the surface, leading to the formation of an amorphous layer with excellent wear resistance. The texturing process can be precisely controlled to create desired surface patterns, enhancing the alloy's tribological properties and making it suitable for high-performance applications (Grabowski et al., 2021).

8.6.3.8 Groove-Type Laser-Engraved Chip Breaker Design

The groove-type laser-engraved chip breaker design, achieved through laser texturization on the substrate, not only enhances machining performance but also addresses concerns regarding tribocorrosion. While analysis of variance suggests that tested texturizations, under specific cutting parameters, do not significantly impact final surface roughness, they play a crucial role in tribocorrosion management. By precisely controlling laser parameters to create grooves that act as chip breakers, this technique effectively reduces frictional forces during machining, thereby mitigating wear-induced tribocorrosion. This approach ensures optimal machining performance without compromising surface finish, making it particularly advantageous for manufacturing processes where both chip control and tribocorrosion resistance are paramount (Fernández-Lucio et al., 2021).

8.6.3.9 Powder Bed-Based Laser Additive Manufacturing

In the fabrication of Ti6Al4V parts through SLM and subsequent annealing, the management of tribocorrosion is a critical consideration. The anisotropy of tensile properties in annealed samples is intricately linked to the formation of texture and microstructure during the SLM process. The layer-by-layer melting and solidification dynamics of SLM contribute to the development of a textured microstructure, influencing the susceptibility to tribocorrosion. Postprocessing annealing plays a vital role in relieving residual stresses and refining the microstructure, thereby enhancing mechanical properties and reducing anisotropy, which are essential for mitigating the effects of tribocorrosion on the longevity and performance of Ti6Al4V implants (Chang et al., 2020).

8.6.3.10 Heat Treatments for Microstructure and Texture Development

In the fabrication of Ti6Al4V samples, diverse heat treatment schedules are utilized to engineer specific microstructures and textures. These procedures involve meticulous control over heating and cooling cycles to tailor grain size and phase distribution. Hardness evaluations reveal a pronounced correlation between surface hardness and grain size, indicating that finer grains yield superior hardness. Beyond enhancing mechanical properties and wear resistance, these treatments also play a pivotal role in mitigating corrosion and tribocorrosion. Controlled microstructural modifications not only bolster the alloy's mechanical strength but also fortify its resistance against corrosive environments. By optimizing grain size and phase distribution, these treatments create barriers against corrosion propagation, ensuring the long-term integrity of the material. Furthermore, the improved wear resistance resulting from finer grains reduces the susceptibility to tribocorrosion-induced degradation, extending the service life of the alloy in high-stress applications. This integrated approach underscores the significance of tailored heat treatments in enhancing both the mechanical performance and corrosion resistance of Ti6Al4V alloys, facilitating their utilization in critical biomedical and engineering applications (Amirnejad et al., 2020).

8.6.3.11 Electron Beam Melting (EBM) and Postheat Treatment

EBM-produced samples undergo conventional annealing followed by a two-step thermal heat processing (THP) method aimed at microstructure refinement. While conventional annealing alleviates internal stresses, THP is implemented to refine the microstructure, which includes microstructure coarsening of both the Widmanstätten and grain boundary alpha (GBα) phases, alongside alterations in the α phase texture. Notably, this treatment induces a 30° rotation of basal planes (0001) around the building direction, significantly enhancing material mechanical properties and anisotropy. Furthermore, the refined microstructure and altered texture contribute to improved resistance against corrosion and tribocorrosion-induced degradation, underscoring the comprehensive benefits of this thermal processing approach in enhancing both mechanical performance and biocompatibility of the implants (Doğu et al., 2020)

8.7 FUTURE SCOPE AND RELEVANT FOCUS AREAS FOR PROGRESSIVE RESEARCH

In terms of orthopedic implants, there is a long way to go in research and find out the best material to be used as implants. It must mainly focus on innovation and safety. Postsurgery studies in the long term are important to reduce bony defects. Developing implants with porosity and a biologically inactive trauma site is needed in the future. These can be brought into play only when we use preemptive computation by FEA modeling and biomechanically fit the implant through surgery can be remarkably improved. In addition to enhancing surgical procedures, implants need to address metal ion sensitivity, infection control, early detection of osseointegration, reduced stiffness, and the possibility of bioresorbability.

Regenerative medicine, like gene therapy, is giving good results in healing bone fractures, but these strategies also need more improvements in devising stem cells or isolated genes to find defective fracture sites. On the other hand, implant sensors are the one that identifies the stress and longevity of the implant that has been placed in the limbs or other bones. The researchers must also concentrate more on 3D-printed implants that could give a push to make the implants cost-effective and more reliable in terms of availability.

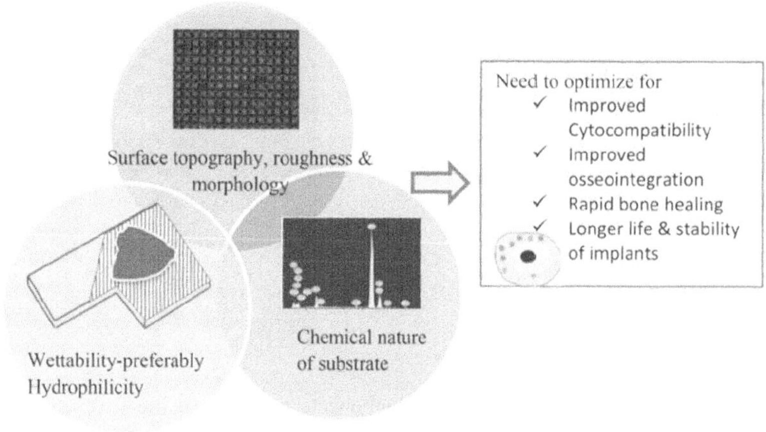

FIGURE 8.21 A diagrammatic representation of surface characteristics of biomaterials for enhancing the stability of implants. (Reproduced with permission from Ref. (Sirdeshmukh and Dongre, 2023). Copyright Elsevier, 2023.)

An improved understanding of the mechanisms of surface processing and characterization of manufacturing methods is necessary for future research. Figure 8.21 shows some of the important areas where we need to improve the research work further.

8.8 APPLICATIONS

8.8.1 General Applications of Ti Alloys

Ti6Al4V alloys have been used in several areas such as aircraft turbines, high-temperature containers, warships, spacecraft, medical implants, and manufacturing industries that require better mechanical and corrosion resistance. Figures 8.22 and 8.23 show some examples of Ti alloy used in the aerospace industry.

There are many applications of Ti alloys in different fields. Some advantages of Ti alloys are mentioned here. (1) Ti has one of the highest strength-density ratios. It is highly preferred in commercial aircraft as landing gear structures where Ti1023 has 1190 MPa, rather than using steel that has an UTS of 1930 MPa, resulting in about 50% reduction in weight because of the better density ratio of Ti alloy. (2) Many structural applications use Al alloys, but in some cases, they are too prone to corrosion. Ti alloys can be used in such cases because of their excellent corrosion resistance with equivalent specific strength. (3) The alpha alloys used in the high-pressure compressor are used at 500°C for Ti6242S, which is 50% lighter compared to steel. It is seen that Ti alloys were used rather than the latter due to their elevated temperature capabilities. (4) Unlike steel, Ti alloys have better ductility and the capability to transition and operate at –250°C, particularly because of Ti's high specific strength and ease of working at lower operating temperatures.

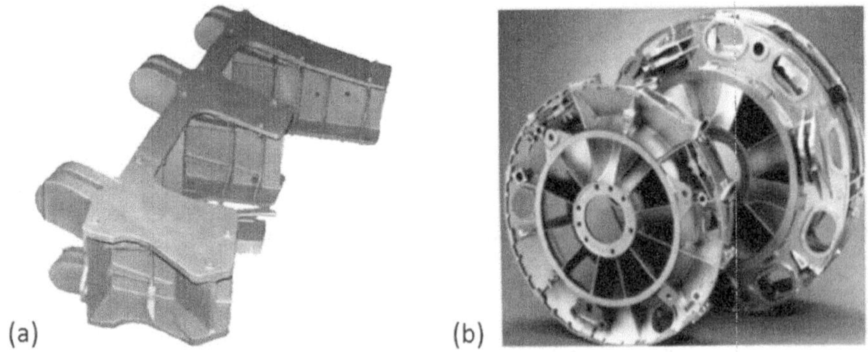

FIGURE 8.22 Examples of aerospace components produced using investment casting include (a) a side body casting for an aircraft (Boyer 1996) and (b) a front frame casting for an aircraft engine. (Reproduced with permission from Ref. (Pasang et al., 2023). Copyright Elsevier, 2023.)

FIGURE 8.23 Forged aerospace component examples. (Reproduced with permission from Ref. (Pasang et al., 2023). Copyright Elsevier, 2023.)

8.8.2 Biomedical Applications of Ti Alloys

The increasing number of severe cases resulting from a global rise in road accidents, coupled with the growing demand for cost-effective surgeries, is driving the development of future biomaterials. Currently, Ti alloys dominate the implant industry market due to their biocompatibility, high strength-to-weight ratio, and superior corrosion resistance compared to other metals and alloys.

Endosseous dental implants have been developed to address various oral health issues with good corrosion resistance. Due to constant exposure to enzymes and acids, the material is expected to exhibit excellent corrosion resistance. Additionally, rapid stabilization of the implant is required for expedited healing in the patient.

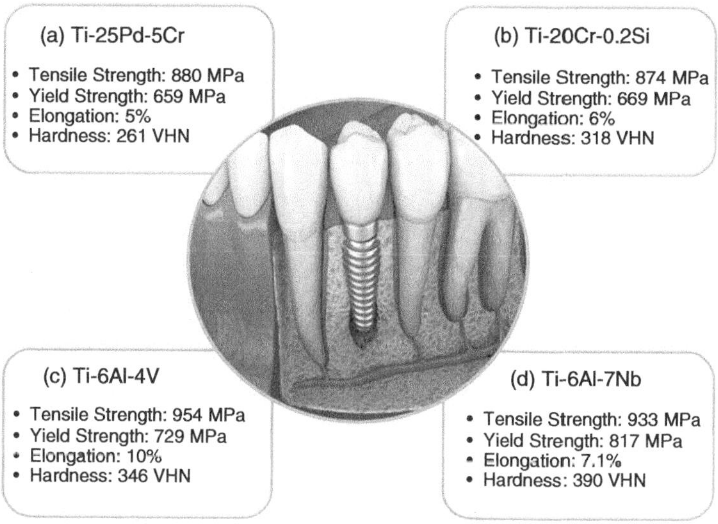

FIGURE 8.24 Mechanical characteristics of different Ti alloys for dental implants. (Reproduced with permission from Ref. (Sarraf et al., 2022). Copyright Springer Nature, 2021.)

Figure 8.24 displays various Ti alloys utilized in dental implants due to their superior mechanical properties. In the subsections below, some of the progress in the orthopedics of Ti alloys is summarized.

8.8.2.1 Spine

Li et al. (2017) presented work on porous Ti6Al4V for implantation and lateral mass screw fixation. The results revealed that there was no initiation of implant displacement or subsidence for 12 months. The custom-designed implants were easy to insert, thus reducing the surgical time and avoiding complex reconstruction. These alloys, with different designs, can prevent stress concentration and implant collapse.

8.8.2.2 Hip

Lima was the first company to develop a porous Ti acetabular cup using AM. Other companies, such as AK Medical and Smith & Nephew, later introduced similar products starting in 2015. The cup is designed with 72% porosity, 1.01 friction, and 546 μm pore size. Commercial AM methods are used to fabricate Ti, which can be easily customized.

8.8.2.3 Ankle

A customized Ti spacer with a better anatomical shape is in development to commercially replace the tantalum spacer. From the perspective of an orthopedic surgeon, many anatomically compatible Ti materials must be manufactured and commercialized. Further research should be conducted to understand the design of biocompatible customized metals and related techniques. Metals should be further developed according to the necessity of the problem, which can be for cortical to cancellous bone. Additional antibacterial properties must be studied for further depth of the Ti alloys to commercialize for the future. Many Ti-based biomaterials research studies are carried out on making nontoxic, non-allergenic material with low modulus and high caliber to work efficiently as implants. This is mainly developed to replace vanadium and aluminum constituents in future (Niinomi, 2003).

8.9 INFERENCES

The research area of corrosion and tribocorrosion related to biomedical implants in combination with cell growth has predominantly improved in the last few decades. The studies on biology-related corrosion are still minimal to different challenges offered by different fields, which include an interdisciplinary perspective on the experimental work. The studies discussed in this review showcase the results of corrosion, tribocorrosion, and manufacturing methods that give an idea about cell growth and cell culture that provides a pathway for future research. Their important value depends on the fact that they represent intersecting points for various research pathways. Crucial points regarding the microstructures and surface characteristics of fabrication on mechanical, corrosion, and biological properties are discussed with the currently used methods. With different surface texturing and different manufacturing methods, we can improve the biomechanical properties. This chapter recommends understanding the material, its behavior in the environment, and the associated failure mechanisms before adopting them for critical applications.

REFERENCES

Acharya GM, Shetty NA (2021a) Exploring the corrosion inhibition properties of an anionic gemini surfactant based on an ethylenediaminetetraacetic acid derivative on AZ31 alloy. *Chemistry Select* 6:8275–8287. https://doi.org/10.1002/slct.202101912

Acharya MG, Shetty AN (2021b) Influence of media pH on corrosion behaviour of AZ31 magnesium alloy in chloride and sulphate media. *Surface Engineering and Applied Electrochemistry* 57:675–688

Alam Md O, Haseeb ASMA (2002) Response of Ti–6Al–4V and Ti–24Al–11Nb alloys to dry sliding wear against hardened steel. *Tribology International* 35:357–362. https://doi.org/10.1016/S0301-679X(02)00015-4

Allahyarzadeh MH, Aliofkhazraei M, Rouhaghdam ARS, Torabinejad V (2017) Electrochemical tailoring of ternary Ni-W-Co (Al_2O_3) nanocomposite using pulse reverse technique. *Journal of Alloys and Compounds* 705:788–800

Amirnejad M, Rajabi M, Jamaati R (2021) The effect of crystallographic texture as a distinct effective parameter on the biocorrosion performance of Ti6Al4V alloy in PBS solution. *Corrosion Science* 179:109100. https://doi.org/10.1016/j.corsci.2020.109100

Amirnejad M, Rajabi M, Jamaati R (2020) Comparative investigation of microstructure and crystallographic texture effect on Ti6Al4V alloy mechanical properties. *Materials Chemistry and Physics* 256:123725. https://doi.org/10.1016/j.matchemphys.2020.123725

Andrade del Olmo J, Alonso JM, Sáez-Martínez V, et al (2023a) Hyaluronic acid-based hydrogel coatings on Ti6Al4V implantable biomaterial with multifunctional antibacterial activity. *Carbohydrate Polymers* 301:120366. https://doi.org/10.1016/j.carbpol.2022.120366

Andrade del Olmo J, Pérez-Álvarez L, Sáez Martínez V, et al (2023b) Multifunctional antibacterial chitosan-based hydrogel coatings on Ti6Al4V biomaterial for biomedical implant applications. *International Journal of Biological Macromolecules* 231:123328. https://doi.org/10.1016/j.ijbiomac.2023.123328

Attar H, Prashanth KG, Chaubey AK, et al (2015) Comparison of wear properties of commercially pure titanium prepared by selective laser melting and casting processes. *Materials Letters* 142:38–41

Aufa AN, Hassan MZ, Ismail Z (2022) Recent advances in Ti-6Al-4V additively manufactured by selective laser melting for biomedical implants: Prospect development. *Journal of Alloys and Compounds* 896:163072. https://doi.org/10.1016/j.jallcom.2021.163072

Bai WQ, Li LL, Xie YJ, et al (2016) Corrosion and tribocorrosion performance of M (MTa, Ti) doped amorphous carbon multilayers in Hank's solution. *Surface and Coatings Technology* 305:11–22

Bai Y, Gai X, Li S, et al (2017) Improved corrosion behaviour of electron beam melted Ti-6Al–4V alloy in phosphate buffered saline. *Corrosion Science* 123:289–296. https://doi.org/10.1016/j.corsci.2017.05.003

Bandyopadhyay A, Mitra I, Goodman SB, et al (2022) Improving biocompatibility for next generation of metallic implants. *Progress in Materials Science* 133:101053

Bäreis J, Semjatov N, Renner J, Ye J, Zongwen F, Körner C, (2023) Electron-optical in-situ crack monitoring during electron beam powder bed fusion of the Ni-Base superalloy CMSX-4. *Progress in Additive Manufacturing* 8(5):801–806.

Bolzoni L, Meléndez IM, Ruiz-Navas EM, Gordo E (2012) Microstructural evolution and mechanical properties of the Ti–6Al–4V alloy produced by vacuum hot-pressing. *Materials Science and Engineering: A* 546:189–197

Boraei NFE, Elshamy IH, Ibrahim MAM (2024) A comparative study on the electrochemical corrosion behaviour of biomedical β-titanium alloy with TiAlV and titanium in Hank's physiological solution and the impact of reactive oxygen species and immersion time. *Journal of Bio- and Tribo-Corrosion* 10:38. https://doi.org/10.1007/s40735-024-00839-z

Borgioli F, Galvanetto E, Iozzelli F, Pradelli G (2005) Improvement of wear resistance of Ti–6Al–4V alloy by means of thermal oxidation. *Materials Letters* 59:2159–2162. https://doi.org/10.1016/j.matlet.2005.02.054

Boyer RR (1996) An overview on the use of titanium in the aerospace industry. *Materials Science and Engineering: A* 213:103–114. https://doi.org/10.1016/0921-5093(96)10233-1

Burstein GT, Liu C (2007) Nucleation of corrosion pits in Ringer's solution containing bovine serum. *Corrosion Science* 49:4296–4306

Çaha I, Alves A, Chirico C, et al (2020) Corrosion and tribocorrosion behavior of Ti-40Nb and Ti-25Nb-5Fe alloys processed by powder metallurgy. *Metallurgical and Materials Transactions A* 51:3256–3267. https://doi.org/10.1007/s11661-020-05757-6

Calabrese L, Khaskhoussi A, Patane S, Proverbio E (2019) Assessment of super-hydrophobic textured coatings on AA6082 aluminum alloy. *Coatings* 9:352

Cerri E, Ghio E, Bolelli G (2022) Effect of surface roughness and industrial heat treatments on the microstructure and mechanical properties of Ti6Al4V alloy manufactured by laser powder bed fusion in different built orientations. *Materials Science and Engineering: A* 851:143635. https://doi.org/10.1016/j.msea.2022.143635

Chang K, Wang X, Liang E, Zhang R (2020) On the texture and mechanical property anisotropy of Ti6Al4V alloy fabricated by powder-bed based laser additive manufacturing. *Vacuum* 181:109732. https://doi.org/10.1016/j.vacuum.2020.109732

Chen J, Li C, Zhou L, et al (2022) The anisotropic of corrosion and tribocorrosion behaviors of Ti15Mo alloy fabricated by selective laser melting. *Materials Characterization* 190:112000. https://doi.org/10.1016/j.matchar.2022.112000

Chen J, Li J, Hu F, et al (2019) Effect of microarc oxidation-treated Ti6Al4V scaffold following low-intensity pulsed ultrasound stimulation on osteogenic cells in vitro. *ACS Biomaterials Science & Engineering* 5:572–581

Chen Q, Thouas GA (2015) Metallic implant biomaterials. *Materials Science and Engineering: R: Reports* 87:1–57. https://doi.org/10.1016/j.mser.2014.10.001

Chopski SG, Fox CS, McKenna KL, et al (2016) Physics-driven impeller designs for a novel intravascular blood pump for patients with congenital heart disease. *Medical Engineering & Physics* 38:622–632

Cinca N, Barbosa M, Dosta S, Guilemany JM (2010) Study of Ti deposition onto Al alloy by cold gas spraying. *Surface and Coatings Technology* 205:1096–1102

Ciszak C, Popa I, Brossard J-M, et al (2016) NaCl induced corrosion of Ti-6Al-4V alloy at high temperature. *Corrosion Science* 110:91–104

Contu F, Elsener B, Böhni H (2005) Corrosion behaviour of CoCrMo implant alloy during fretting in bovine serum. *Corrosion Science* 47:1863–1875

Cosyn J, De Rouck T (2009) Aesthetic outcome of single-tooth implant restorations following early implant placement and guided bone regeneration: Crown and soft tissue dimensions compared with contralateral teeth. *Clinical Oral Implants Research* 20:1063–1069. https://doi.org/10.1111/j.1600-0501.2009.01746.x

Dai J, Sun C, Wang A, et al (2021) High temperature oxidation and hot corrosion behaviors of Ti2AlNb alloy at 923 K and 1023 K. *Corrosion Science* 184:109336

Dai N, Zhang L-C, Zhang J, et al (2016) Corrosion behavior of selective laser melted Ti-6Al-4V alloy in NaCl solution. *Corrosion Science* 102:484–489

Davoren B, Ferg EE, Hattingh D (2017) Tribocorrosion of friction-stir-welded Ti6AL4V in 3.5% NaCl aqueous solution. *Wear* 390–391:246–252. https://doi.org/10.1016/j.wear.2017.08.004

Deng H, Chen Y, Jia Y, et al (2021) Microstructure and mechanical properties of dissimilar NiTi/Ti6Al4V joints via back-heating assisted friction stir welding. *Journal of Manufacturing Processes* 64:379–391. https://doi.org/10.1016/j.jmapro.2021.01.024

Detor AJ, Schuh CA (2007) Tailoring and patterning the grain size of nanocrystalline alloys. *Acta Materialia* 55:371–379

Dimah MK, Albeza FD, Borrás VA, Muñoz AI (2012) Study of the biotribocorrosion behaviour of titanium biomedical alloys in simulated body fluids by electrochemical techniques. *Wear* 294:409–418

Ding J, Zou S, Choi J, et al (2020) A laser texturing study on multi-crystalline silicon solar cells. *Solar Energy Materials and Solar Cells* 214:110587

Diomidis N, Mischler S, More NS, Roy M (2012) Tribo-electrochemical characterization of metallic biomaterials for total joint replacement. *Acta Biomaterialia* 8:852–859

Doğu MN, Esen Z, Davut K, et al (2020) Microstructural and texture evolution during thermo-hydrogen processing of Ti6Al4V alloys produced by electron beam melting. *Materials Characterization* 168:110549. https://doi.org/10.1016/j.matchar.2020.110549

Doni Z, Alves AC, Toptan F, et al (2013) Dry sliding and tribocorrosion behaviour of hot pressed CoCrMo biomedical alloy as compared with the cast CoCrMo and Ti6Al4V alloys. *Materials & Design (1980–2015)* 52:47–57. https://doi.org/10.1016/j.matdes.2013.05.032

Durdu S, Tosun S, Yalcin E, et al (2022) Characterization and investigation of properties of copper nanoparticle coated TiO_2 nanotube surfaces on Ti6Al4V alloy. *Materials Chemistry and Physics* 292:126741. https://doi.org/10.1016/j.matchemphys.2022.126741

Escano LI, Clark SJ, Chuang AC, et al (2022) An electron beam melting system for in-situ synchrotron X-ray monitoring. *Additive Manufacturing Letters* 3:100094

Fan L, Liu L, Lv Y, et al (2020) Effects of pre-oxidation on the corrosion behavior of pure Ti under coexistence of solid NaCl deposit and humid oxygen at 600° C: The diffusion of chlorine. *Scientific Reports* 10:1–13

Fernández-Lucio P, Villarón-Osorno I, Pereira Neto O, et al (2021) Effects of laser-textured on rake face in turning PCD tools for Ti6Al4V. *Journal of Materials Research and Technology* 15:177–188. https://doi.org/10.1016/j.jmrt.2021.08.004

Fu Y, Zhou F, Feng J, et al (2021) Long-term corrosion resistance and friction-induced antibacterial enhancement of CrMoSiCN coating on Ti6Al4V alloy. *Corrosion Science* 186:109478. https://doi.org/10.1016/j.corsci.2021.109478

Gangwar K, Ramulu M (2018) Friction stir welding of titanium alloys: A review. *Materials & Design* 141:230–255

García-Rueda AK, Guzmán-Castillo D, García-González L, et al (2022) Surface modification of a Ti6Al4V alloy by thermal oxidation to improve its tribological properties. *Materials Letters* 317:132082. https://doi.org/10.1016/j.matlet.2022.132082

Gärtner F, Stoltenhoff T, Schmidt T, Kreye H (2006) The cold spray process and its potential for industrial applications. *Journal of Thermal Spray Technology* 15:223–232. https://doi.org/10.1361/105996306X108110

Ge J, Huang Q, Wang Y, et al (2023) Microstructural optimization and mechanical enhancement of SLM Ti6Al4V TPMS scaffolds through vacuum annealing treatment. *Journal of Alloys and Compounds* 934:167524. https://doi.org/10.1016/j.jallcom.2022.167524

Ghelichi R, Guagliano M (2009) Coating by the cold spray process: A state of the art. *Frattura ed Integrità Strutturale* 3:30–44

Ghods S, Schultz E, Wisdom C, et al (2020) Electron beam additive manufacturing of Ti6Al4V: Evolution of powder morphology and part microstructure with powder reuse. *Materialia* 9:100631

Goldbaum D, Chromik RR, Yue S, et al (2011) Mechanical property mapping of cold sprayed Ti splats and coatings. *Journal of Thermal Spray Technology* 20:486–496

Grabowski A, Florian T, Wieczorek J, Adamiak M (2021) Structuring of the Ti6Al4V alloy surface by pulsed laser remelting. *Applied Surface Science* 535:147618. https://doi.org/10.1016/j.apsusc.2020.147618

Gulati K, Ivanovski S (2017) Dental implants modified with drug releasing titania nanotubes: therapeutic potential and developmental challenges. *Expert Opinion on Drug Delivery* 14:1009–1024. https://doi.org/10.1080/17425247.2017.1266332

Haseeb A, Blanpain B, Wouters G, et al (1993) Electrochemical deposition: A method for the production of artificially structured materials. *Materials Science and Engineering: A* 168:137–140

Hoomehr B, Raeissi K, Ashrafizadeh F, et al (2022) Corrosion performance and biological properties of electrophoretically deposited bioactive glass-zirconia core-shell composite coating on Ti6Al4V substrate. *Surface and Coatings Technology* 434:128209. https://doi.org/10.1016/j.surfcoat.2022.128209

Hua N, Wang W, Wang Q, et al (2021) Mechanical, corrosion, and wear properties of biomedical Ti–Zr–Nb–Ta–Mo high entropy alloys. *Journal of Alloys and Compounds* 861:157997. https://doi.org/10.1016/j.jallcom.2020.157997

Huang J, Huang Y, Yu X, et al (2022) Corrosion characterization of Ti-6Al-4V coating layer by the alternating current assisted GTAW method. *Corrosion Science* 197:110066. https://doi.org/10.1016/j.corsci.2021.110066

Huang J, Zhang W, Xu H, Fang W (2024) Effect of heat treatment on corrosion properties of Ti-6Al-4V titanium alloy produced by electron powder bed fusion. *JOM* 76:3039–3049. https://doi.org/10.1007/s11837-024-06537-y

Hussain T, McCartney DG, Shipway PH, Marrocco T (2011) Corrosion behavior of cold sprayed titanium coatings and free standing deposits. *Journal of Thermal Spray Technology* 20:260–274. https://doi.org/10.1007/s11666-010-9540-x

Jain A, Kumari N, Jagadevan S, Bajpai V (2020) Surface properties and bacterial behavior of micro conical dimple textured Ti6Al4V surface through micro-milling. *Surfaces and Interfaces* 21:100714. https://doi.org/10.1016/j.surfin.2020.100714

Jeong Y-H, Choe H-C, Brantley WA, Sohn I-B (2013) Hydroxyapatite thin film coatings on nanotube-formed Ti–35Nb–10Zr alloys after femtosecond laser texturing. *Surface and Coatings Technology* 217:13–22. https://doi.org/10.1016/j.surfcoat.2012.11.066

Ju J, Zhao C, Kang M, et al (2021) Effect of heat treatment on microstructure and tribological behavior of Ti–6Al–4V alloys fabricated by selective laser melting. *Tribology International* 159:106996

Kaliaraj GS, Siva T, Ramadoss A (2021) Surface functionalized bioceramics coated on metallic implants for biomedical and anticorrosion performance–a review. *Journal of Materials Chemistry B* 9:9433–9460

Kamkar S, Mohammadi M, Karimi M, Salehi M (2022) Electrochemical and biological properties of mono- and bilayer nitride coatings deposited on Ti–6%Al–4%V alloy. *Materials Chemistry and Physics* 286:126185. https://doi.org/10.1016/j.matchemphys.2022.126185

Kanishka K, Acherjee B (2023) A systematic review of additive manufacturing-based remanufacturing techniques for component repair and restoration. *Journal of Manufacturing Processes* 89:220–283. https://doi.org/10.1016/j.jmapro.2023.01.034

Ke D, Vu AA, Bandyopadhyay A, Bose S (2019) Compositionally graded doped hydroxyapatite coating on titanium using laser and plasma spray deposition for bone implants. *Acta Biomaterialia* 84:414–423. https://doi.org/10.1016/j.actbio.2018.11.041

Khodaei T, Schmitzer E, Suresh AP, Acharya AP (2023) Immune response differences in degradable and non-degradable alloy implants. *Bioactive Materials* 24:153–170

Kim K, Watanabe M, Kawakita J, Kuroda S (2008) Grain refinement in a single titanium powder particle impacted at high velocity. *Scripta Materialia* 59:768–771

Kityk A, Švec P, Šoltys J, et al (2023) Deep inside of the mechanism of electrochemical surface etching of α+β Ti6Al4V alloy in room-temperature deep eutectic solvent Ethaline. *Journal of Molecular Liquids* 121316. https://doi.org/10.1016/j.molliq.2023.121316

Komotori J, Hisamori N, Ohmori Y (2007) The corrosion/wear mechanisms of Ti–6Al–4V alloy for different scratching rates. *Wear* 263:412–418

Körner C (2016) Additive manufacturing of metallic components by selective electron beam melting—a review. *International Materials Reviews* 61:361–377

Lazić MM, Mitić D, Radović K, et al (2024) Corrosion behavior of nickel–titanium continuous-casted alloys. *Metals* 14. https://doi.org/10.3390/met14010088

Lee H-K, Lee H-Y, Jeon J-M (2007) Codeposition of micro-and nano-sized SiC particles in the nickel matrix composite coatings obtained by electroplating. *Surface and Coatings Technology* 201:4711–4717

Lek JY, Bhowmik A, Tan AW- Y, et al (2018) Understanding the microstructural evolution of cold sprayed Ti-6Al-4V coatings on Ti-6Al-4V substrates. *Applied Surface Science* 459:492–504

Lenis JA, Romero EC, Macías AH, et al (2022) Mechanical, structural, and biological evaluation of multilayer HA-Ag/TiO_2/TiN/Ti coatings on Ti6Al4V obtained by magnetron sputtering for implant applications. *Surface and Coatings Technology* 449:128925. https://doi.org/10.1016/j.surfcoat.2022.128925

Li J, Lin X, Yang Y, et al (2021) Distinction in electrochemical behaviour of Ti6Al4V alloy produced by direct energy deposition and forging. *Journal of Alloys and Compounds* 860:157912. https://doi.org/10.1016/j.jallcom.2020.157912

Li W, Cao C, Yin S (2020) Solid-state cold spraying of Ti and its alloys: A literature review. *Progress in Materials Science* 110:100633

Li X, Wang Y, Zhao Y, et al (2017) Multilevel 3D printing implant for reconstructing cervical spine with metastatic papillary thyroid carcinoma. *Spine* 42:E1326–E1330

Li XT, An Q, Huang LJ, et al (2022) The hot corrosion behaviors and mechanisms of hot-dipped Ti-Al-Si coatings on TiBw/Ti6Al4V composite against NaCl and Na_2SO_4. *Journal of Alloys and Compounds* 921:166151. https://doi.org/10.1016/j.jallcom.2022.166151

Licausi M-P, Muñoz AI, Borrás VA (2013) Tribocorrosion mechanisms of Ti6Al4V biomedical alloys in artificial saliva with different pHs. *Journal of Physics D: Applied Physics* 46:404003

Ling W, Wang X, Wang L, et al (2023) Defect formation mechanism of laser additive manufacturing joint for large-thickness Ti6Al4V titanium alloy with Y2O3 nanoparticles. *Optics & Laser Technology* 157:108648. https://doi.org/10.1016/j.optlastec.2022.108648

Liu Y, Su J, Tan C, et al (2021) Effect of laser texturing on mechanical strength and microstructural properties of hot-pressing joining of carbon fiber reinforced plastic to Ti6Al4V. *Journal of Manufacturing Processes* 65:30–41

Lodes MA, Guschlbauer R, Körner C (2015) Process development for the manufacturing of 99.94% pure copper via selective electron beam melting. *Materials Letters* 143:298–301. https://doi.org/10.1016/j.matlet.2014.12.105

Lopez F, Witherell P, Lane B (2016) Identifying uncertainty in laser powder bed fusion additive manufacturing models. *Journal of Mechanical Design* 138:114502.

Luo X, Ren X, Qu H, et al (2022) Research on influence of deep cryogenic treatment and ultrasonic rolling on improving surface integrity of Ti6Al4V alloy. *Materials Science and Engineering: A* 843:143142. https://doi.org/10.1016/j.msea.2022.143142

Mahlobo MGR, Chikosha L, Olubambi PA (2022) Study of the corrosion properties of powder rolled Ti-6Al-4V alloy applied in the biomedical implants. *Journal of Materials Research and Technology* 18:3631–3639. https://doi.org/10.1016/j.jmrt.2022.04.004

Maity T, Balcı Ö, Gammer C, et al (2020) High pressure torsion induced lowering of Young's modulus in high strength TNZT alloy for bio-implant applications. *Journal of the Mechanical Behavior of Biomedical Materials* 108:103839. https://doi.org/10.1016/j.jmbbm.2020.103839

Manivasagam G, Dhinasekaran D, Rajamanickam A (2010) Biomedical implants: Corrosion and its prevention – A review. *Recent Patents on Corrosion Science* 2:40–54. https://doi.org/10.2174/1877610801002010040

Ming Q, Yong-zhen Z, Jian-heng Y, Jun Z (2006) Microstructure and tribological characteristics of Ti–6Al–4V alloy against GCr15 under high speed and dry sliding. *Materials Science and Engineering: A* 434:71–75. https://doi.org/10.1016/j.msea.2006.07.043

Miranda G, Carvalho O, Silva FS, Soares D (2013) Effect of sintering stage in NiTi short-fibre-reinforced aluminium–silicon composites interface properties. *Journal of Composite Materials* 47:1625–1631

Moghadasi K, Isa MSM, Ariffin MA, et al (2022) A review on biomedical implant materials and the effect of friction stir based techniques on their mechanical and tribological properties. *Journal of Materials Research and Technology* 17:1054–1121. https://doi.org/10.1016/j.jmrt.2022.01.050

Moon B-H, Choe H-C, Brantley WA (2012) Surface characteristics of TiN/ZrN coated nanotubular structure on the Ti–35Ta–xHf alloy for bio-implant applications. *Applied Surface Science* 258:2088–2092. https://doi.org/10.1016/j.apsusc.2011.04.050

Moradi M, Saidi R, Hoomehr B, Raeissi K (2022) The effect of bioactive glass nanoparticles on corrosion barrier performance and bioactivity of zinc oxide coatings electrodeposited on Ti6Al4V substrate. *Ceramics International*. https://doi.org/10.1016/j.ceramint.2022.11.088

More NS, Diomidis N, Paul SN, et al (2011) Tribocorrosion behavior of β titanium alloys in physiological solutions containing synovial components. *Materials Science and Engineering:* C 31:400–408

Moura CG, Carvalho O, Gonçalves LMV, et al (2019) Laser surface texturing of Ti-6Al-4V by nanosecond laser: Surface characterization, Ti-oxide layer analysis and its electrical insulation performance. *Materials Science and Engineering:* C 104:109901. https://doi.org/10.1016/j.msec.2019.109901

Murphy CM, Matsiko A, Haugh MG, et al (2012) Mesenchymal stem cell fate is regulated by the composition and mechanical properties of collagen–glycosaminoglycan scaffolds. *Journal of the Mechanical Behavior of Biomedical Materials* 11:53–62

Murr LE, Gaytan SM, Ramirez DA, et al (2012) Metal fabrication by additive manufacturing using laser and electron beam melting technologies. *Journal of Materials Science & Technology* 28:1–14. https://doi.org/10.1016/S1005-0302(12)60016-4

Ni X, Kong D, Zhou Y, et al (2023) Heat treatment effects on the electrochemical behavior of pore-free Ti6Al4V ELI fabricated by laser powder bed fusion in fluoride-containing chloride solution. *Journal of Alloys and Compounds* 937:168439. https://doi.org/10.1016/j.jallcom.2022.168439

Niinomi M (2003) Recent research and development in titanium alloys for biomedical applications and healthcare goods. *Science and Technology of Advanced Materials* 4:445–454. https://doi.org/10.1016/j.stam.2003.09.002

Nune KC, Misra RDK, Li SJ, et al (2017) Osteoblast cellular activity on low elastic modulus Ti–24Nb–4Zr–8Sn alloy. *Dental Materials* 33:152–165. https://doi.org/10.1016/j.dental.2016.11.005

Pacheco JT, Cardoso EL (2022) Simulation of the thermomechanical behavior of discrete particles in the laser directed energy deposition process. *Powder Technology* 405:117568. https://doi.org/10.1016/j.powtec.2022.117568

Pang Z, Wang S, Yin X, et al (2022) Effect of spindle speed during ultrasonic rolling on surface integrity and fatigue performance of Ti6Al4V alloy. *International Journal of Fatigue* 159:106794. https://doi.org/10.1016/j.ijfatigue.2022.106794

Pasang T, Budiman AS, Wang JC, et al (2023) Additive manufacturing of titanium alloys – Enabling re-manufacturing of aerospace and biomedical components. *Microelectronic Engineering* 270:111935. https://doi.org/10.1016/j.mee.2022.111935

Pattison J, Celotto S, Morgan R, et al (2007) Cold gas dynamic manufacturing: A non-thermal approach to freeform fabrication. *International Journal of Machine Tools and Manufacture* 47:627–634

Pellier J, Geringer J, Forest B (2011) Fretting-corrosion between 316L SS and PMMA: Influence of ionic strength, protein and electrochemical conditions on material wear. Application to orthopaedic implants. *Wear* 271:1563–1571

Perez-Andrade LI, Bhattiprolu VS, Schuette WM, Brewer LN (2022) Influence of powder properties and processing gas on the microstructural evolution of Armstrong CP-titanium and Ti6Al4V powders processed by cold spray. *Surface and Coatings Technology* 431:128011. https://doi.org/10.1016/j.surfcoat.2021.128011

Pfleging W, Kumari R, Besser H, et al (2015) Laser surface textured titanium alloy (Ti–6Al–4V): Part 1–Surface characterization. *Applied Surface Science* 355:104–111

Qin P, Chen LY, Zhao CH, et al (2021) Corrosion behavior and mechanism of selective laser melted Ti35Nb alloy produced using pre-alloyed and mixed powder in Hank's solution. *Corrosion Science* 189:109609. https://doi.org/10.1016/j.corsci.2021.109609

Qu D, Dan LIU, Wang X, et al (2022) Corrosion and wear properties of TB2 titanium alloy borided by pack boriding with La_2O_3. *Transactions of Nonferrous Metals Society of China* 32:868–881

Rafaja D, Schucknecht T, Klemm V, et al (2009) Microstructural characterisation of titanium coatings deposited using cold gas spraying on Al2O3 substrates. *Surface and Coatings Technology* 203:3206–3213

Rajesh K, Ghosh S, Islam A, et al (2022) Multilayered porous hydroxyapatite coating on Ti6Al4V implant with enhanced drug delivery and antimicrobial properties. *Journal of Drug Delivery Science and Technology* 70:103155. https://doi.org/10.1016/j.jddst.2022.103155

Raoelison RN, Verdy C, Liao H (2017) Cold gas dynamic spray additive manufacturing today: Deposit possibilities, technological solutions and viable applications. *Materials & Design* 133:266–287

Ren B, Wan Y, Liu C, et al (2021) Improved osseointegration of 3D printed Ti-6Al-4V implant with a hierarchical micro/nano surface topography: An in vitro and in vivo study. *Materials Science and Engineering:* C 118:111505. https://doi.org/10.1016/j.msec.2020.111505

Ren X, Peng H, Li J, et al (2022) Selective electron beam melting (SEBM) of pure tungsten: Metallurgical defects, microstructure, texture and mechanical properties. *Materials* 15:1172

Revathi A, Borrás AD, Muñoz AI, et al (2017) Degradation mechanisms and future challenges of titanium and its alloys for dental implant applications in oral environment. *Materials Science and Engineering:* C 76:1354–1368

Ríos JM, Quintero D, Castaño JG, et al (2022) Effect of EDTA addition on the biotribological properties of coatings obtained from PEO on the Ti6Al4V alloy in a phosphate-based solution. *Surfaces and Interfaces* 30:101857. https://doi.org/10.1016/j.surfin.2022.101857

Rupert TJ, Schuh CA (2010) Sliding wear of nanocrystalline Ni–W: Structural evolution and the apparent breakdown of Archard scaling. *Acta Materialia* 58:4137–4148

Sadeghi M, Kharaziha M, Salimijazi HR (2019) Double layer graphene oxide-PVP coatings on the textured Ti6Al4V for improvement of frictional and biological behavior. *Surface and Coatings Technology* 374:656–665. https://doi.org/10.1016/j.surfcoat.2019.06.048

Sampaio M, Buciumeanu M, Henriques B, et al (2016) Tribocorrosion behavior of veneering biomedical PEEK to Ti6Al4V structures. *Journal of the Mechanical Behavior of Biomedical Materials* 54:123–130. https://doi.org/10.1016/j.jmbbm.2015.09.010

Sarraf M, Rezvani Ghomi E, Alipour S, et al (2022) A state-of-the-art review of the fabrication and characteristics of titanium and its alloys for biomedical applications. *Bio-Design and Manufacturing* 5:371–395. https://doi.org/10.1007/s42242-021-00170-3

Schmidt KTA, Chen ZW (2022) Effect of surface morphology of Ti6Al4V alloy parts manufactured by electron beam powder bed fusion on bonding strength of adhesive joints. *Additive Manufacturing Letters* 3:100059. https://doi.org/10.1016/j.addlet.2022.100059

Shi G, Ren L, Wang L, et al (2009) H2O2/HCl and heat-treated Ti-6Al-4V stimulates pre-osteoblast proliferation and differentiation. *Oral Surgery, Oral Medicine, Oral Pathology, Oral Radiology, and Endodontology* 108:368–375

Siddiqui MA, Ren L, Macdonald DD, Yang K (2021) Effect of Cu on the passivity of Ti–xCu (x=0, 3 and 5 wt%) alloy in phosphate-buffered saline solution within the framework of PDM-II. *Electrochimica Acta* 386:138466. https://doi.org/10.1016/j.electacta.2021.138466

Sirdeshmukh N, Dongre G (2023) Achieving controlled topography and wettability through laser surface texturing of Ti6Al4V for bioengineering applications. *Results in Engineering* 17:100898. https://doi.org/10.1016/j.rineng.2023.100898

Sochalski-Kolbus LM, Payzant EA, Cornwell PA, et al (2015) Comparison of residual stresses in Inconel 718 simple parts made by electron beam melting and direct laser metal sintering. *Metallurgical and Materials Transactions A* 46:1419–1432

Solorio VM, Vergara-Hernández HJ, Olmos L, et al (2022) Effect of the Ag addition on the compressibility, sintering and properties of Ti6Al4V/xAg composites processed by powder metallurgy. *Journal of Alloys and Compounds* 890:161813. https://doi.org/10.1016/j.jallcom.2021.161813

Song B, Dong S, Zhang B, et al (2012) Effects of processing parameters on microstructure and mechanical property of selective laser melted {Ti6Al4V}. *Materials & Design* 35:120–125. https://doi.org/10.1016/j.matdes.2011.09.051

Talha M, Ma Y, Kumar P, et al (2019) Role of protein adsorption in the bio corrosion of metallic implants–A review. *Colloids and Surfaces B: Biointerfaces* 176:494–506

Tang Y-C, Katsuma S, Fujimoto S, Hiromoto S (2006) Electrochemical study of type 304 and 316L stainless steels in simulated body fluids and cell cultures. *Acta Biomaterialia* 2:709–715

Tarafder P, Rock C, Horn T (2021) Quasi-static tensile properties of unalloyed copper produced by electron beam powder bed fusion additive manufacturing. *Materials* 14:2932

Tareq MS, Rahman T, Hossain M, Dorrington P (2021) Additive manufacturing and the COVID-19 challenges: An in-depth study. *Journal of Manufacturing Systems* 60:787–798

Toh WQ, Wang P, Tan X, et al (2016) Microstructure and wear properties of electron beam melted Ti-6Al-4V parts: A comparison study against as-cast form. *Metals* 6:284

Vidaller MV, List A, Gaertner F, et al (2015) Single impact bonding of cold sprayed Ti-6Al-4V powders on different substrates. *Journal of Thermal Spray Technology* 24:644–658. https://doi.org/10.1007/s11666-014-0200-4

Vieira AC, Rocha LA, Papageorgiou N, Mischler S (2012) Mechanical and electrochemical deterioration mechanisms in the tribocorrosion of Al alloys in NaCl and in $NaNO_3$ solutions. *Corrosion Science* 54:26–35

Vilaro T, Colin C, Bartout J-D (2011) As-fabricated and heat-treated microstructures of the Ti-6Al-4V alloy processed by selective laser melting. *Metallurgical and Materials Transactions A* 42:3190–3199

Villafuerte J (2015) *Modern Cold Spray: Materials, Process, and Applications*. Springer

Vrancken B, Thijs L, Kruth J-P, Van Humbeeck J (2012) Heat treatment of Ti6Al4V produced by selective laser melting: Microstructure and mechanical properties. *Journal of Alloys and Compounds* 541:177–185

Wang G, Zhao Y, Hao Y (2018) Friction stir welding of high-strength aerospace aluminum alloy and application in rocket tank manufacturing. *Journal of Materials Science & Technology* 34:73–91

Wang K, Zou X, Wang H, et al (2023a) Development of bone adapted 3D printed Ti6Al4V scaffold via unit optimization and acid etching-MAO double treatment. *Materials Letters* 331:133500. https://doi.org/10.1016/j.matlet.2022.133500

Wang R, Li F, Wei D, et al (2023b) Scanning electron beam rotary polishing of column-faced Ti6Al4V. *Materials Letters* 330:133228. https://doi.org/10.1016/j.matlet.2022.133228

Wang T, Zhu YY, Zhang SQ, et al (2015) Grain morphology evolution behavior of titanium alloy components during laser melting deposition additive manufacturing. *Journal of Alloys and Compounds* 632:505–513

Wang W, Yang K, Wang Q, et al (2022) Novel Ti-Zr-Hf-Nb-Fe refractory high-entropy alloys for potential biomedical applications. *Journal of Alloys and Compounds* 906:164383. https://doi.org/10.1016/j.jallcom.2022.164383

Wu H, Gong P, Hu S, et al (2021) Hot corrosion damage mechanism of Ti6Al4V alloy in marine environment. *Vacuum* 191:110350. https://doi.org/10.1016/j.vacuum.2021.110350

Xiong Y, Zhu S, Wang F (2008) Synergistic corrosion behavior of coated Ti60 alloys with NaCl deposit in moist air at elevated temperature. *Corrosion Science* 50:15–22

Yang G, Yang P, Yang K, et al (2019) Effect of processing parameters on the density, microstructure and strength of pure tungsten fabricated by selective electron beam melting. *International Journal of Refractory Metals and Hard Materials* 84:105040. https://doi.org/10.1016/j.ijrmhm.2019.105040

Yang W, Liu Y, Hua N, et al (2021) Formation and properties of biocompatible Ti-based bulk metallic glasses in the Ti–Cu–Zr–Fe–Sn–Si–Ag system. *Journal of Non-Crystalline Solids* 571:121060. https://doi.org/10.1016/j.jnoncrysol.2021.121060

Yang W, Pang S, Liu Y, et al (2022) Design and properties of novel Ti–Zr–Hf–Nb–Ta high-entropy alloys for biomedical applications. *Intermetallics* 141:107421. https://doi.org/10.1016/j.intermet.2021.107421

Yin C, Zhang T, Wei Q, et al (2022) Surface treatment of 3D printed porous Ti6Al4V implants by ultraviolet photofunctionalization for improved osseointegration. *Bioactive Materials* 7:26–38. https://doi.org/10.1016/j.bioactmat.2021.05.043

Yin S, Cavaliere P, Aldwell B, et al (2018) Cold spray additive manufacturing and repair: Fundamentals and applications. *Additive Manufacturing* 21:628–650

Yuan S, Lin N, Zou J, et al (2020a) In-situ fabrication of gradient titanium oxide ceramic coating on laser surface textured Ti6Al4V alloy with improved mechanical property and wear performance. *Vacuum* 176:109327. https://doi.org/10.1016/j.vacuum.2020.109327

Yuan S, Lin N, Zou J, et al (2020b) Manipulation tribological behavior of Ti6Al4V alloy via a duplex treatment of double glow plasma surface molybdenizing-laser surface texturing (LST). *Journal of Materials Research and Technology* 9:6360–6375. https://doi.org/10.1016/j.jmrt.2020.03.061

Zhang G, Hao Z, Wang M, et al (2023) Globularization mechanism and near isotropic properties in subcritical heat-treated Ti6Al4V fabricated by directed energy deposition. *Acta Metallurgica Sinica (English Letters)* 36:937–948. https://doi.org/10.1007/s40195-022-01516-1

Zhang H, Man C, Dong C, et al (2021) The corrosion behavior of Ti6Al4V fabricated by selective laser melting in the artificial saliva with different fluoride concentrations and pH values. *Corrosion Science* 179:109097. https://doi.org/10.1016/j.corsci.2020.109097

Zhang J, Liu J, Wang C, et al (2020) A comparative study of the osteogenic performance between the hierarchical micro/submicro-textured 3D-printed Ti6Al4V surface and the SLA surface. *Bioactive Materials* 5:9–16. https://doi.org/10.1016/j.bioactmat.2019.12.008

Zhang L, Attar H (2016) Selective laser melting of titanium alloys and titanium matrix composites for biomedical applications: a review. *Advanced Engineering Materials* 18:463–475

Zhang Q, Chen J, Lin X, et al (2016) Grain morphology control and texture characterization of laser solid formed Ti6Al2Sn2Zr3Mo1. 5Cr2Nb titanium alloy. *Journal of Materials Processing Technology* 238:202–211

Zhang Z, Liu D, Zhang Y, et al (2022) Fabrication and droplet impact performance of superhydrophobic Ti6Al4V surface by laser induced plasma micro-machining. *Applied Surface Science* 605:154661. https://doi.org/10.1016/j.apsusc.2022.154661

Zhu G, Wang G (2021) Advances in implant surface modifications to improve osseointegration. *Materials Advances* 6901–6927. https://doi.org/10.1039/d1ma00675d

Zivic F, Babic M, Mitrovic S, Vencl A (2011) Continuous control as alternative route for wear monitoring by measuring penetration depth during linear reciprocating sliding of Ti6Al4V alloy. *Journal of Alloys and Compounds* 509:5748–5754

Zsurzsa S, Péter L, Kiss LF, Bakonyi I (2017) Magnetic and magnetoresistance studies of nanometric electrodeposited Co films and Co/Cu layered structures: Influence of magnetic layer thickness. *Journal of Magnetism and Magnetic Materials* 421:194–206

Index

Note: **Bold** page numbers refer to tables; *italic* page numbers refer to figures.

N

O

P

Q

R

S

T

U

V

W

X

Y

Z